# The biochemistry of natural pigments

*Cambridge Texts in Chemistry and Biochemistry*

# The biochemistry of natural pigments

G. BRITTON

CAMBRIDGE UNIVERSITY PRESS
Cambridge
London   New York   New Rochelle
Melbourne   Sydney

CAMBRIDGE UNIVERSITY PRESS
Cambridge, New York, Melbourne, Madrid, Cape Town, Singapore, São Paulo, Delhi

Cambridge University Press
The Edinburgh Building, Cambridge CB2 8RU, UK

Published in the United States of America by Cambridge University Press, New York

www.cambridge.org
Information on this title: www.cambridge.org/9780521105316

First published 1983
This digitally printed version 2009

A catalogue record for this publication is available from the British Library

Library of Congress Catalogue Card Number: 82-9512

ISBN 978-0-521-24892-1 hardback
ISBN 978-0-521-10531-6 paperback

# Contents

# Preface

As I write these words, the leaves on the trees in my garden are changing from their summer green to the yellow, red and brown of autumn, roses and other flowers are still blooming brightly and wading birds on the shore close by are mostly in their winter plumage. These simple observations provide good examples of the manifestations of colour and pattern in the living world, often in ways that are so familiar to us that we take them for granted. Colour and pattern are important for camouflage, to enable animals to escape the notice of predators, they are important in advertising the presence of an animal to potential mates, etc. and bright colours are important in drawing flowers and fruits to the attention of pollinating and seed-dispersing creatures. In our own everyday world, millions of gardeners grow flowers of many vivid or subtle hues to delight the eye, and brightly coloured fruits are displayed in the shops to attract customers. We should not be surprised, therefore, that the interest of scientists turned very early towards investigating the nature of these plant and animal colours and identifying the underlying mechanisms of colour production and display. It is now well known that there are two fundamentally different mechanisms for natural colour production: the physical or optical phenomena based upon the structures of the cells and tissues and giving rise to structural colours, and the presence of light-absorbing substances, pigments, responsible for the pigmentary colours. This book is concerned with the biochemistry of these natural pigments, the molecules responsible for so much of the colour in the living world. But it is not only for their colour that many of these molecules are important; the property of absorbing visible light renders them useful in many ways, for example in such vital processes as light harvesting in photosynthesis, light detection and colour discrimination in vision, and many other light-mediated responses and regulatory mechanisms. All these topics must be included in a book on natural pigments.

This book is divided into two sections. The first section describes the main features of the chemistry and biochemistry of the main groups of natural pigments; the second section is concerned with biological aspects, dealing with the main functional roles of pigments in Nature. The approach is descriptive and concentrates on the main features and principles. It cannot be comprehensive; this would lead to each chapter expanding into a several-volume series. The aim is rather to give an overall picture, to draw attention to the main points of interest, to stimulate the appetite and send the reader off in search of the key references quoted. I have had to be very selective about which topics went into the book and in how much detail. Readers may not agree with my choice or may think that I have the emphasis and balance wrong, but this is an overview of the subject as I see it. The writing and preparation of this book have been a new challenge, often enjoyable, sometimes frustrating and demanding time and attention that should really have been employed differently. During the preparation, however, I have read much and learned a lot about natural pigments. This has been very rewarding, and I hope that I have been able to pass on to the reader some of the knowledge and understanding gained and some of the great interest that the subject holds for me.

Finally, and with much pleasure, I must acknowledge the great debt of gratitude that I owe to so many people. First I wish to express my thanks publicly for the first time to my parents for their sacrifices and support during the years of my formal education which allowed me to spend these later years happily studying the world of natural pigments. My thanks are due also to Dr E. Haslam and Professor T. W. Goodwin who stimulated and encouraged my interest in the subject, and from whom I have learned so much. I acknowledge the forbearance of members of my research group over the years when I have devoted to the book time and attention which they could justly claim should have been accorded to them. I also wish to thank Dr Ernest Kirkwood, Mrs Marion Jowett and others at Cambridge University Press for their work in converting my typescript into a book.

My greatest debt of gratitude is, of course, to my family, for the many occasions when I have given in to the demands that the writing and preparation of the book made on my time and energy, when perhaps I should have put them first. My wife, Pat, has borne this with perseverance and patience and given me the added encouragement of producing a virtually perfect typescript from my imperfect and sometimes illegible handwriting. My children, Rebecca and Jonathan, have at times been deprived of the companionship and fatherly guidance to which they are entitled and which I should like to have given. It is to them that this book is dedicated, in the hope that they may derive as much pleasure as I from the world of Nature in which colour plays such a large part.

# SECTION I
# CHEMICAL AND BIOCHEMICAL ASPECTS

# 1      Light and colour

## 1.1     Introduction
### 1.1.1   Solar electromagnetic radiation

All living processes on Earth ultimately depend upon that portion of the vast resources of the Sun's energy which eventually reaches the surface of our planet. The Sun emits a wide range of electromagnetic radiations, from long-wavelength infrared (i.r.) and radio frequencies to very short-wavelength ultraviolet (u.v.) and $\gamma$-rays (fig. 1.1). However, the Earth's atmosphere as it is today effectively and efficiently filters out much of this radiation, particularly the high-energy u.v., X-rays, and $\gamma$-rays that can have a disastrous effect on living tissues.

### 1.1.2   Visible light

Amongst the radiations that do reach the surface of the Earth, those with wavelengths between approximately 380 and 750 nanometers (nm, $\equiv 10^{-9}$ m) penetrate the atmosphere most readily, *i.e.* suffer least restriction on their passage. This wavelength range, 380–750 nm, is of fundamental importance in maintaining life. It is also the range which we recognise as 'visible light'. Animals, including ourselves, have developed very sophisticated photoreceptor systems for the detection of this light and also for accurate

Fig. 1.1. The electromagnetic spectrum.

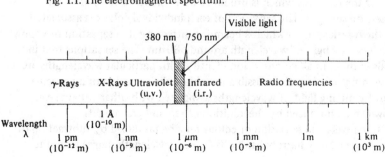

discrimination of different wavelengths within this region in the processes of colour vision. Colour, and the property of being coloured, thus become very important in the living world.

It is the same range of light energy which is harnessed by plants and microbes in the process of photosynthesis by which atmospheric carbon dioxide is fixed into a chemical form that is not only used by the plant but also provides the primary food source for the rest of the natural world. Variations in the amount of available visible light, for example variations in length of day and night, are also monitored by various photoreceptors. This provides the basis of extremely important mechanisms for regulating growth and development.

All these properties and processes – being coloured, detecting light and colour, photosynthesis, photoregulation – require mechanisms for detecting or absorbing light from the visible range. Molecules which have the special property of absorbing light of wavelengths in the 380–750 nm range are therefore of fundamental importance. Such compounds are the **natural pigments** or **biochromes**. It is the purpose of this book to review the main features of the chemistry and biochemistry of groups of natural pigments, and to describe, as far as possible, how these pigments function at the molecular level.

## 1.2    Colour and colour perception
### *1.2.1    Colour*

Simultaneous perception of radiations over the entire range, 380–750 nm, produces (in man) the sensation that we recognise as white light. Other animals are able to perceive radiations of wavelengths outside this range, *e.g.* bees can 'see' u.v. wavelengths invisible to us.

The sensation of colour is given if radiations are received from only part of the visible range. 'White light' is a continuum of electromagnetic radiations covering the wavelength range 380–750 nm. When this continuum is separated by passage (refraction) through a prism, then a series of beams is obtained, each consisting of a much narrower range of wavelengths. We see these beams as a series of colours, the familiar red, orange, yellow, green, blue (indigo), violet of the rainbow, which is produced by prismatic effects of water droplets on sunlight. The sensation of each individual colour is associated with the wavelength on which a beam is centred, *e.g.* the sensation of yellow is produced by light of wavelength around 580 nm. The sensations that individuals with 'normal' colour vision identify with particular wavelengths are shown in fig. 1.2. It is also possible to achieve the sensation of a particular colour by mixing light of wavelengths associated with other colours, *e.g.* yellow can be produced by the addition of red and green light.

Alternatively the sensation of colour may be produced by subtraction of what can be a fairly narrow band (20–30 nm wavelength range) from the white-light continuum. In this case what is 'seen' is the colour complementary to that of the missing waveband. Thus, if white light is passed through a filter or substance which absorbs blue light, i.e. 480 ± 30 nm, the emergent beam is seen as the colour complementary to blue, *i.e.* yellow. The complementary or subtraction colours observed when light of a particular colour or wavelength range is subtracted from the white-light continuum are also shown in fig. 1.2.

### 1.2.2 Light perception and colour discrimination

The above brief introduction to light and colour has frequently alluded to our ability to 'see colours'. Although the identification and description of colours is to a considerable degree subjective, there must be an underlying fundamental physiological mechanism which is not only capable of detecting electromagnetic radiations in the wavelength range 380–750 nm, but is also able to produce different sensations in response to radiations of different wavelengths within this range. Plants and microbes do not have vision as we know it, although they are able to use the energy of light of specific wavelengths in, for example, photosynthesis (chapter 10), and they may have the ability to move or grow towards or away from a light source (chapter 11). It is only in the animal kingdom that the mechanisms of light detection and colour discrimination have developed into the accurate and sensitive powers of vision that we know and enjoy.

### 1.2.3 The eye and colour vision

The processes of light detection and colour recognition by animals take place in the eye. In man and many other animals there are specific rod and cone cells in the retina of the eye; these cells contain the photoreceptors, or visual pigments. Mammalian retinal rod cells are responsible for the detection of low-intensity light. They contain pigments, **scotopsins**, which are sensitive to very low levels of light. The sensitivity maximum of the human scotopsin, **rhodopsin**, is at about 520 nm, although light of quite a wide

Fig. 1.2. The visible spectrum, showing the colours which individuals with 'normal' colour vision identify with particular wavelengths and also the complementary or subtraction colours observed when light of a particular colour or wavelength range is subtracted from the white light continuum.

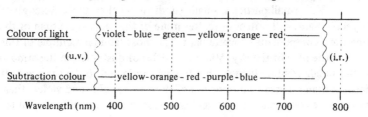

range of wavelengths around this value can be detected. In the visual process, light of the appropriate wavelength is absorbed by the visual pigment. This leads in turn to the generation of an electrical stimulus and a neural impulse. The same response is produced by light of all wavelengths that can be absorbed by the scotopsin, *i.e.* there is no differential response to light of different wavelengths.

The retinal cone cells, on the other hand, contain a set of visual pigments, **photopsins**, that are sensitive to different wavelength ranges. In man there are three such pigments sensitive to blue, green and red light, respectively. These three photoreceptors cover almost the entire range of the visible spectrum and provide a colour-discrimination mechanism sensitive enough to distinguish very subtle variations in colour, shade or hue. This trichromatic system and the pigments involved are described in more detail in chapter 9. Modern colour television also employs a trichromatic system to produce any colour, shade or hue by mixing red, green and blue light.

## 1.3    Colour in living organisms

When most living organisms or tissues receive white light, *e.g.* from the sun, they pass on to the eye of the observer light of only part of the visible range. In other words they appear to possess colour. This may be **structural colour**, produced as a consequence of the physical nature of the surface of the tissue. Alternatively the colour may be due to the presence of chemical compounds (**pigments** or **biochromes**) which absorb specifically some of the wavelengths of visible light.

### 1.3.1    *Structural colours*

In the animal kingdom there are many examples in which the observed colour is the result of optical phenomena such as light scattering, interference or diffraction by microscopic structures present in the tissues. Colours produced in this way are known as structural colours. The subject of structural colours is a large and important one, but detailed descriptions of the characteristics of structural colours and of the optical phenomena that produce them are not really within the scope of this book. Only a very brief account will be given.

### 1.3.2    *Light scattering – Tyndall blue*

Very small particles, smaller in diameter than the wavelength of red or yellow light, will reflect or scatter more of the short-wave than of the long-wave components of white light. The most familiar example of this effect is the blue of the sky. Minute particles of dust, *etc.*, in the atmosphere scatter incident white light so that the light reflected to the surface of the Earth contains a greater proportion of short-wave (blue and violet) than of longer-wave (red–yellow) light, and is thus seen as the familiar sky-blue colour. This process is often called Rayleigh or Tyndall scattering, and the colour produced is known as Tyndall blue.

Most non-iridescent blue colours in animals are Tyndall blues. Thus the blue colour of human eyes is due to the scattering of white light by minute protein particles in the iris. In the blue feathers of many birds, *e.g.* blue tit, budgerigar, parrot, light-scattering particles in the form of minute air-filled lamellae are present within the keratin of the feather barbs.

Tyndall blues are identified as structural colours by virtue of the fact that no blue pigment can be isolated from the tissues and because the blue colour is not evident when the tissues are viewed by transmitted white light. They are characterised by a matt, non-iridescent appearance, and by exhibiting the same colour when viewed from almost all angles.

Green colour, especially in feathers, is often due to superimposition of structural blue and a yellow pigment.

### 1.3.3   Iridescent colours

Among the most striking visual effects produced in Nature are the glittering, iridescent structural colours frequently encountered in the animal kingdom, particularly in birds, insects and fishes. It is a characteristic of iridescent colours that the observed hues change according to the angle of viewing. Two optical phenomena are involved, interference and diffraction.

*Interference.* The property of interference is perhaps known best from the example of a thin film of oil on the surface of water. Light reflected from the lower surface (oil–water interface) of the film travels a small but finite distance farther than light reflected from the upper, oil–air, surface. When the difference in distance travelled is equivalent to half the wavelength of the light, then the two light rays reflected from the upper and lower surfaces will effectively be 'out of phase' and will cancel each other out. Thus light of this particular wavelength will not be present in the reflected light observed. The reflected beam will therefore appear coloured. With a more acute viewing angle the travelling distance between the upper and lower surfaces is greater. The interference will therefore occur in a different part of the spectrum (longer wavelength) and hence a different colour will be seen.

There are many examples of interference colours in animals. The transparent wing structure of many insects serves as a thin film producing a range of interference colours when seen from different angles. Many butterflies have in the surface of their wing scales laminae with minute air spaces between them. The intralamellar distance is approximately constant so that an almost constant colour may be given over a reasonably wide range of viewing angles.

Interference colours are commonly found in birds, *e.g.* the peacock. The flattened feather barbules that contain the laminar structures which constitute the interference film are twisted so that their flat surface faces the viewer. In many cases the brilliance of the iridescent colours is enhanced by the presence of an underlying black surface (melanin) which absorbs all other light. A metallic lustre often results.

*Diffraction.* Rather similar iridescent colour effects may be produced by diffraction, although this is much less common than interference in natural tissues. The artificial diffraction grating used in certain optical instruments consists of a series of very close, equidistant, parallel lines scratched on to a polished surface. There are some examples of iridescent colours being produced by natural lamellar structures that behave as diffraction gratings, *e.g.* mother of pearl.

### 1.3.4    Structural white

The concept of structural white may be illustrated by the example of snow, which owes its brilliant white appearance to the reflection of incident white light from countless small crystal surfaces. Similar effects are given by reflection of white light by other solid or liquid particles or surfaces maintained in media of different refractive index. The particles must not be so small as to effect differential scattering of different wavelengths, *i.e.* Tyndall scattering. Some examples of structural whites in Nature are white hair (reflection from air bubbles trapped in a translucent solid), white feathers (reflection from numerous small colourless barbules), milk (reflection from droplets in an emulsion of two liquids of different refractive indices), white moths and butterflies (reflection from ribbed and reticulated scale surfaces) and white and silver fish (reflection from guanine crystals).

### 1.3.5    Chemical colour – natural pigments

Most natural colour results from the preferential absorption of some of the wavelengths of visible light by chemical substances present in the tissue. The chemical compounds responsible for this light absorption are the **natural pigments**. It is with the biochemistry of these natural pigments or **biochromes** that this book is concerned. The natural pigments are, almost without exception, organic molecules and differ greatly from the industrial pigments widely used in paints and dyes. Before details of the various groups of natural pigments are presented, however, it is necessary to consider in general terms those molecular properties which are responsible for the absorption of visible light.

## 1.4    Light absorption
### 1.4.1    Energy and wavelength

When an atom or molecule absorbs a quantum of electromagnetic radiation energy, the absorption results in a transition from one energy state to another. Any particle may occupy only certain discrete energy levels, and absorption is possible only when the radiation energy $E$ is equal to the energy difference between two of these energy states. However, according to 'selection rules' not all these transitions are 'allowed'; some are 'forbidden'. The absorption depends strictly on the energy of the quanta, and hence on the wavelength of the radiation, since

$$E = h\nu$$

where $E$ is the energy of the quantum, $h$ is Planck's constant and $\nu$ is the frequency of the radiation, which is related to the wavelength, $\lambda$, by the expression

$$\nu = c/\lambda$$

where $c$ is the velocity of light. There is thus an inverse relationship between energy and wavelength, in other words the greater the quantum energy difference between the two energy states, then the shorter will be the wavelength of the radiation required to bring about the transition. Light in the u.v. and visible regions of the spectrum is of sufficiently high energy to bring about electronic transitions, *i.e.* to promote an electron from a lower to a higher energy state. In addition, each electronic energy state of even the simplest diatomic molecule may occupy various vibrational and rotational energy levels. The vibrational and rotational energy differences are very much smaller and correspond to i.r. radiation quanta.

### 1.4.2    Radiation absorption by atoms
Absorption of radiation by an atom is relatively simple, and causes an electron to be raised from its normal lowest energy level (ground state) to a higher-energy excited state by a quantum whose energy is exactly equal to the energy difference between the two electronic energy levels. Because of the relationship between energy, frequency and wavelength, it follows that for a simple electronic transition such as this the radiation or light absorbed can only be of a single wavelength, and a single absorption line will be observed in the absorption spectrum.

### 1.4.3    Radiation absorption by molecules
In the case of molecules things are not so simple. Even diatomic molecules are quite large compared with atoms, and can no longer be considered as rigid particles. Molecular rotation and vibration of the nuclei occur, and the rotational and vibrational energy levels are again quantified (fig. 1.3). Thus for any electron in a molecule the electronic ground state may exist in several vibrational energy levels; for each of these in turn several rotational energy levels are possible. The same is true of the electronically excited state. Although the vibrational and rotational energy differences are small compared

to differences in electronic energy levels they must be taken into account in a discussion of electronic transitions. An electronic transition therefore does not have a single, sharply defined, energy requirement, but may be brought about by a range of energy quanta corresponding to the energy differences between the electronic ground state and excited state in the various vibrational and rotational energy levels. Electronic excitation may also be accompanied by vibrational and rotational excitation. The overall effect is for electronic absorption to occur over a range of radiation energies or wavelengths, so that for molecules the spectroscopic absorption lines are broadened into absorption bands, centred on the wavelength of maximal absorption ($\lambda_{max}$), and usually with a band-width of 50–100 nm. Normally it is not possible to resolve the absorption bands well enough to see the vibrational and rotational fine structure.

Fig. 1.3. Diagram illustrating electronic (and vibrational and rotational) energy levels and some possible electronic transitions for (*a*) an atom and (*b*) a diatomic or polyatomic molecule.

(*a*) An atom                    (*b*) A di–or polyatomic molecule

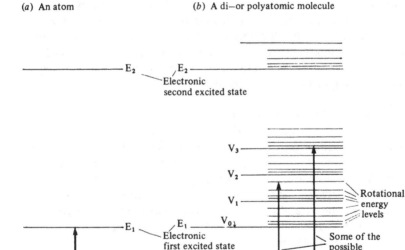

The excitation of an electron to a higher energy level takes place extremely rapidly (*ca* $10^{-15}$ s), and the much heavier atomic nuclei cannot move appreciably in this time. Electronic transitions therefore take place with essentially no change in the nuclear positions or internuclear distances (Franck–Condon principle).

Molecular vibrations are of a more or less harmonic form, so the vibrating nuclei will spend the greatest time at the extreme positions where the motion is slowest. Electronic transitions are therefore most likely to occur when internuclear distances are at their extreme maxima and minima, so quanta of the corresponding energy are those most likely to be absorbed, and light absorption at the corresponding wavelength will be the most intense.

Overall, the *position* of an absorption band $\lambda_{max}$ is determined by the energy required to bring about the most probable electronic transition, and the *intensity* of the absorption reflects the probability of that transition taking place.

### 1.4.4    Absorption properties of some simple molecules

Any electron in a simple molecule is capable of excitation. The energy required will depend primarily on the type of orbital occupied (fig. 1.4). Thus excitation of an electron in a $\sigma$-orbital requires the highest energy. Saturated hydrocarbons are therefore the most difficult simple organic molecules to excite electronically. The only accessible electrons are those in

Fig. 1.4. The five electronic energy levels normally encountered with organic molecules, and the most common electronic transitions that can be brought about by absorption of u.v. and visible light.

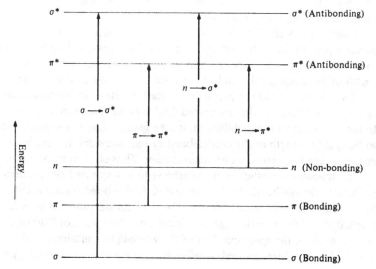

the σ-bonds, and a large absorption of energy is required to promote a
σ-bonding electron to the high-energy antibonding $\sigma^*$-orbital. Thus in a
simple molecule such as methane, $CH_4$, an energy input of around 600 kJ/mole
is required for electronic excitation. This degree of energy can be supplied
by electromagnetic radiation in the u.v. region, in the case of methane
(vapour) at *ca* 122 nm.

A carbon–carbon double bond, C=C, may be considered as a σ-bond and
a π-bond. Excitation of an electron from a bonding π-orbital to an anti-
bonding $\pi^*$-orbital is much easier to achieve than a $\sigma \rightarrow \sigma^*$ transition. The
smaller amount of energy required can be supplied by u.v. radiation of
somewhat longer wavelength (162 nm for ethylene, $CH_2=CH_2$).

When a heteroatom such as oxygen or nitrogen is present in either a
saturated or unsaturated molecule, a non-bonding, 'lone-pair', p electron
(designated $n$) on the heteroatom may be promoted to a $\sigma^*$-antibonding
orbital. These $n \rightarrow \sigma^*$ transitions require an even smaller energy input and
can be achieved by the absorption of electromagnetic radiation of even longer
wavelength (183 nm for methanol, $CH_3OH$).

The transition which requires the least energy is the $n \rightarrow \pi^*$ transition in
molecules containing C=O unsaturation. Thus acetone, $(CH_3)_2C=O$, has an
$n \rightarrow \pi^*$ absorption maximum at 280 nm.

### 1.4.5     Absorption properties of conjugated polyenes

In the case of complex molecules, just as for the simple examples
discussed above, the easier it is for an electron to be excited, then the smaller
will be the amount of energy required and the longer will be the wavelength
of the light that can effect the electronic promotion. This can be illustrated
conveniently by a simple qualitative treatment of the light-absorption
properties of conjugated polyenes.

If, instead of a single molecular double bond, a series of conjugated double
bonds is present, then the π-electrons of the double bonds may be considered
to be delocalised over the entire conjugated double-bond system. As the
length of the conjugated double-bond system increases, so does the extent of
stabilisation due to the resonance delocalisation, although for maximum
resonance stabilisation the conjugated double-bond system must be planar.
In the excited state this stabilisation is even greater than in the ground state
so that, as the length of the double-bond system increases, the energy difference between the two states becomes smaller. Thus electronic excitation
($\pi \rightarrow \pi^*$ transition) is much easier to achieve in a conjugated system than with
an isolated double bond. As the conjugated double-bond system is extended,
the excitation energy requirement will become smaller, and a stage will be
reached at which excitation can be achieved by absorption of light of the
visible region of the spectrum. Part of the white-light continuum will be
absorbed, the rest transmitted or reflected, and the compound will appear

coloured. As the length of the conjugated double-bond system increases, the wavelength of maximal absorption increases, and the observed colour changes: yellow → orange → red → purple. This effect is beautifully illustrated by the carotenoid group of natural pigments (see chapter 2).

The structural feature mainly responsible for the absorption of light is called the **chromophore**. In the case of the polyenes outlined above the chromophore is the conjugated double-bond system. Other functional groups or substituents in the molecule may modify (especially increase) the absorption maximum; such groups have been called **auxochromes**. These may either serve to extend the length of the chromophore, or they may make electronic transitions rather easier by contributing to the stability of the excited state.

### 1.4.6    Light absorption by some groups of natural pigments

The carotenoids (see chapter 2) are polyenes which have an extended conjugated double-bond chromophore. Other groups of natural pigments owe their colour to other chromophores. In most cases a conjugated or aromatic $\pi$-electron system is involved, and additional electron-donating or -attracting groups are present; nitrogen and oxygen atoms are especially important. With

Fig. 1.5. Some ways in which energy can be lost from the first excited singlet state. Vibrational relaxation, fluorescence, intersystem crossing, and phosphorescence are shown.

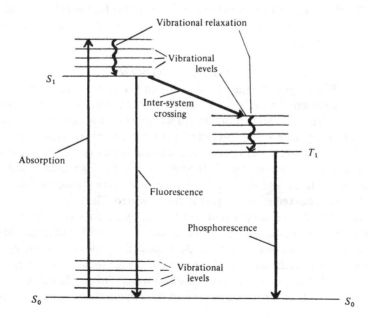

molecules of this kind, charge-separated species can make a substantial contribution to the overall resonance structure, resulting in a high degree of stabilisation, especially of the excited state. Excitation energy requirements are therefore low and light in the visible region of the spectrum is absorbed. Quinones and analogous systems are good examples of this effect and form the basic structures of many natural pigments. In other classes the contribution of non-bonding electrons of oxygen and nitrogen atoms to a heteroaromatic system is the most significant feature.

Particularly intense absorption is given when a molecule has a dipole moment which is different in the ground and electronically excited states, *i.e.* excitation results in a redistribution of charge. The greater this redistribution or transition dipole moment, the more intense is the light absorption band produced. The transition dipole moments are vectorial. For a complex molecule, several different electronic excitations may be possible, with different transition dipole moments, and different spatial orientations.

When pigment chromophores are in close proximity, electronic (exciton) interactions are possible between them due to coupling of the transition dipole moment of one molecule with the corresponding transition moment of the other similar or identical molecules. This results in the production of different excited electronic energy levels and splitting of the absorption bands. Similar exciton effects on circular dichroism bands are also observed. The detection of exciton splitting can give useful information about the relative orientations of chromophore molecules, *e.g.* evidence of molecular stacking.

The electronic and structural features responsible for the light-absorption properties of the main groups of natural pigments will be discussed in the relevant chapters.

### 1.4.7    *Triplet states, fluorescence and phosphorescence*

When pigment molecules absorb light energy they are promoted to higher-energy excited electronic states. These excited states with electrons in antibonding orbitals are unstable and will give up their excess energy to return to the ground state. This can take place in several ways (fig. 1.5). First a very rapid ($10^{-12}$ s) vibrational relaxation occurs as the electronically excited molecule loses excess vibrational energy to the surroundings and comes down to a lower vibrational state. There are several processes that this vibrationally relaxed excited state can then undergo. The simplest to envisage is a return to the electronic ground state either by emission of radiation as **fluorescence**, or by passing the excitation energy on to some other molecule, which therefore becomes electronically excited. Since vibrational energy has already been lost, the amount of energy available for fluorescence emission or transfer is always smaller than that which was absorbed originally. Fluorescence therefore occurs at a longer wavelength than absorption. Likewise, energy can only be passed on to a molecule with absorption maxima at longer wavelength than those of the originally excited molecule, *e.g.* from carotenoid ($\lambda_{max}$ 450 nm) to chlorophyll ($\lambda_{max}$ about 680 nm) in photosynthesis.

A second possibility is decay through a lower-energy triplet state. Excited states produced by absorption of light usually have all electrons paired and are described as **singlet states**. A **triplet state** has two unpaired electrons and is therefore normally more stable than the corresponding singlet state (Hund's rule). Conversion of the lowest singlet excited state ($S_1$) into the triplet state ($T_1$) is therefore energetically favourable, though it is of low probability and hence usually occurs rather slowly. However, if the singlet state is sufficiently long lived, the singlet–triplet change ($S_1 \rightarrow T_1$), often called **intersystem crossing**, may occur for a very considerable proportion of the excited singlet molecules. Triplet states are usually much longer lived than the original singlet excited states.

In some cases the triplet state, like the excited singlet state, returns to the ground state ($S_0$) by a radiative transition ($T_1 \rightarrow S_0$) even though this process has low probability. Because of the smaller energy change in such a transition, the light emitted is of considerably longer wavelength than either that absorbed originally or that emitted by fluorescence, and is called **phosphorescence**. As the $T_1 \rightarrow S_0$ transition has a low probability, and the triplet state may be long lived, phosphorescence may persist for several seconds, as opposed to fluorescence, which normally takes place within $10^{-9}$–$10^{-7}$ s of the original absorption.

In biological systems, loss of excess energy by processes that do not involve the emission of radiation is especially important. The excess energy may be transferred from either the singlet excited state, $S_1$, or the longer-lived triplet state, $T_1$, to other molecules close by, and may be used to bring about chemical reactions. It is such radiationless or non-radiative processes that are responsible for the conversion of harvested light into chemical energy in photosynthesis and for many other important processes in biological tissues (see chapters 9–11).

## 1.5    Spectroscopic methods in natural pigment research

It is clear from the above brief discussion that light absorption is of fundamental importance in studies of natural pigments. Electronic absorption spectroscopy, which monitors the absorption of light of u.v. and visible wavelengths, is the obvious basic spectroscopic method to use both for the characterisation of pigments and for their quantitative analysis. However, their special light-absorbing properties make many pigments also particularly amenable to study by other methods, notably resonance Raman spectroscopy and circular dichroism. In addition, infrared (i.r.) and nuclear magnetic resonance (n.m.r.) spectroscopic methods and mass spectrometry are extremely useful, as they are for all organic molecules.

### 1.5.1    Visible-light absorption spectroscopy

Natural pigments, by definition, absorb light in the visible region of the electromagnetic spectrum, between 380 and 750 nm wavelength. The visible-light absorption spectrum will therefore show at least one absorption maximum at a wavelength ($\lambda_{max}$) which is characteristic of the chromophore of the molecule. This, and the overall shape of the spectrum, give useful information about the molecular structure, and are commonly used as a first means of identifying the pigment. The $\lambda_{max}$ is strongly dependent upon the solvent used and in some groups of pigments also upon pH. The absorption spectra of pigments *in vivo* are frequently influenced by the immediate micro-environment of the molecule.

Some details of the light absorption spectroscopic properties of the various groups of pigments, and of some individual examples, will be given in the relevant chapters of this book.

In research into the functioning of pigments, use is made of very sophisticated developments of the basic light absorption spectroscopic method. These make it possible to study events which take place very rapidly (picoseconds – nanoseconds). The system is pulsed with a very short, intense flash of light, and changes in the light absorption spectrum which occur as a consequence of the flash treatment are followed. Such methods have proved extremely valuable in studies of the primary reactions of photosynthesis.

### 1.5.2    Quantitative spectrophotometric analysis

Light absorption spectra are also extremely valuable for the accurate, sensitive and reproducible quantitative analysis of pigments. The intensity of an absorption band at any wavelength is recorded experimentally as the absorbance, extinction, absorption, or optical density of a solution, and is directly proportional both to the concentration of the pigment in the solution and to the distance that the light path travels through the solution (Beer and Lambert laws).

Two coefficients are used in this analysis. The molar absorbance or extinction coefficient is defined as the absorbance of a 1M solution of a compound in a 1 cm light path, whereas the specific absorbance or extinction coefficient, $A_{1\,cm}^{1\%}$ or $E_{1\,cm}^{1\%}$, is the absorbance of a 1% (w/v) solution in a 1 cm light path. Reference tables of these coefficients are available for most groups of natural pigments.

The concentration of any solution can be obtained from its absorbance and the standard absorbance coefficient by simple proportionality. Thus the amount $(x \, \mathrm{g})$ of pigment in $y$ ml of solution is given by

$$x = (A \times y)/(A_{1\,\mathrm{cm}}^{1\%} \times 100)$$

where $A$ is the recorded absorbance of the solution.

### 1.5.3  Difference spectra

Two samples of the same material maintained under different conditions may show slight displacements of absorption wavelengths and small changes in absorption intensities, though it may be difficult to see these differences when the spectra are examined separately. The variations are, however, much more readily apparent when one sample is used as the reference against which the spectrum of the other sample is determined. A difference spectrum obtained in this way is a very sensitive means of detecting small variations in light-absorption properties. For example, light–dark difference spectra which compare the light absorption of an illuminated sample with that of a sample maintained in the dark have been extremely valuable in identifying minor absorbance changes which occur on illumination of photosynthetic tissues or particles. Also oxidised–reduced difference spectra have been used to give information about the involvement of cytochromes in electron transport chains, and about the predominant redox state of individual cytochromes under specified conditions. Much information about the physical states of pigments and their functioning in photosynthesis and electron transport has been obtained by this basic technique and the many sophisticated developments from it.

### 1.5.4  Raman spectroscopy

In Raman spectroscopy it is vibrational and rotational energy changes that are detected, though u.v. and visible-light absorption are used to excite these transitions. When monochromatic light is passed through a sample some light will be scattered at right angles to the incident beam. When this process is examined it is found that the wavelength of the scattered light is not identical to that of the incident light because of changes which occur in the molecular vibrational and rotational energies during the very brief time that the incident light photons are 'captured' by the molecules of the sample. Since these wavelength differences are due to vibrational and rotational energy changes, their values correspond to the absorption or emission of i.r. radiation. A spectrum is obtained, containing numerous absorption bands (Raman lines), each corresponding to a particular vibrational mode in the molecule, *e.g.* C=C stretching at around $1500 \, \mathrm{cm}^{-1}$. It is significant, however, that i.r. absorption and Raman spectra are by no means always identical. The differences may in fact give valuable information about molecular symmetry, since a symmetrical molecule or vibration does not give an i.r. absorption band but may give a strong Raman band.

In work with pigments the technique usually used is resonance Raman spectroscopy. When the wavelength of the incident or exciting light approaches a light absorption maximum of the sample, the momentary capture of the light quanta is much more likely, so the scattering of light is very greatly enhanced. This gives rise to Raman lines of greatly increased intensity. In a sample containing a mixture of compounds, resonance enhancement is obtained only for the Raman lines which arise from the molecules that are excited by the incident u.v./visible light. Information is therefore obtained about the light-absorbing molecules selectively; other molecules which do not absorb light of the exciting wavelength do not give resonance-enhanced Raman lines. The resonance Raman method is thus particularly useful for investigating pigments *in situ*. Pigments in shell, skin, *etc.*, can be detected and analysed quantitatively in this way, without the need for extraction from the tissue and extensive purification.

Another field of application is in studies of pigment chromophore–protein interactions, since information about the chromophore molecule can be obtained selectively. In particular it is possible to detect perturbations to the ground state of the chromophore molecule which occur when this is bound to the protein or as a result of other changes in the microenvironment. The properties of the chromophore in its natural environment can thus be studied. This approach has been extremely useful in studies of the mechanism of retinaldehyde–opsin binding in the visual pigments (see chapter 9).

*1.5.5    Linear dichroism*

All the above considerations of light absorption have been concerned with natural, unpolarised light. Various phenomena associated with the absorption of polarised light can also give valuable information.

According to the electromagnetic theory, a light wave consists of electric and magnetic vector components which are at right angles to each other and to the direction of propagation. The frequency of oscillation is the frequency of the radiation. In a beam of light from a natural or ordinary incandescent source all possible planes of polarisation are present, but, if the beam passes through a polariser the only components transmitted are those in the allowed electric or magnetic vector planes. In general, a pigment sample in which the chromophore groups are disposed randomly will absorb light of a particular wavelength to the same extent whether the light is polarised or not. If asymmetry is introduced, by virtue of regular orientation of the chromophores in a natural structure, then the absorbance will depend upon the plane of polarisation of the light beam. Two planes of polarisiation, at right angles to each other, will be found which give the maximum and minimum absorbance values, respectively, and a dichroic ratio can be obtained. This phenomenon is known as **linear dichroism**. Linear dichroism studies have proved very useful in investigations of the orientation of pigment chromophores in ordered biological structures, especially in photosynthetic pigment–protein complexes.

### 1.5.6    *Circular dichroism*

Related to linear dichroism is the extremely valuable **circular dichroism** (c.d.), which is useful not only for determining molecular asymmetry in the pigment molecules themselves but also for detecting asymmetry induced in the chromophore by, for example, binding to protein. When two plane-polarised light beams of the same wavelength and amplitude, but differing in phase by $\pi/2$ (or $3\pi/2$ to produce the opposite rotation), and with their planes at right angles are combined, the resulting light beam is circularly polarised in a left- or right-handed sense. An asymmetric or optically active pigment will not absorb right- and left-circularly polarised light to the same extent. This effect (circular dichroism) varies with the light wavelength so that a c.d. spectrum can be obtained. Circular dichroism is observed only in those regions of the spectrum where there are absorption bands. Somewhat similar information is given about the optical activity of pigment molecules by the technique of optical rotatory dispersion (o.r.d.), which measures the variation of rotation of plane-polarised light with wavelength. These techniques are extremely useful in organic chemistry for the determination of relative and absolute configurations of chiral molecules.

Circular dichroism, however, is not only useful for detecting optical activity in chiral molecules. It is particularly valuable for revealing asymmetry that is induced when a normally symmetrical or achiral molecule is incorporated into an organised structure, *e.g.* bound to protein, in an asymmetrical way. The asymmetry induced may result in a differential absorption of the right- and left-circularly polarised light, *i.e.* c.d. may then be seen in the region of the main absorption bands of the chromophore. This can reveal the presence of different forms of a pigment *in situ*, and may provide information about the conformational distortion of the chromophore molecule due to the binding or structural organisation.

### 1.6    **The significance of colour in Nature**

Many natural pigments are involved in important metabolic or physiological processes. The functioning of chlorophyll and other pigments in photosynthesis and the role of haemoglobin as an oxygen carrier are particularly well known. In many cases, however, the only known function of a pigment may be to bestow colour upon the organism or part of the organism that contains it.

In the plant kingdom, flowers and fruits of brilliant colours that contrast with the background green of the foliage advertise their presence to insects and other animals. From this the plant gains an advantage in pollination and seed dispersal. In the animal kingdom, pigmentary and structural colours may serve either to advertise or to conceal.

The significance of colour in Nature will be discussed in more detail in chapter 8.


### 1.7    Conclusions and comments

Any attempt to explain the many ways in which living organisms react and respond to, or use, light requires an understanding of the interactions between light and molecules. Those molecules with the special property of absorbing visible light are the natural pigments. Without these pigments, and these interactions, life as we know it would not be possible. Colour is important. This chapter is intended to give a very brief outline of the physical basis of colour and of the physical and chemical means by which natural colour can be produced and detected.

It is relatively simple to determine whether a given colour effect is chemical (*i.e.* pigmentary) or physical (structural) in origin. The identification and characterisation of a pigment is usually a standard organic chemistry exercise. The main features of the chemistry of the largest natural pigment groups are given in the following chapters of section I of this book. A much more challenging problem concerns the mutual interactions between the pigment molecules and their immediate microenvironment, *e.g.* association with proteins and in membranes. Refinement and sophisticated application of physico-chemical techniques such as resonance Raman spectroscopy, linear and circular dichroism, and nuclear magnetic resonance should allow this challenge to be taken up, and should also give information about molecular changes which are involved in the functioning of some pigments. Section II of this book is devoted to a survey of the functions of the natural pigments, both as colouring agents and in much more complex processes such as photosynthesis, vision and other photoresponses, where a time-scale of picoseconds may be involved.

Although a great deal is known about pigmentary colours, very few examples of structural colours have been studied in detail. Systematic investigation of the microscopic structures involved in the production of structural colours should be fascinating and revealing.


### 1.8    Suggested further reading

Most readers of a biochemistry book such as this will be satisfied with the general descriptive accounts of molecular light-absorption properties that can be found in most textbooks of organic chemistry and in monographs on photobiology, such as the two volumes by Clayton (1971). Some readers, however, may wish to have a deeper understanding of the physical processes involved and of the theoretical concepts which allow more exact definition and interpretation of these light-absorption properties. Such a theoretical treatment is provided by Murrell (1963).

There are many books and reviews which deal with the widely varied aspects of the chemistry, biochemistry and biology of the natural pigments as a whole and of individual pigment groups. This discussion includes only a small number of them, primarily the most extensive recent works which in turn give references to older publications. There are, however, some older works which make fascinating reading and give a feeling of the way that the study of natural pigments has developed. Much of the pioneering work on plant pigments was done by chemists, but the study of animal colours evolved from the work of the classical naturalist-collectors. Two articles by MacMunn (1883, 1890) and a book by Newbigin (1898) contain early extensive surveys of natural colours in animals. Increased knowledge of the chemistry of the animal pigments then gave Denis Fox (1953) the opportunity to produce an exhaustive monograph on the subject of animal colours ('biochromes'). This book has recently (1976) been brought more up to date by the addition of an appendix, and provides a very useful source of information on the distribution of animal pigments. A third book by the same author (Fox, 1979) gives an entertaining account of natural colours for the non-specialist. Another very readable book by H. M. Fox and Vevers (1960) also gives an entertaining broad account of the subject. A great deal of information on the physiological aspects of 'zoochromes' can be obtained from a substantial volume by Needham (1974). More limited in scope but nonetheless containing useful material is a monograph by Vuillaume (1969) on pigments in invertebrate animals.

Especially useful as a source of information on plant pigments is an advanced two-volume set edited by Goodwin (1976) which contains detailed accounts of the chemistry, distribution, biosynthesis and functions of the main classes of plant pigments, and invaluable chapters on the experimental methods used for their study. The second edition of a book edited by Czygan (1980) also deals with plant pigments in general. Other books and articles dealing with more specialised aspects, for example with each individual group of pigments, will be noted in later chapters.

Structural colours are not so well served as pigmentary colours. The general books by Fox (1976) and Fox and Vevers (1960) outline the main characteristics of structural colours and introduce the physical phenomena which produce them. A monograph by Simon (1971) also provides an introduction to the subject, and some recent detailed papers are worth consulting, *e.g.* Huxley (1975). However, for a proper understanding of the physical phenomena involved the reader is recommended to consult a textbook of physics or optics.

Spectroscopic methods now form an essential part of the training of an organic chemist, and consequently several textbooks have been devoted to these methods. As an example, Banwell (1972) includes sections on electronic absorption and Raman spectroscopy. An organic chemistry library will include other equally useful examples. For more information on circular dichroism the reader is referred to a monograph by Snatzke (1967), and for resonance Raman spectroscopy of biological molecules to reviews by Warshel (1977) and Carey (1978). Junge (1976) gives an excellent account of flash kinetic spectroscopy, especially as applied to photosynthesis. Details of the spectroscopic properties of different groups of pigments, and the application of spectroscopic methods to their study can be obtained from references quoted in the following chapters.

## 1.9    Selected bibliography

Banwell, C. N. (1972) *Fundamentals of molecular spectroscopy*, 2nd edition. London: McGraw-Hill.

Carey, P. R. (1978) Resonance Raman spectroscopy in biochemistry and biology, *Quart. Rev. Biophys.*, **11**, 309.

Clayton, R. K. (1971) *Light and living matter*, vols 1 and 2. New York: McGraw-Hill.

Czygan, F.-C. (Ed.) (1980) *Pigments in plants*, 2nd edition. Stuttgart and New York: Gustav Fischer.

Fox, D. L. (1953) *Animal biochromes and structural colours*. Cambridge University Press.

Fox, D. L. (1976) *Animal biochromes and structural colors*, 2nd edition. Berkeley, Los Angeles and London: University of California Press.

Fox, D. L. (1979) *Biochromy: natural coloration of living things*. Berkeley, Los Angeles and London: University of California Press.

Fox, H. M. and Vevers, G. (1960) *The nature of animal colours*. London: Sidgwick and Jackson.

Goodwin, T. W. (Ed.) (1976) *Chemistry and biochemistry of plant pigments*, 2nd edition, vols. 1 and 2. London, New York and San Francisco: Academic Press.

Huxley, J. (1975) The basis of structural colour variation in two species of *Papilio*, *J. Entomol.*, **50A**, 9.

Junge, W. (1976) Flash kinetic spectrophotometry in the study of plant pigments, in *Chemistry and biochemistry of plant pigments*, 2nd edition, vol. 2, ed. T. W. Goodwin, p. 233. London, New York and San Francisco: Academic Press.

MacMunn, C. A. (1883) Studies on animal chromatology, *Proc. Birmingham Nat. Hist. Soc.*, **3**, 351.

MacMunn, C. A. (1890) Contributions to animal chromatology, *Quart. J. Microsc. Sci.*, **30**, 51.

Murrell, J. N. (1963) *The theory of the electronic spectra of organic molecules*. London: Methuen.

Needham, A. E. (1974) *The significance of zoochromes*. Berlin, Heidelberg and New York: Springer-Verlag.

Newbigin, M. I. (1898) *Colour in nature*. London: John Murray.

Simon, H. (1971) *The splendor of iridescence: structural colors in the animal world*. New York: Dodd, Mead.

Snatzke, G. (Ed.) (1967) *Optical rotatory dispersion and circular dichroism in organic chemistry*. London: Heyden.

Vuillaume, M. (1969) *Les pigments des invertébrés*. Paris: Masson.

Warshel, A. (1977) Interpretation of resonance Raman spectra of biological molecules, *Ann. Rev. Biophys. Bioeng.*, **6**, 273.

# 2    Carotenoids

## 2.1    Introduction
Of all the various classes of natural pigments, carotenoids are probably the most widely distributed and are certainly among the most important.
They are found throughout the plant kingdom in both photosynthetic and
non-photosynthetic tissues, they are frequently encountered as microbial
pigments, and they are responsible, wholly or in part, for the colours of many
animals, notably birds, fish, and insects and other invertebrates. Carotenoids
and their derivatives are also of great importance in animals as the basis of the
visual pigments responsible for light detection and colour discrimination.

## 2.2    Structures and nomenclature
### 2.2.1    Basic structure
Almost all carotenoids are, or are derived from, **tetraterpenes**, $C_{40}$
compounds with a carbon skeleton built up from eight $C_5$ isoprene units
(2.1). The basic skeleton is symmetrical, consisting of two $C_{20}$ halves, and is
illustrated by lycopene (2.2), familiar as the red pigment of tomatoes.

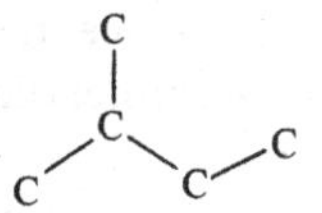

(2.1) Isoprene unit

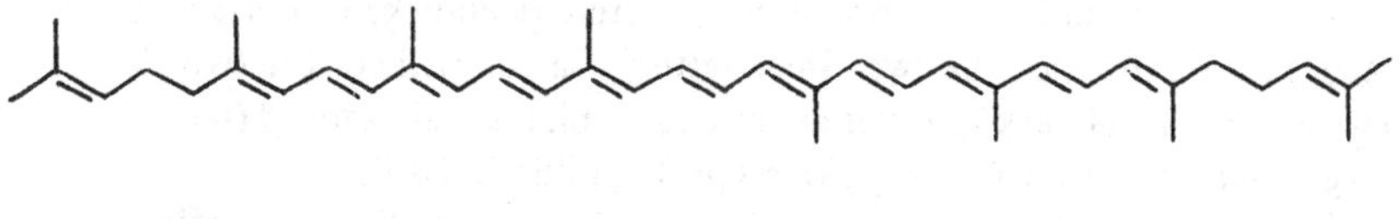

(2.2) Lycopene

This fundamental structure may be modified by the presence of a six-membered (or occasionally five-membered) ring at one or both ends of the molecule, *e.g.* β-carotene (2.3). This compound is the orange pigment of carrot roots, and is generally considered as the 'parent' of all the carotenoid group.

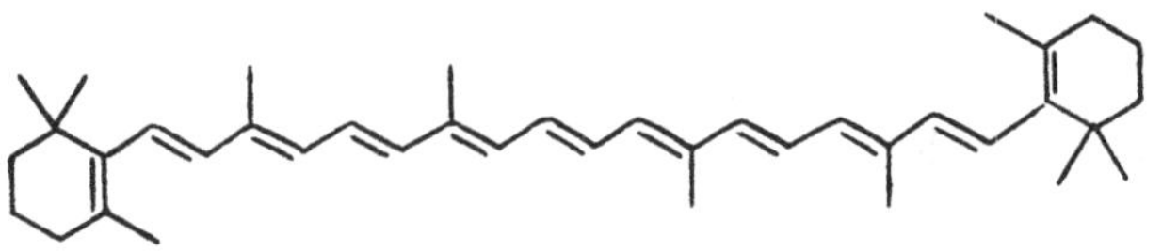

(2.3) β-Carotene

The carotenoid hydrocarbons are known as **carotenes**. All derivatives containing oxygen functions are **xanthophylls**. Most of the common oxygen-containing functional groups may be found in carotenoids, *e.g.* hydroxy-, methoxy-, epoxy-, oxo-, aldehyde and carboxylic acid; appropriate groups may be esterified or glycosylated. Substituents are usually located in the $C_9$ end-groups (see below). To date no carotenoids containing nitrogen, sulphur or halogen substituents have been found in Nature.

### 2.2.2     Nomenclature

Some 500 naturally occurring carotenoids have been identified and characterised. Many of these, especially some of the most important ones that have been known for many years, have well-established trivial names, usually derived from the biological source from which they were first isolated. In recent years a new, semi-systematic nomenclature has been introduced in an attempt to clarify the haphazard and often confusing trivial nomenclature.

In this book, well-known carotenoids will generally be referred to by their trivial names but, in line with current practice, the semi-systematic name of each carotenoid will be given when the carotenoid is first mentioned.

### 2.2.3     IUPAC rules for carotenoid nomenclature

All specific names are based on the stem name 'carotene', which corresponds to the structure and numbering shown in fig. 2.1. The several possible structures of the $C_9$ end-group are also illustrated. The name of a specific carotenoid hydrocarbon is constructed by adding, as prefixes to the stem name 'carotene', the two Greek letters characteristic of the two $C_9$ end-groups. The Greek prefixes are cited in alphabetical order: β (beta), γ (gamma), ε (epsilon), κ (kappa), φ (phi), χ (chi), ψ (psi).

Substituent groups are designated by prefixes and suffixes according to the standard rules of organic chemistry.

The basic system of numbering is that shown in fig. 2.1. If the two end-groups are dissimilar, lower (unprimed) numbers are given to that end of the molecule which is associated with the Greek letter prefix cited first in the name. Formulae should be drawn so that unprimed numbers are on the left-hand side.

Fig. 2.1. Basic carotenoid structure, $C_9$ end-group types, and numbering scheme according to IUPAC rules.

$C_9$ end-group types

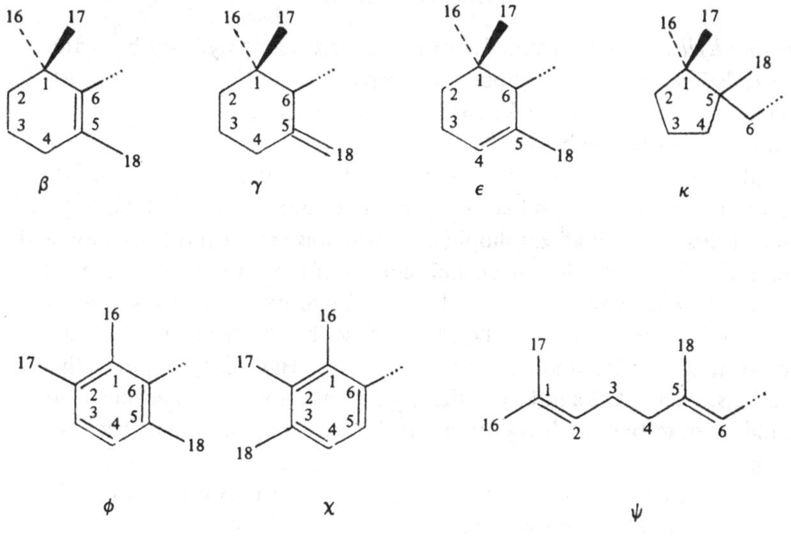

### 2.2.4 Some examples of carotenoid structures and nomenclature

*Carotenes.* The best known of all carotenoids, β-carotene (**2.3**), is described by the IUPAC system as β,β-carotene. The isomer α-carotene (**2.4**) becomes β,ε-carotene, and the acyclic lycopene (**2.2**) is named ψ,ψ-carotene.

Carotenoid hydrocarbons that differ in hydrogenation level from the parent carotenes as defined above are important biosynthetic intermediates.

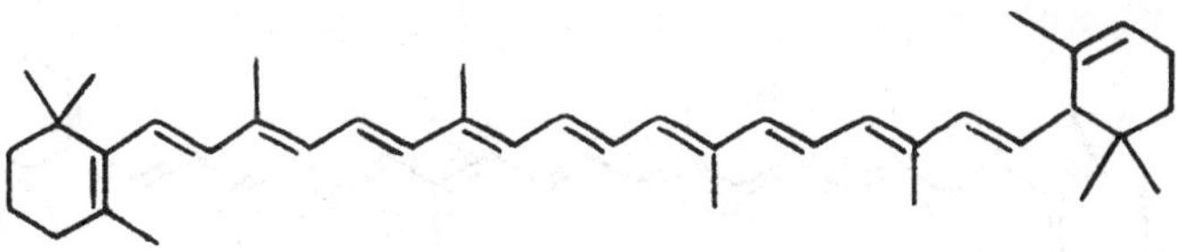

(**2.4**) α-Carotene

They are named from the parent carotene by use of the prefixes 'dehydro' and 'hydro' (with the appropriate multiplier) together with the locants (numbers) specifying the carbon atoms where hydrogen atoms have been added or removed. As an example, β-zeacarotene (2.5) becomes 7',8'-dihydro-β,ψ-carotene.

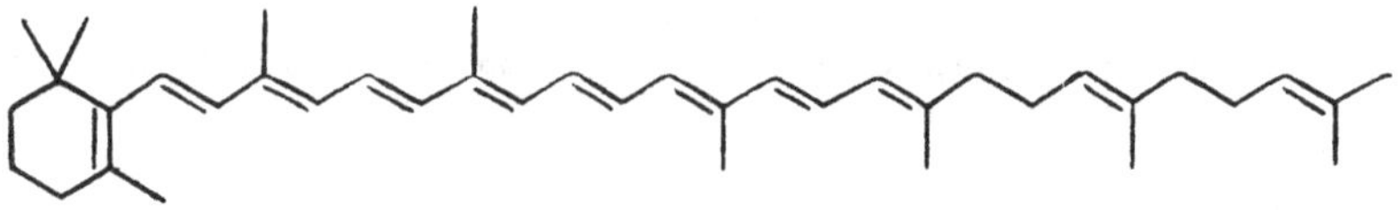

(2.5) β-Zeacarotene

*Xanthophylls.* All oxygenated derivatives of carotenoid hydrocarbons are now called xanthophylls, although the name 'xanthophyll' has in the past been used more specifically for carotenols and even for one particular compound, lutein (see below).

Only a few examples of the structures and nomenclature of some of the most important of the hundreds of naturally occurring xanthophylls will be given at this stage. Of all xanthophylls, carotenols are the most common and important. The most widespread and familiar of these are the leaf pigments lutein (2.6) and zeaxanthin (2.7) which are dihydroxy-derivatives of α-carotene and β-carotene, respectively, and are described by the new system as β,ε-carotene-3,3'-diol and β,β-carotene-3,3'-diol. Hydroxy-groups at other positions, such as C-2 and C-4 of the ring, are also found, and acyclic compounds with tertiary hydroxy-groups at C-1 are common in some bacteria (see §2.4.4).

Cyclic carotenoids may have epoxy-groups at C-5,6. A typical example is the chloroplast pigment violaxanthin (5,6,5',6'-diepoxy-5,6,5',6'-tetrahydro-β,β-carotene-3,3'-diol, 2.8).

Keto-groups in cyclic carotenoids are usually located at C-4, in conjugation with the polyene system. Astaxanthin (3,3'-dihydroxy-β,β-carotene-4,4'-

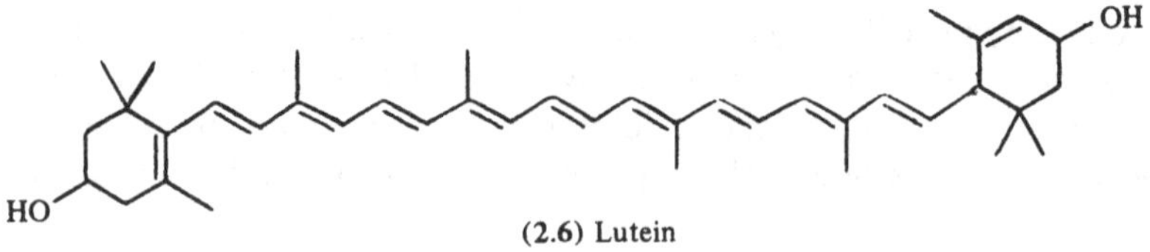

(2.6) Lutein

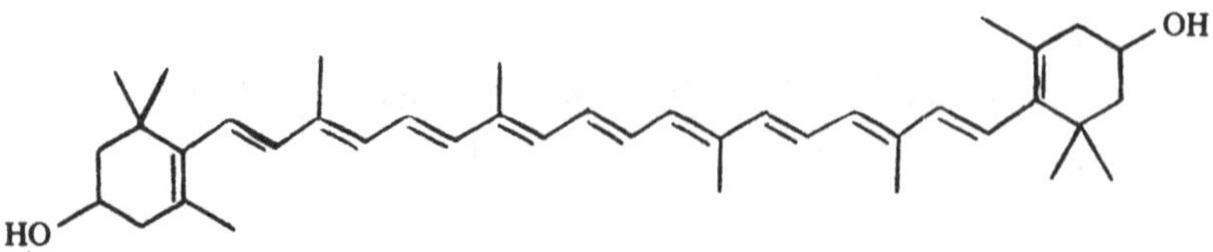

(2.7) Zeaxanthin

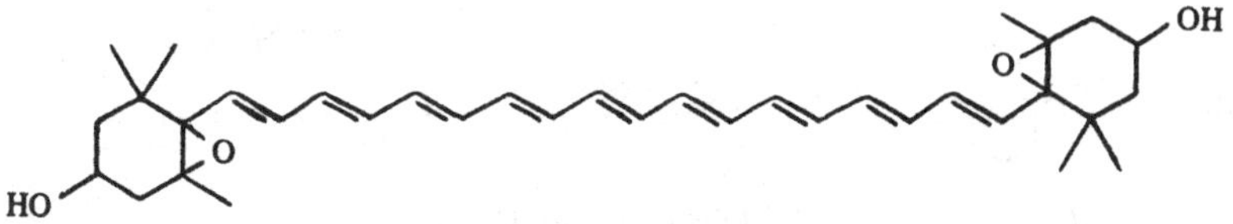

(2.8) Violaxanthin

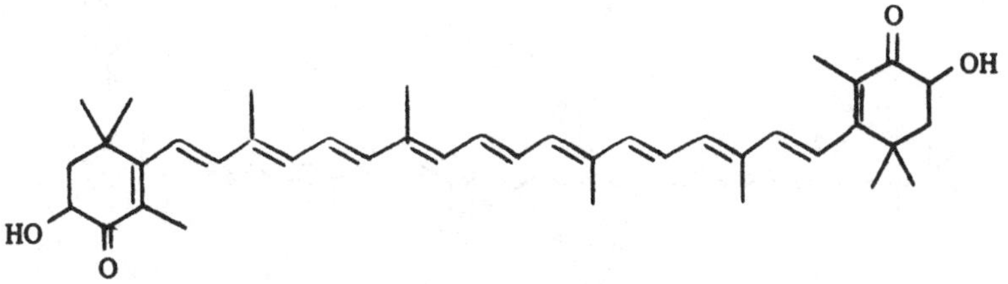

(2.9) Astaxanthin

dione, **2.9**), the characteristic carotenoid of many marine animals, is a good example.

Many other examples of xanthophyll structures and nomenclature will be encountered in later sections of this chapter.

*Retro-carotenoids.* The term *retro*-carotenoid is used to describe a structure in which a formal shift has occurred, by one position, of the single and double bonds of the conjugated polyene system. An example of a naturally occurring *retro*-carotenoid is rhodoxanthin (**2.10**) (4',5'-didehydro-4,5'-*retro*-β,β-carotene-3,3'-dione), which gives the red colour to the arils ('berries') of the yew tree. *N.B.* According to the new nomenclature, a pair of locants precedes the prefix '*retro*', the first locant being that of the carbon atom which has formally lost a proton, the second that of the carbon atom which has gained one.

*Homo-, apo-, and nor-carotenoids.* Although most carotenoids are $C_{40}$ compounds, there are some which have more or fewer than 40 carbon atoms. The $C_{45}$ and $C_{50}$ carotenoids which are found in some bacteria are often referred to as higher or 'homo-carotenoids'. Structurally they consist simply of a normal $C_{40}$ carotenoid skeleton with one or two additional $C_5$ substituents at C-2, C-2'. They may be cyclic or acyclic, for example decaprenoxanthin (2,2'-bis(4-hydroxy-3-methylbut-2-enyl)-ε,ε-carotene, **2.11**), and bacterioruberin (2,2'-bis(3-hydroxy-3-methylbutyl)-3,4,3',4'-tetradehydro-1,2,1',2'-tetrahydro-ψ,ψ-carotene-1,1'-diol, **2.12**).

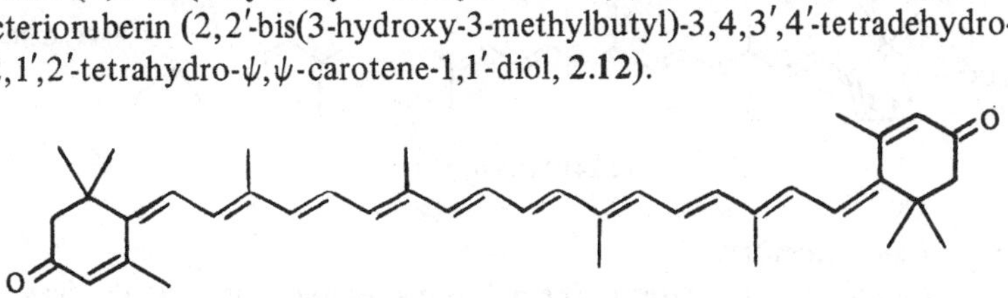

(2.10) Rhodoxanthin

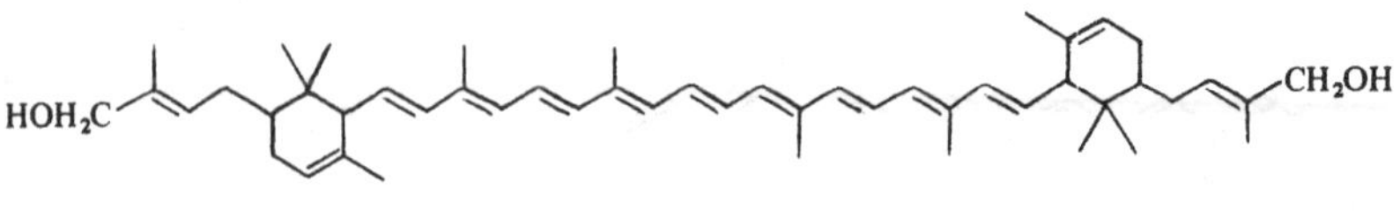

(2.11) Decaprenoxanthin

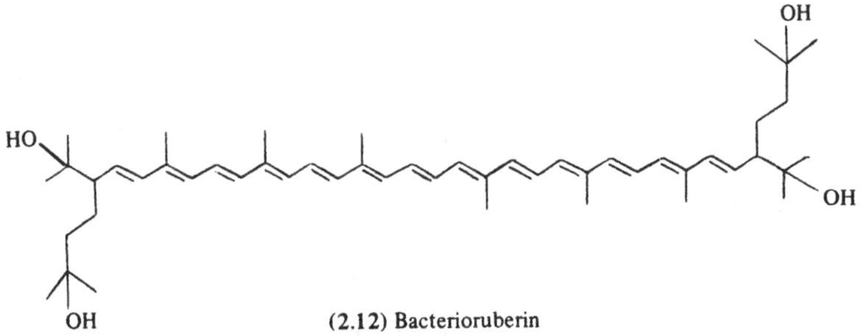

(2.12) Bacterioruberin

Carotenoids with fewer than 40 carbon atoms fall into two categories. A molecule in which the carbon skeleton has been shortened by the formal removal of fragments from one or both ends of a $C_{40}$ carotenoid is referred to as an apo- (or diapo-) carotenoid. An example is the $C_{30}$ compound β-citraurin (3-hydroxy-8'-apo-β-caroten-8'-al) (2.13). This and related apo-carotenoids are responsible for the colour of oranges and other citrus fruits.

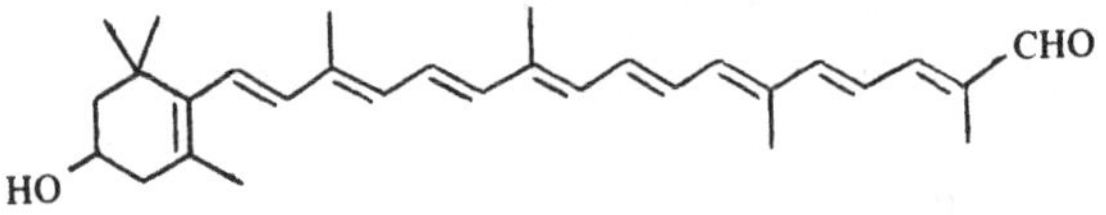

(2.13) β-Citraurin

A nor-carotenoid, on the other hand, is one which has lost a carbon atom or small group of carbon atoms not from the end of the molecule but from some internal portion of the carbon chain. For example, the sea anemone *Actinia equina* contains substantial amounts of a purple pigment, actinio-erythrin, which has lost the C-2 and C-2' carbon atoms and is therefore 3,3'-dihydroxy-2,2'-dinor-β,β-carotene-4,4'-dione 3,3'-diacylate (2.14).

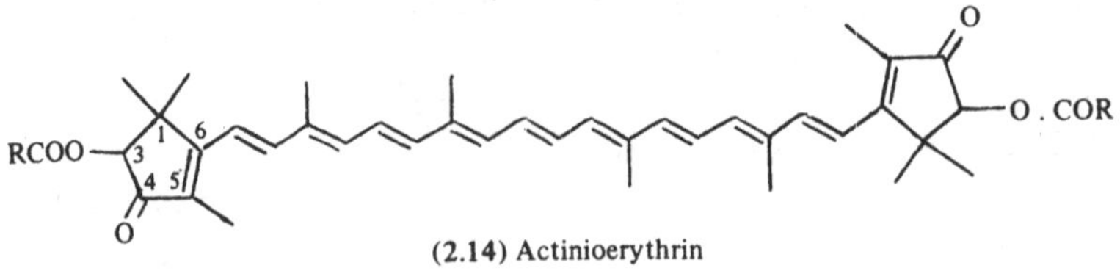

(2.14) Actinioerythrin

## 2.2.5    Stereochemistry

*Geometrical isomerism.* The polyene system of carotenoids gives extensive scope for the existence of large numbers of geometrical (*cis–trans*) isomers. Thus lycopene (**2.2**), a symmetrical molecule with 11 conjugated double bonds, is theoretically capable of existing in 1056 forms, and many more isomers are possible for unsymmetrical carotenoids. Fortunately many of the possible isomers are sterically hindered and therefore difficult to produce, but nevertheless there are still 72 possible 'sterically unhindered' isomers of lycopene to contend with (fig. 2.2). By convention the terms *cis* and *trans* are used in the carotenoid field to specify the relative disposition of those two double-bond substituents which form part of the main chain of carbon atoms.

Fig. 2.2. Sterically-hindered and '-unhindered' *cis* double bonds.

Hindered                'Unhindered'

In Nature, most carotenoids exist entirely or very largely in the all-*trans* (all-*E*) form. There are, however, some notable exceptions. Phytoene ($7,8,11,12,7',8',11',12'$-octahydro-$\psi,\psi$-carotene), generally considered to be the first $C_{40}$ hydrocarbon intermediate in carotenoid biosynthesis (see §*2.6.3*), is usually isolated as the 15-*cis* (*Z*) isomer (**2.15**). Perhaps the most remarkable *cis*-carotenoid is prolycopene, a poly-*cis* isomer of lycopene occurring naturally in several plant systems. The structure of prolycopene has recently been established as $7,9,7',9'$-tetra*cis*-lycopene (**2.16**).

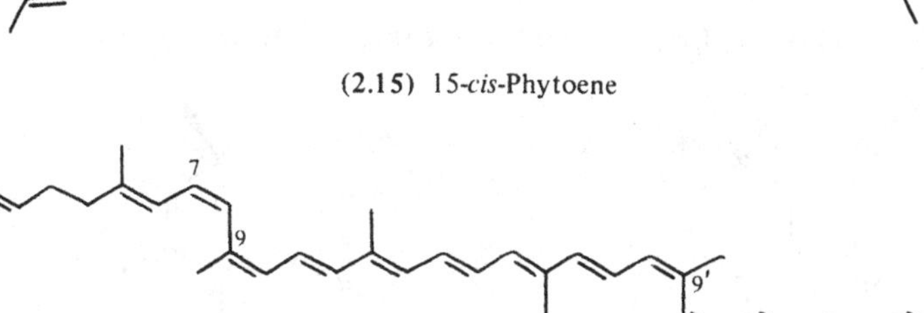

(2.15) 15-*cis*-Phytoene

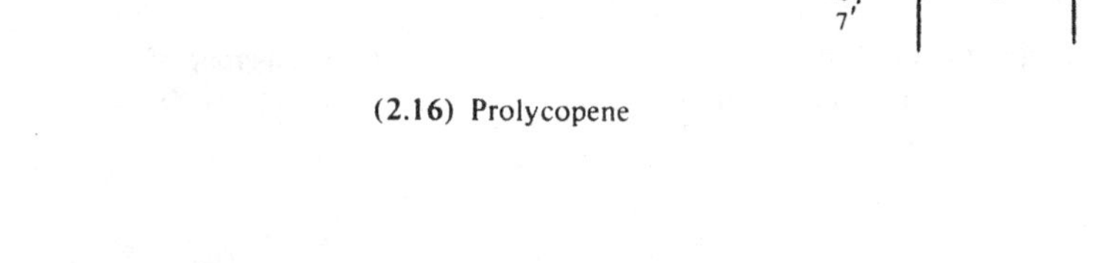

(2.16) Prolycopene

In general, traces of *cis* isomers present in natural extracts are usually considered to be artefacts produced by stereomutation of natural all-*trans* carotenoids.

The conformation of the constituent single bonds of the polyene chromophore is *s-trans*. In the cyclic carotenoids the C-6,7 bond usually adopts the *s-cis* conformation, both in the crystalline state and in solution. Other details of carotenoid conformation will be discussed later in relation to light-absorption properties (see §*2.3.3*).

*Absolute configuration.* Many organic molecules are chiral, that is they exist as only one of two possible enantiomers, stereoisomers which bear a mirror-image relationship but which are not superimposable. The most common example of chirality is the presence of an asymmetrically substituted carbon atom which theoretically can exist in two spatial configurations. Chirality is frequently observed in carotenoids, and can usually be attributed to the presence of an asymmetrically substituted carbon atom in the molecule. Examples of such chiral centres are C-6 of an ε-ring carotenoid (**2.17**) and C-3 of the common cyclic caroten-3-ols (**2.18**).

The absolute configurations of many chiral carotenoids have been determined, largely by chiroptical methods (optical rotatory dispersion, circular dichroism) and nuclear magnetic resonance (n.m.r.) studies. Recently examples have been found of the natural occurrence of different optical isomers of a carotenoid in different living organisms, *e.g.* the yeast *Phaffia rhodozyma* produces (3*R*,3′*R*)-astaxanthin (**2.19**), whereas the lobster accumulates mainly the (3*S*,3′*S*)-isomer (**2.20**), together with smaller amounts of the (3*R*,3′*R*) and *meso*-(3*R*,3′*S*) forms.

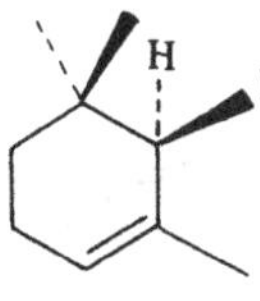

(2.17) (6*R*)-ε-Ring

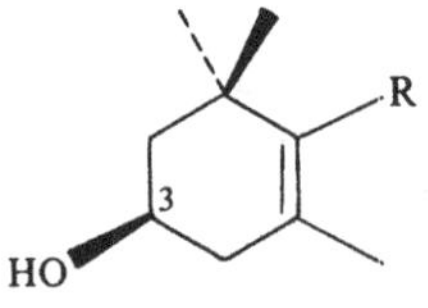

(2.18) (3*R*)-3-Hydroxy-β-Ring

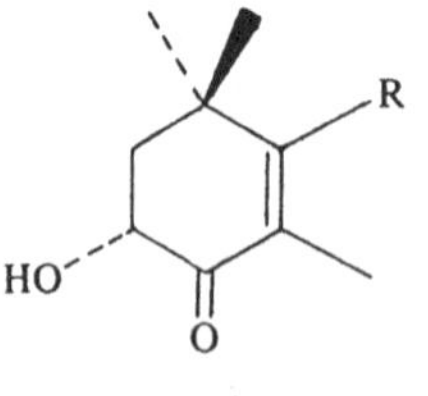

(2.19) (3*R*)-End-group of
astaxanthin (2.9)

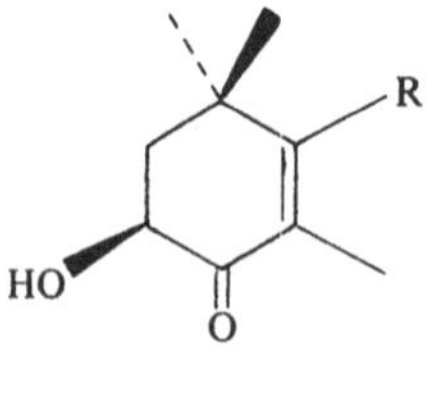

(2.20) (3*S*)-End-group of
astaxanthin (2.9)

## 2.3    Properties

### 2.3.1    *General physical properties*

Carotenoids are lipids. They are soluble in organic solvents and can be extracted from natural tissues by polar solvents such as acetone and alcohols. Even xanthophylls with four or more hydroxy-groups are virtually insoluble in water, but water-solubility can be achieved by glycosylation or by complexing with protein. *In vivo*, carotenoids are generally located in lipophilic, hydrophobic regions in the cell, such as lipid globules, crystalline structures and, in association with protein, in membranes.

Apart from the biosynthetic intermediates phytoene, phytofluene and usually ζ-carotene, the carotenoids are all solids at room temperature, and most can be crystallised from appropriate solvent mixtures.

### 2.3.2    *Stability*

Isolated carotenoids, either dry or in solution, are very sensitive to light and heat (which cause *cis–trans* isomerisation), to acid (which causes *cis–trans* isomerisation and especially converts 5,6-epoxides into 5,8-furanoid oxides, fig. 2.3) and in some cases to base (which causes autooxidation of 3-hydroxy-4-oxo end-groups, as for example in astaxanthin, fig. 2.4). The conjugated double-bond system renders carotenoids particularly susceptible to oxidative bleaching brought about by oxygen in the air.

*In vivo*, carotenoids are usually an integral part of the structure of the cell or organelle, and are present in association either with protein or with other cellular lipids. Under such conditions the carotenoids are stabilised and protected from the harmful influences listed above.

Carotenoproteins, especially the astaxanthin–protein complexes of invertebrate animals (see §2.5), are also much more stable *in vitro* than the free carotenoids.

Fig. 2.3. Acid-catalysed isomerisation of a carotenoid-5,6-epoxide into the 5,8-furanoid oxide.

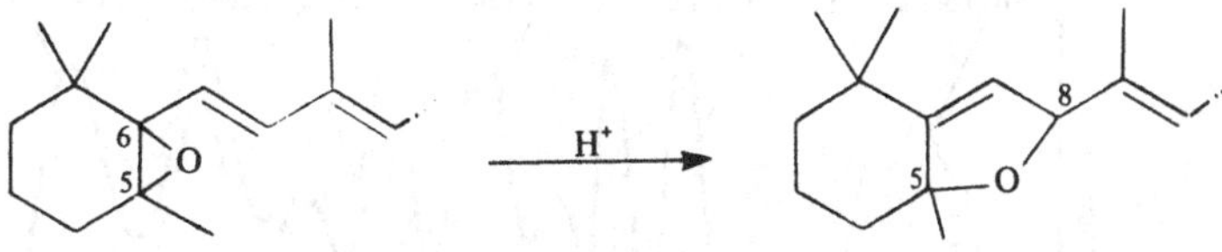

Fig. 2.4. Oxidation of a carotenoid 3-hydroxy-4-oxo end-group.

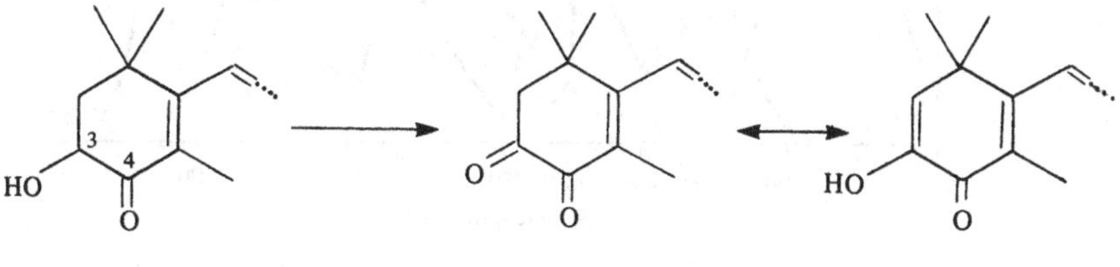

### 2.3.3    Light absorption

The chromophore responsible for the absorption of visible light by carotenoids is the conjugated double-bond system (see chapter 1). Increasing the length of the polyene $\pi$-electron system confers greater stability on the first excited state, so that electronic excitation is easier, requires less energy, and is brought about by light of longer wavelength.

This progressive **bathochromic effect** (movement to higher or longer wavelength) is illustrated by the absorption spectra of the biosynthetic series (see §2.6.4) of acyclic carotenoids of increasing chromophore length (fig. 2.5). Thus phytoene (**2.15**) (3 conjugated double bonds, c.d.b.) and phytofluene (7,8,11,12,7',8'-hexahydro-$\psi$,$\psi$-carotene, **2.21**) (5 c.d.b.) are colourless with $\lambda_{max}$ at 275, 285, 296 and 331, 348, 367 nm, respectively, ζ-carotene (7,8,7',8'-tetrahydro-$\psi$,$\psi$-carotene, **2.22**) (7 c.d.b.) has some absorption in the blue region of the spectrum ($\lambda_{max}$ 378, 400, 425 nm) and hence appears slightly yellow, and neurosporene (7,8-dihydro-$\psi$,$\psi$-carotene, **2.23**) (9 c.d.b.) and lycopene (**2.2**) (11 c.d.b.) absorb strongly in the visible region at 414, 439, 467 nm and 444, 470, 502 nm, respectively, and exhibit characteristic yellow and orange colours. These compounds all show the typical three-peaked carotenoid absorption spectrum with well-defined maxima and minima (fine structure or persistence).

(**2.21**) Phytofluene

Fig. 2.5. Light absorption spectra of acyclic carotenoids of increasing chromophore length; phytoene (**2.15**), phytofluene (**2.21**), ζ-carotene (**2.22**), neurosporene (**2.23**), lycopene (**2.2**) (in light petroleum).

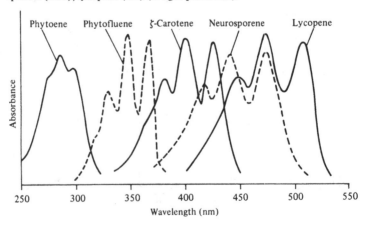

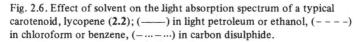

(2.22) ζ-Carotene

(2.23) Neurosporene

The absorption maxima and sometimes the degree of fine structure in the
spectrum are influenced considerably by the solvent used. Absorption
maxima are observed at longer wavelengths when spectra are determined
in aromatic or halogenated solvents or $CS_2$, than when light petroleum,
hexane or ethanol are used (fig. 2.6).

Cyclisation of lycopene to $\epsilon,\epsilon$-carotene (2.24) or $\gamma,\gamma$-carotene (2.25)
removes two double bonds from conjugation, so that the absorption spectra
of these cyclic carotenoids are very similar to that of neurosporene. In the
case of β-ring carotenoids such as β-carotene, the endocyclic double bond is
conjugated with the main polyene chain. However, because of steric hindrance
between the ring methyl substituents and the main polyene chain, the molecule is twisted about the C-6,7 single bond, so that the π-orbital of the ring
5,6-double bond is not coplanar with the polyene π-electron system. This

Fig. 2.6. Effect of solvent on the light absorption spectrum of a typical
carotenoid, lycopene (2.2); (———) in light petroleum or ethanol, (– – – –)
in chloroform or benzene, (– ··· – ···) in carbon disulphide.

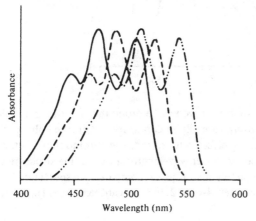

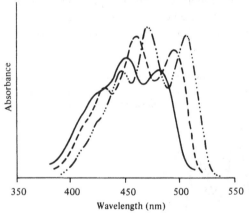

(2.24) ε, ε-Carotene

(2.25) γ, γ-Carotene

decreased π-orbital overlap means that the effective chromophore is less than
a full 11 conjugated double bonds, so that the absorption maxima occur at
shorter wavelengths than those of lycopene. The molecular distortion also
results in reduced fine structure, *i.e.* less pronounced maxima and minima
in the spectrum. In fig. 2.7 the spectra of lycopene, γ-carotene (β,ψ-carotene)
and β-carotene are compared.

Fig. 2.7. Effect of β-rings on the light absorption spectra of carotenoids;
spectra of β-carotene (2.3) (———), γ-carotene (2.78) (– – – –) and lycopene
(2.2) (– ··· – ···) (in light petroleum).

The effect on fine structure is even more pronounced with carotenoids
containing a carbonyl group formally in conjugation with the polyene system.
The conjugated C=O group effectively extends the chromophore so that
absorption maxima occur at longer wavelengths, but the fine structure of the
spectrum is almost completely lost. Fig. 2.8 compares the spectra of β-carotene, echinenone (β,β-caroten-4-one, 2.26) and canthaxanthin (β,β-carotene-
4,4'-dione, 2.27).

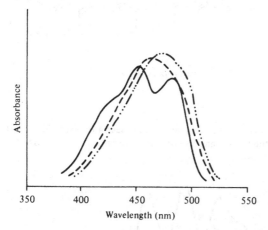

(2.26) Echinenone

(2.27) Canthaxanthin

Fig. 2.8. Effect of conjugated carbonyl groups on the light absorption spectra of carotenoids; (———) β-carotene (2.3), (– – – –) echinenone (2.26, 1 x C=O), (– ··· – ···) canthaxanthin (2.27, 2 x C=O) (spectra in light petroleum).

Other substituent groups, *e.g.* OH, generally have little or no effect on the absorption spectrum. Thus the spectra of β-carotene, β-cryptoxanthin (β,β-caroten-3-ol, 2.28), isocryptoxanthin (β,β-caroten-4-ol, 2.29), zeaxanthin (2.7) and isozeaxanthin (β,β-carotene-4,4′-diol, 2.30) are virtually identical in $\lambda_{max}$ and shape.

*Cis*-isomers exhibit absorption maxima of rather lower intensity and at somewhat shorter wavelengths than do the corresponding all-*trans* compounds (fig. 2.9). This occurs because the polyene system is distorted from

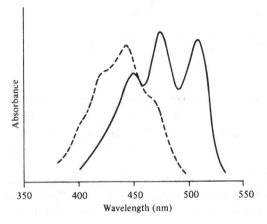

**(2.28)** β-Cryptoxanthin

**(2.29)** Isocryptoxanthin

**(2.30)** Isozeaxanthin

Fig. 2.9. Comparison of the light absorption spectra (in light petroleum) of all-*trans*-lycopene (———) and its poly-*cis* isomer prolycopene **(2.16)** (– – – –).

(2.31) Auroxanthin

coplanarity by partial rotation about a single bond adjacent to the *cis* double bond to relieve steric interference between respectively two hydrogen atoms ('unhindered *cis*') or between a hydrogen atom and a methyl group ('hindered *cis*') (fig. 2.2).

Shifts in absorption maxima as a result of chemical reactions performed on a very small scale in the spectrophotometer cuvette are very useful diagnostic tests. Thus treatment of a ketocarotenoid such as canthaxanthin (2.27) with sodium borohydride results in a **hypsochromic shift** in the spectrum (*i.e.* to lower or shorter wavelength) and an increase in fine structure as the canthaxanthin is reduced to isozeaxanthin (2.30) (fig. 2.10). Treatment of a 5,6-epoxycarotenoid such as violaxanthin (2.8) with traces of acid causes a spectral shift of 40 nm to lower wavelengths (20 nm for a monoepoxide) due to formation of the furanoid oxide auroxanthin (5,8,5′,8′-diepoxy-5,8,5′,8′-tetrahydro-$\beta,\beta$-carotene-3,3′-diol, 2.31) (fig. 2.11).

Carotenoids *in vivo* commonly exhibit absorption maxima of approximately 10 nm longer wavelength than in hexane or ethanol, as a result of their occurrence in association with a lipid or protein environment. The absorption

Fig. 2.10. Spectroscopic test for conjugated oxo-groups in a carotenoid. Light absorption spectra, in ethanol, of canthaxanthin ($\beta,\beta$-carotene-4,4′-dione, 2.27) (———) and its NaBH$_4$ reduction product isozeaxanthin (2.30) (- - - -).

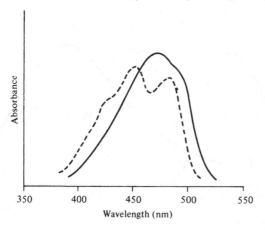

spectra *in vivo* often resemble those of artificially produced micellar suspensions of carotenoids with other lipids in aqueous medium.

The stoichiometric carotenoprotein complexes present in many invertebrate marine animals (see §2.5) show enormous shifts (*e.g. ca* 100 nm) in absorption maxima, and are thus often purple or blue in colour ($\lambda_{max}$ 550–630 nm) in contrast to the yellow-orange of the free carotenoid ($\lambda_{max}$ *ca* 470 nm) (fig. 2.12). No satisfactory explanation of this phenomenon has yet been advanced.

Fig. 2.11. Spectroscopic test for 5,6-epoxy-groups in a cyclic carotenoid. Light absorption spectra, in ethanol, of the diepoxycarotenoid violaxanthin (**2.8**) (——), and of auroxanthin (**2.31**) (– – – –) the furanoid oxide formed by addition of a drop of dilute HCl.

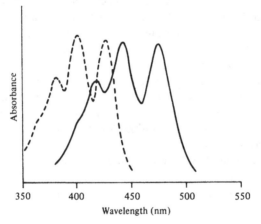

Fig. 2.12. Light absorption spectra of the carotenoprotein crustacyanin from lobster (——) (aqueous solution) and of the free carotenoid prosthetic group astaxanthin (**2.19**) (– – – –) (in ethanol).

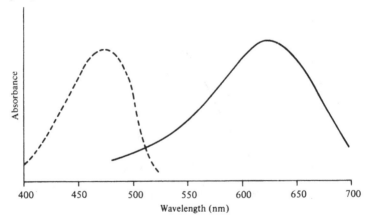

## 2.4    Distribution

Carotenoids have perhaps the widest distribution of all classes of natural pigments. They occur throughout the plant kingdom and are also commonly encountered as animal and microbial pigments.

### 2.4.1    Higher plants

*Photosynthetic tissues.* All green tissues of higher plants contain the same major carotenoids, which are located, probably exclusively, in the chloroplasts. These pigments are β-carotene (**2.3**), lutein (**2.6**), violaxanthin (**2.8**), and neoxanthin (5′,6′-epoxy-6,7-didehydro-5,6,5′,6′-tetrahydro-β,β-carotene-3,5,3′-triol, **2.32**) often accompanied by lesser amounts of α-carotene (**2.4**), zeaxanthin (**2.7**), β-cryptoxanthin (**2.28**), and antheraxanthin (5,6-epoxy-5,6-dihydro-β,β-carotene-3,3′-diol, **2.33**). The carotenoids are probably located in the chloroplast grana, as chromoproteins. Some, if not all, of the chloroplast carotenoids may be present *in vivo* in a number of different forms, at different sites within the chloroplast. The significance of this, and the possible functions of carotenoids in chloroplasts will be discussed in chapter **10**.

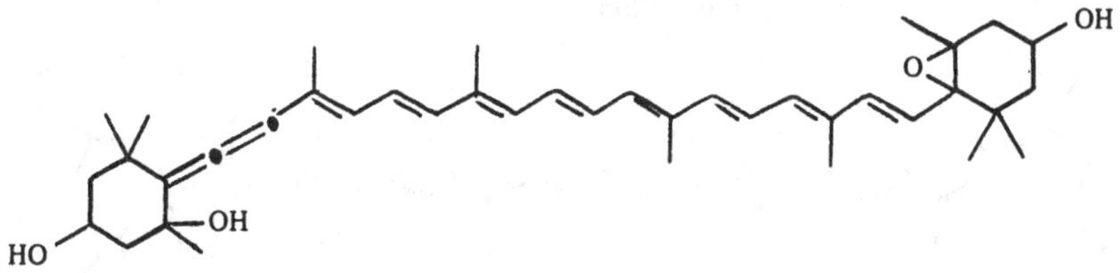

(2.32) Neoxanthin

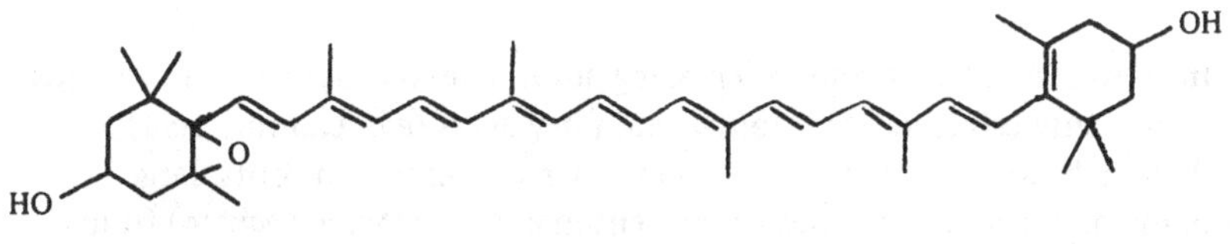

(2.33) Antheraxanthin

*Non-photosynthetic tissues.* Many yellow flowers and orange-red fruits owe their colour to carotenoids, which are normally located in chromoplast structures. Yellow flowers, *e.g.* daffodil, dandelion, often contain large amounts of carotenoid epoxides, such as violaxanthin (**2.8**), whereas orange (*e.g.* apricot) and red (*e.g.* tomato) fruits, respectively, are often coloured by β-carotene and lycopene or their simple hydroxy-derivatives. In several cases, however, pigmentation is due to the presence of carotenoids that occur in only one or very few species. Thus the blooms of the Californian poppy (*Eschscholtzia californica*) contain the *retro*-carotenoid eschscholtzxanthin

(4′,5′-didehydro-4,5′-*retro*-β,β-carotene-3,3′-diol, **2.34**), yew (*Taxus baccata*) arils ('berries') contain the *retro*-carotenoid ketone rhodoxanthin (**2.10**) and the familiar red of red peppers (*Capsicum annuum*) is derived from the cyclopentanone carotenoids capsanthin (3,3′-dihydroxy-β,κ-caroten-6′-one, **2.35**) and capsorubin (3,3′-dihydroxy-κ,κ-carotene-6,6′-dione, **2.36**).

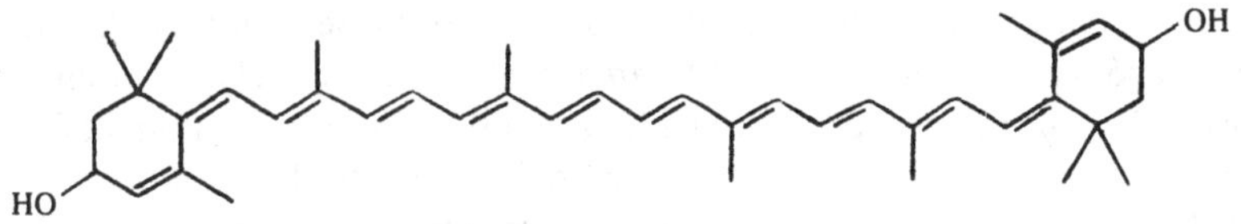

(2.34) Eschscholtzxanthin

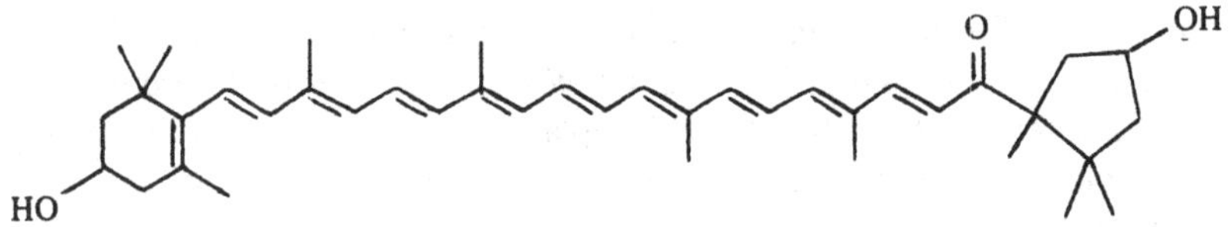

(2.35) Capsanthin

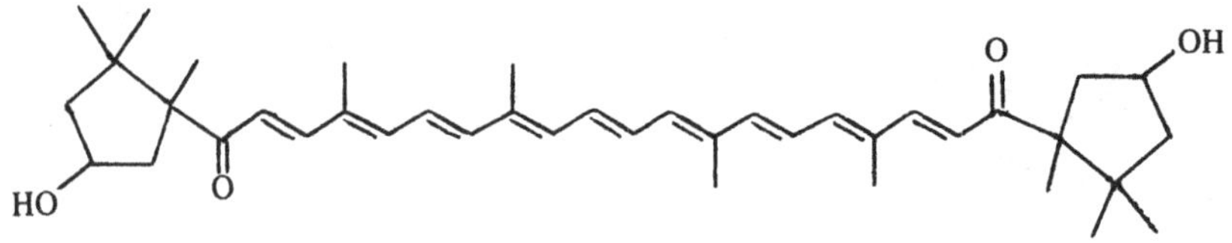

(2.36) Capsorubin

In flowers and fruits, and in senescing autumn leaves, hydroxycarotenoids are commonly esterified with long-chain fatty acids (*e.g.* palmitic acid).

Although carotenoids are not commonly encountered in roots, large amounts of β- and α-carotenes are present in carrots (*Daucus carota*) from which this class of pigments derives its name.

### 2.4.2    Algae

Carotenoids are normally found in the chloroplasts of algae, including the familiar seaweeds. The chloroplast carotenoids of the various classes of algae show considerable qualitative differences, and carotenoid compositions have been used in attempts to solve some problems of algal taxonomy and evolution. Carotenoids are sometimes present also outside the chloroplast, *e.g.* in the light-sensitive 'eye-spot' of *Euglena* (see chapter 11) and in reproductive areas of colonial species, such as *Ulva*, the green 'sea lettuce' commonly found in inshore waters and rock pools. Ketocarotenoids, including astaxanthin (**2.9**), may accumulate outside the chloroplast in some green algae (Chlorophyta) under adverse cultural conditions, notably mineral or nitrogen deficiency.

The chloroplast carotenoid compositions of green algae are generally similar to those of higher plant chloroplasts, suggesting a strong evolutionary link. Most other algal classes produce acetylenic or allenic carotenoids; the annual natural production of fucoxanthin (5,6-epoxy-3,3′,5′-trihydroxy-6′7′-didehydro-5,6,7,8,5′,6′-hexahydro-$\beta$,$\beta$-caroten-8-one 3′-acetate, **2.37**) the characteristic carotenoid of the brown seaweeds (Phaeophyta) has been estimated at several million tons.

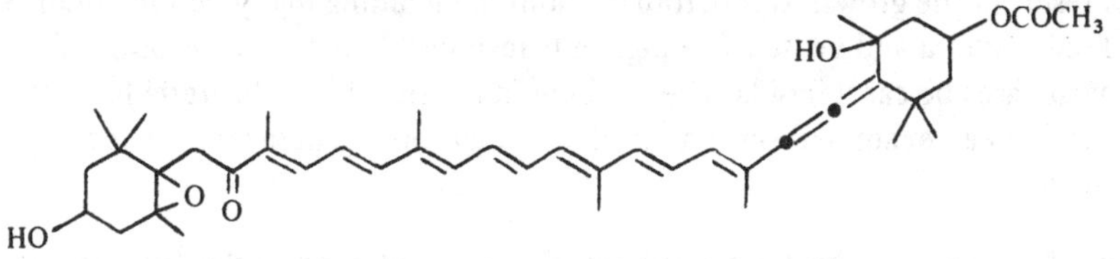

(**2.37**) Fucoxanthin

The primitive prokaryotic blue-green algae or bacteria (Cyanophyta or Cyanobacteria) produce $\beta$-carotene and several of its simple hydroxy- and keto-derivatives but many species also accumulate myxoxanthophyll (2′-($\beta$-L-rhamnopyranosyloxy)-3′,4′-didehydro-1′,2′-dihydro-$\beta$,$\psi$-carotene-3,1′-diol, **2.38**) and other glycosidic carotenoids which, structurally, are more typical of non-photosynthetic bacteria.

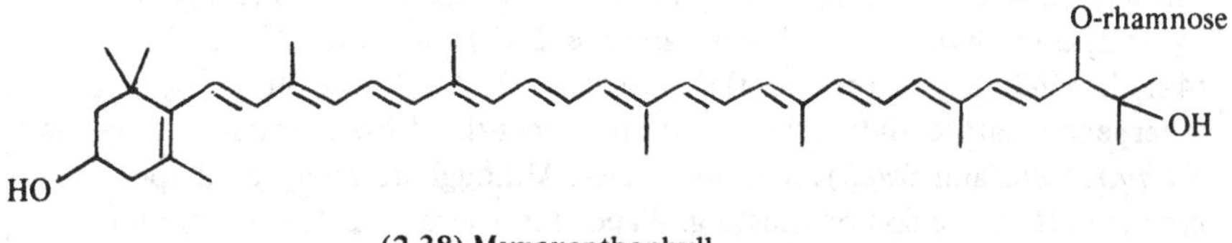

(**2.38**) Myxoxanthophyll

### 2.4.3    Fungi

Carotenoids are found widely, but by no means universally, in fungi. Some species, notably *Blakeslea trispora*, synthesise $\beta$-carotene in the mycelium in such quantities that commercial production by fermentation is possible. Most carotenogenic fungi accumulate only carotenes, especially $\beta$-carotene and $\gamma$-carotene; xanthophylls are rare. Very few of the macro-fungi, *i.e.* mushrooms and toadstools, are coloured by carotenoids, but the edible chanterelle mushroom *Cantharellus cibarius* is coloured yellow by canthaxanthin (**2.27**). The red yeasts (*Rhodotorula* spp.), characteristically

**(2.39)** Torularhodin

produce the carotenoid acid torularhodin ($3',4'$-didehydro-$\beta$,$\psi$-caroten-$16'$-oic acid, **2.39**).

### 2.4.4 Bacteria

Even brief exposure of a nutrient agar plate to the open atmosphere will result in the growth of microbial colonies, including many yellow, orange and red bacteria and yeasts. The pigments responsible for these colours will in most cases be carotenoids. The carotenoid distribution in bacteria has not, however, been examined systematically, though some generalisations may be made.

*Non-photosynthetic bacteria*. Carotenoids are found in examples from many classes and families of non-photosynthetic bacteria, usually located in the cell membranes or wall. Frequently the carotenoids are synthesised in response to light. Many species accumulate simple $C_{40}$ carotenoids such as $\beta$-carotene and $\gamma$-carotene and their derivatives. Other bacteria, however, elaborate $C_{45}$ and $C_{50}$ structures. The characteristic carotenoid of the salt-loving Halobacteria is the acyclic $C_{50}$ pigment bacterioruberin (**2.12**), whereas some *Flavobacterium*, *Sarcina* and *Corynebacterium* species, including some common aerial contaminants and some plant and animal pathogens, contain cyclic $C_{50}$ carotenoids such as decaprenoxanthin (**2.11**), sarcinaxanthin ($2,2'$-bis-(4-hydroxy-3-methylbut-2-enyl)-$\gamma$,$\gamma$-carotene, **2.40**) and 'C.p.450' ($2,2'$-bis-(4-hydroxy-3-methylbut-2-enyl)-$\beta$,$\beta$-carotene, **2.41**). A number of $C_{30}$ triterpenoid carotenoids have recently been identified from certain pathogenic *Streptococcus* and *Staphylococcus* species. Although described as 'diapocarotenoids', these compounds, *e.g.* diaponeurosporene (**2.42**) are genuine triterpenes, derived from farnesyl pyrophosphate *via* a $C_{30}$ analogue of phytoene (see §*2.6.3*). Glycosides of $C_{30}$, $C_{40}$ and $C_{50}$ carotenoids are common in non-photosynthetic bacteria.

*Photosynthetic bacteria*. Carotenoids are found universally in photosynthetic bacteria, where they are an important part of the photosynthetic apparatus (chapter **10**). Purple, non-sulphur bacteria (Rhodospirillaceae) are characterised by the presence of acyclic pigments with tertiary hydroxy- and methoxy-groups at C-1 and C-1', *e.g.* spirilloxanthin ($1,1'$-dimethoxy-$3,4,3',4'$-tetra-dehydro-$1,2,1',2'$-tetrahydro-$\psi$,$\psi$-carotene, **2.43**) from *Rhodospirillum rubrum*, and hydroxyspheroidene ($1'$-methoxy-$3',4'$-didehydro-$1,2,7,8,1',2'$-hexahydro-$\psi$,$\psi$-caroten-1-ol, **2.44**) from *Rhodopseudomonas sphaeroides*.

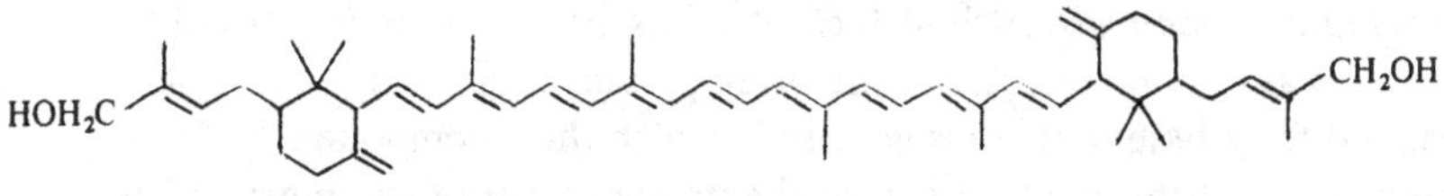

(2.40) Sarcinaxanthin

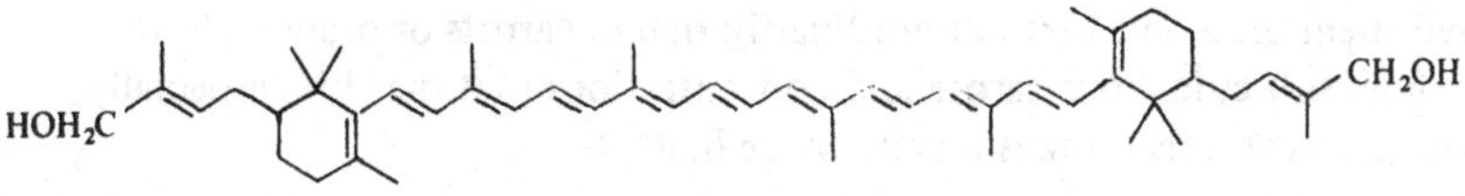

(2.41) 'C.p. 450'

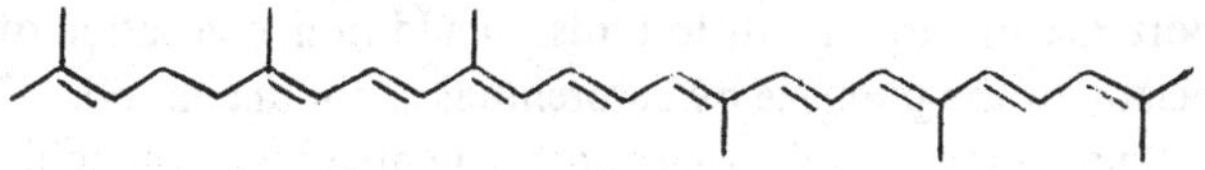

(2.42) 'Diaponeurosporene'

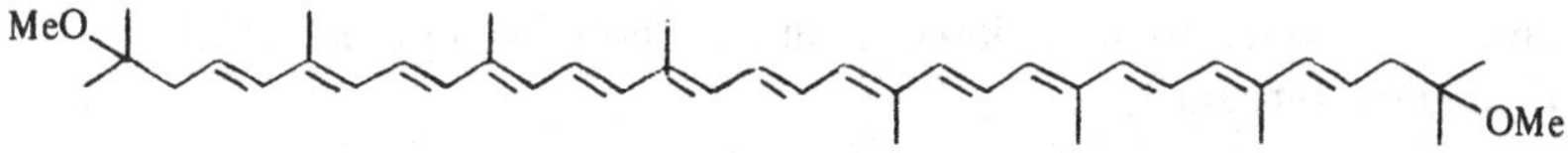

(2.43) Spirilloxanthin

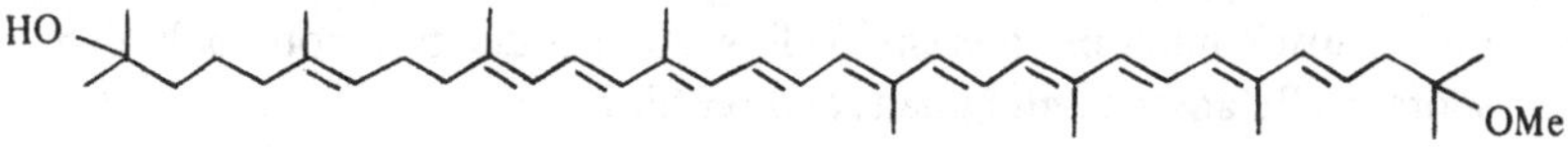

(2.44) Hydroxyspheroidene

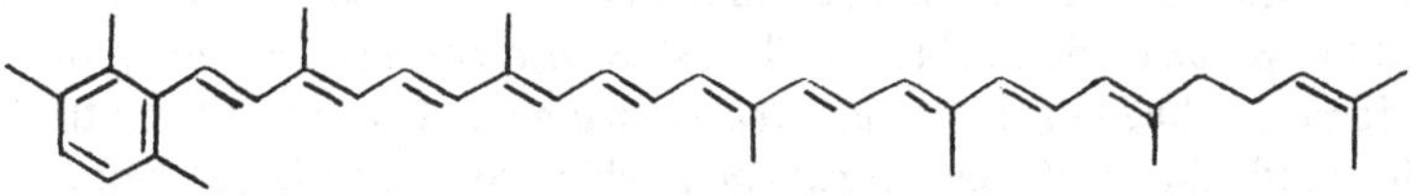

(2.45) Chlorobactene

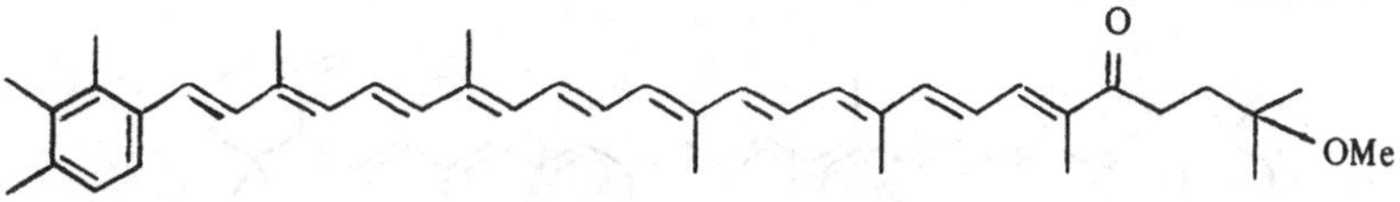

(2.46) Okenone

In purple and green sulphur bacteria (Chromatiaceae and Chlorobiaceae) aromatic carotenoids such as chlorobactene ($\phi,\psi$-carotene, 2.45) and okenone (1′-methoxy-1′,2′-dihydro-$\chi,\psi$-caroten-4′-one, 2.46) are commonly present.

## 2.4.5    Animals

Carotenoids are responsible for integumental colours in many animals of all classes except mammals.

*Vertebrates – mammals.* Although carotenoids are obviously important to mammals as provitamins A, and are often present in small amounts in the liver and fatty tissues, there is no case in which these compounds make a significant contribution to the normal external colour of any mammalian species. However, cases are known in which human patients, who have subjected themselves to a diet extraordinarily rich in carrots or oranges, have such high sub-cutaneous carotenoid concentrations that the skin, especially of the hands and feet, takes on an orange hue.

*Birds.* The familiar deep-yellow colour of egg-yolk bears witness to the universal importance of carotenoids to birds. In addition, coloration of the skin and especially of the plumage by carotenoids is not uncommon. The yellow-red feathers of many bird species are pigmented by carotenoids; the pink-red coloration of the flamingo tribe by ketocarotenoids, mainly cantha-xanthin (2.27) is perhaps the best-known example. Also many green feather colours result from a combination of a blue structural colour and a yellow carotenoid background.

*Fishes, amphibians and reptiles.* Carotenoids may colour the skin (*e.g.* gold-fish) or the flesh (*e.g.* salmon) of fish. Astaxanthin and its esters are the carotenoids most commonly present. Yellow colours due to carotenoids also occur sporadically among amphibians and reptiles.

*Invertebrates – insects.* Carotenoids are responsible for the colours of some, but by no means all, yellow, orange and red insects (see also chapters **3** and **6**). Well-known examples include the locust, *Schistocerca* (β-carotene), the colorado potato beetle *Leptinotarsa decemlineata* (canthaxanthin) and the ladybird beetle *Coccinella septempunctata*, which contains a large number of carotenes, notably the rare β,γ-carotene (**2.47**) and related γ-ring compounds.

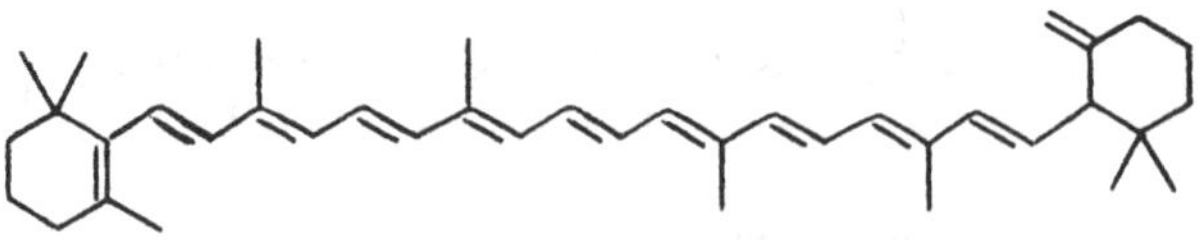

(2.47) β, γ-Carotene

*Marine invertebrates.* Pigmentation by carotenoids is especially common and important in invertebrate marine animals of almost every class. The typical carotenoids present are keto compounds, such as canthaxanthin and asta-xanthin. Other, unusual carotenoids are sometimes found. In sponges (Porifera), which are found in great proliferation and variety of colours in tropical seas, aryl carotenes, such as renieratene (φ,χ-carotene, **2.48**) are

**(2.48)** Renieratene

prevalent. The beadlet sea anemone *Actinia equina*, a common inhabitant of rock pools, is coloured purple by the nor-carotenoid ester actinioerythrin (2.14).

In many marine invertebrates the main carotenoid is present not in the free form but as a stoicheiometric carotenoid–protein complex (see §2.5).

Carotenoids and carotenoproteins are most commonly found in the epidermis or the shell of invertebrate animals, but also, sometimes in high concentrations, in reproductive organs and eggs, though the significance of their presence here is unknown.

## 2.5 Carotenoproteins

In marine invertebrate animals, ketocarotenoids are commonly present as stable, water-soluble carotenoprotein complexes. These complexes are stoicheiometric associations between the carotenoid and a protein, lipo-protein or glycoprotein. In appropriate cases, the carotenoprotein may be firmly associated with the structural material of the integument, *e.g.* chitin, calcium carbonate. Some carotenoproteins exhibit complex sub-unit structures, *e.g.* the lobster shell pigment crustacyanin, whereas others, *e.g.* the violet carotenoprotein of the starfish *Asterias rubens*, are much simpler.

Carotenoprotein complex formation usually results in a large batho-chromic shift in the absorption spectrum (fig. 2.12) so that the complexes are often purple, blue or green in colour, in contrast to the yellow or orange of the free carotenoid. The keto-groups of the carotenoid prosthetic group, which is usually astaxanthin or canthaxanthin, appear to be essential for production of the spectral shift, but the nature of the carotenoid–protein binding is not yet understood. Covalent linkages are not involved, since denaturation by heat or organic solvents readily liberates the free carotenoid. This process is sometimes reversible.

Perhaps the most familiar example of coloration by a carotenoprotein complex is the lobster *Homarus vulgaris*. The grey-blue colour of the lobster shell is due to the carotenoprotein crustacyanin. Cooking denatures the carotenoprotein and liberates the free carotenoid, astaxanthin, and its auto-oxidation product astacene. Boiled lobster is therefore red.

Although readily denatured by excessive heat or organic solvents, the carotenoprotein complexes in aqueous solution are considerably more stable towards light and oxygen than are the free carotenoids.

## 2.6    Biosynthesis

### 2.6.1    *Introduction*

Only plants and microorganisms have been shown to be capable of biosynthesising carotenoids. Many animals can accumulate dietary carotenoids, and may even modify them structurally (see §2.8), but the biosynthesis of carotenoids *de novo* has never been demonstrated unequivocally in any animal system. The biosynthetic pathway to be discussed below is a generalisation based upon investigations with many different carotenogenic systems, notably leaves and chloroplasts, tomato fruits and chromoplasts, fungi, and bacteria, as well as enzyme preparations derived from them.

Carotenoids are tetraterpenes and are biosynthesised by the normal isoprenoid pathway which gives rise also to other important natural products such as rubber, steroids, the mono-, sesqui- and diterpenes present in many essential oils, and the sidechains of the electron transport quinones.

The carotenoid biosynthetic pathway may be considered in several stages:

(i)    formation of the $C_{20}$ intermediate, geranylgeranyl pyrophosphate;

(ii)   formation of phytoene, the first $C_{40}$ carotene;

(iii)  a series of desaturation reactions;

(iv)   cyclisation and related reactions at the C-1,2 double bond;

(v)    final modifications.

Of these, stage (i), formation of geranylgeranyl pyrophosphate, follows the general isoprenoid biosynthetic pathway, whereas stages (ii)–(v) are peculiar to the biosynthesis of carotenoids.

### 2.6.2    *Formation of geranylgeranyl pyrophosphate*

Most of the available experimental data on these early stages have come from studies of the biosynthesis of cholesterol rather than of carotenoids, but there is no reason to believe that the reactions involved may be different.

The first general isoprenoid precursor is usually considered to be acetate, as acetyl-coenzyme A. The biosynthetic pathway from acetyl-CoA to geranylgeranyl pyrophosphate (GGPP) is summarised in fig. 2.13. Acetyl-CoA (3 molecules) is converted, *via* acetoacetyl-CoA (**2.49**) into 3-hydroxy-3-methylglutaryl-CoA (HMG-CoA, **2.50**). There are conflicting reports about whether this sequence involves a malonyl-CoA intermediate as in fatty acid biosynthesis.

HMG-CoA undergoes a two-step reduction to mevalonic acid (MVA, **2.51**). This HMG-CoA reductase reaction is a major control point in cholesterol biosynthesis; it remains to be seen if this is also the case in carotenogenesis. MVA is the first compound that serves solely as an isoprenoid intermediate, and many species of this compound labelled with radioactive ($^{3}$H and $^{14}$C) and stable ($^2$H and $^{13}$C) isotopes have been prepared and used extensively as substrates in studies of carotenoid biosynthesis.

In the next steps, MVA is phosphorylated twice to give mevalonic acid-5-phosphate (2.52) and -5-pyrophosphate (2.53) by kinase enzymes and ATP. The MVA-5-pyrophosphate is then decarboxylated to give the 'isoprene unit'

Fig. 2.13. Formation of geranylgeranyl pyrophosphate (GGPP) by the basic isoprenoid biosynthetic pathway from acetyl-CoA.

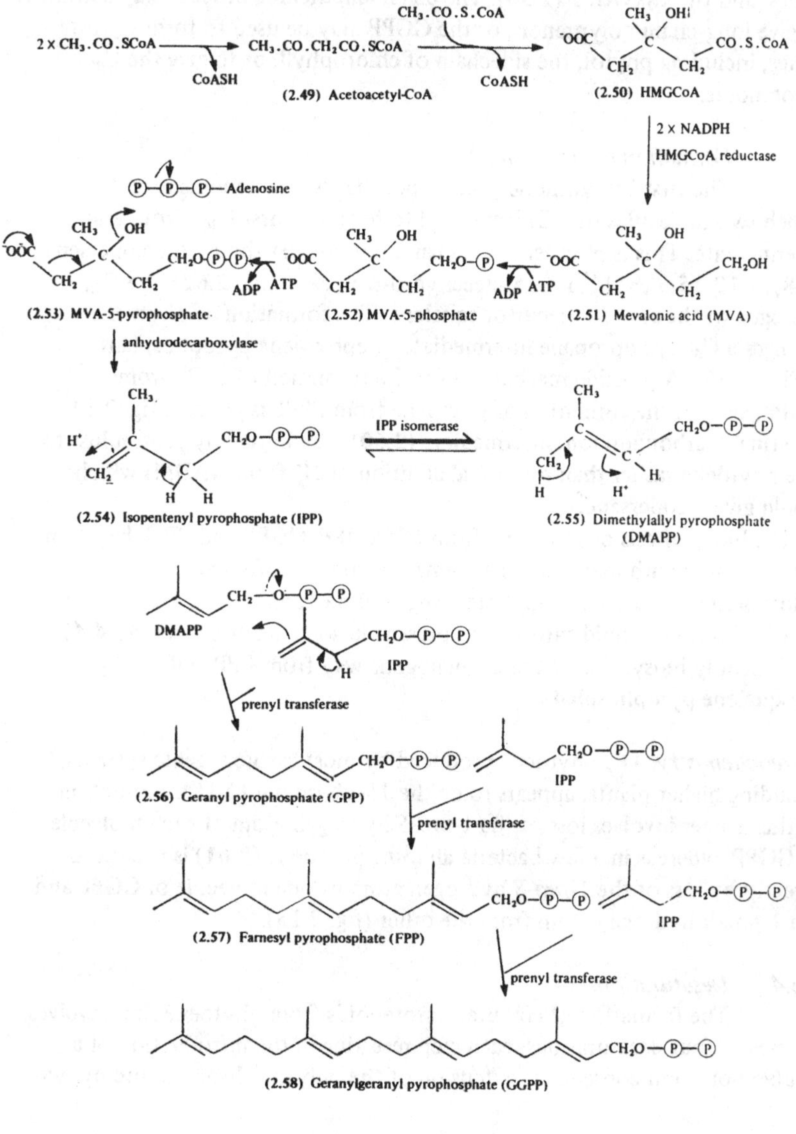

$2 \times CH_3.CO.SCoA \longrightarrow CH_3.CO.CH_2CO.SCoA \longrightarrow$

CoASH

(2.49) Acetoacetyl-CoA

$CH_3.CO.S.CoA$

CoASH

(2.50) HMGCoA

$2 \times$ NADPH

HMGCoA reductase

$P$—$P$—$P$—Adenosine

(2.53) MVA-5-pyrophosphate

ADP  ATP

(2.52) MVA-5-phosphate

ADP  ATP

(2.51) Mevalonic acid (MVA)

anhydrodecarboxylase

(2.54) Isopentenyl pyrophosphate (IPP)

IPP isomerase

(2.55) Dimethylallyl pyrophosphate (DMAPP)

DMAPP

IPP

prenyl transferase

(2.56) Geranyl pyrophosphate (GPP)

IPP

prenyl transferase

(2.57) Farnesyl pyrophosphate (FPP)

IPP

prenyl transferase

(2.58) Geranylgeranyl pyrophosphate (GGPP)

intermediate isopentenyl pyrophosphate, IPP (2.54). An isomerase enzyme catalyses the reversible isomerisation of IPP and dimethylallyl pyrophosphate (DMAPP, 2.55). These two molecules are the first substrates of the prenyl transferase enzymes by which isoprenoid chains are built up. DMAPP acts as a 'primer' molecule and undergoes condensation with IPP to give the $C_{10}$ intermediate geranyl pyrophosphate, GPP (2.56), precursor of monoterpenes. Successive addition of two further molecules of IPP gives the $C_{15}$ farnesyl pyrophosphate, FPP (2.57), precursor to sesquiterpenes, steroids and triterpenes, and the $C_{20}$ GGPP (2.58). The chain-lengthening process may continue, to give long-chain polyprenols, or the GGPP may be used to form $C_{20}$ diterpenes, including phytol, the sidechain of chlorophyll, or to give the $C_{40}$ carotenoids.

### 2.6.3   Formation of phytoene

The first biosynthetic process peculiar to carotenoids is that in which two molecules of GGPP are used to form the first $C_{40}$ carotenoid intermediate. This is phytoene (2.61) and not, as first thought, lycopersene (7,8,11,12,15,7',8',11',12',15'-decahydro-$\psi$,$\psi$-carotene, 2.62), the $C_{40}$ analogue of the steroid precursor squalene. The formation of phytoene involves a $C_{40}$ cyclopropane intermediate, prephytoene pyrophosphate (PPPP, 2.59). A possible mechanism for the formation of PPPP from GGPP, and for the formation of phytoene from PPPP is given in fig. 2.14. The final 'carbonium ion' intermediate (2.60) is stabilised by proton loss to give phytoene rather than by formal addition of $H^-$ from NADPH which would give lycopersene.

The biosynthesis of phytoene from MVA, IPP, GGPP and PPPP has been demonstrated with many crude enzyme preparations, for example from chloroplasts, tomato chromoplasts, fungi and bacteria.

The $C_{30}$ triterpenoid carotenoids present in some bacteria (see §2.4.4) are probably biosynthesised in an analogous way from FPP and the $C_{30}$ presqualene pyrophosphate.

*Stereochemistry.* The phytoene produced by most carotenogenic systems, including higher plants, appears to be the 15-*cis* isomer (2.15). Formation of this isomer involves loss of the 1-*pro-S* hydrogen atom of each molecule of GGPP, whereas in a few bacteria all-*trans*-phytoene (2.61) is produced directly by loss of the 1-*pro-S* hydrogen atom of one molecule of GGPP and the 1-*pro-R* hydrogen atom from the other (fig. 2.15).

### 2.6.4   Desaturation

The formation of coloured carotenoids from phytoene first involves a series of four desaturations, each step resulting in the introduction of a double bond and consequent extension of the polyene chromophore by two

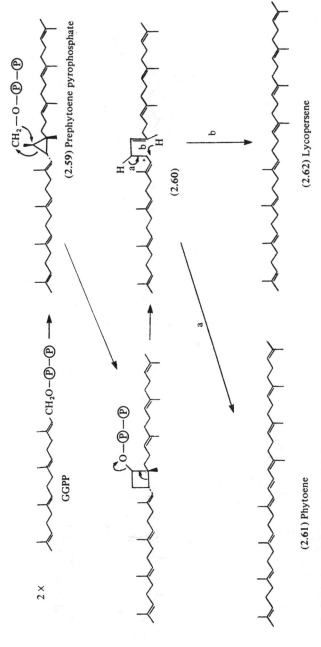

Fig. 2.14. A possible mechanism for the formation of phytoene.

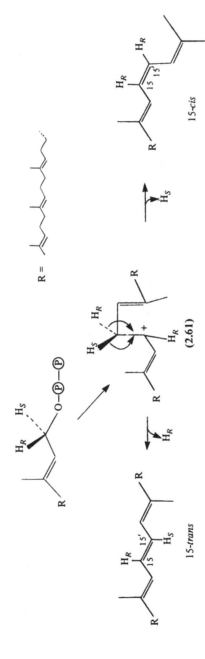

Fig. 2.15. Stereochemistry of hydrogen loss in the biosynthesis of all-*trans*- and 15-*cis*-phytoene.

conjugated double bonds (fig. 2.16). The intermediates in this sequence are phytofluene (**2.63**), ζ-carotene (**2.64**) and neurosporene (**2.66**) and the final product of the desaturations is lycopene (**2.67**). In many microorganisms, ζ-carotene is replaced, entirely or in part, by the unsymmetrical isomer 7,8,11,12-tetrahydro-ψ,ψ-carotene (**2.65**). Conversions of labelled phytoene,

Fig. 2.16. The stepwise desaturation of phytoene to lycopene.

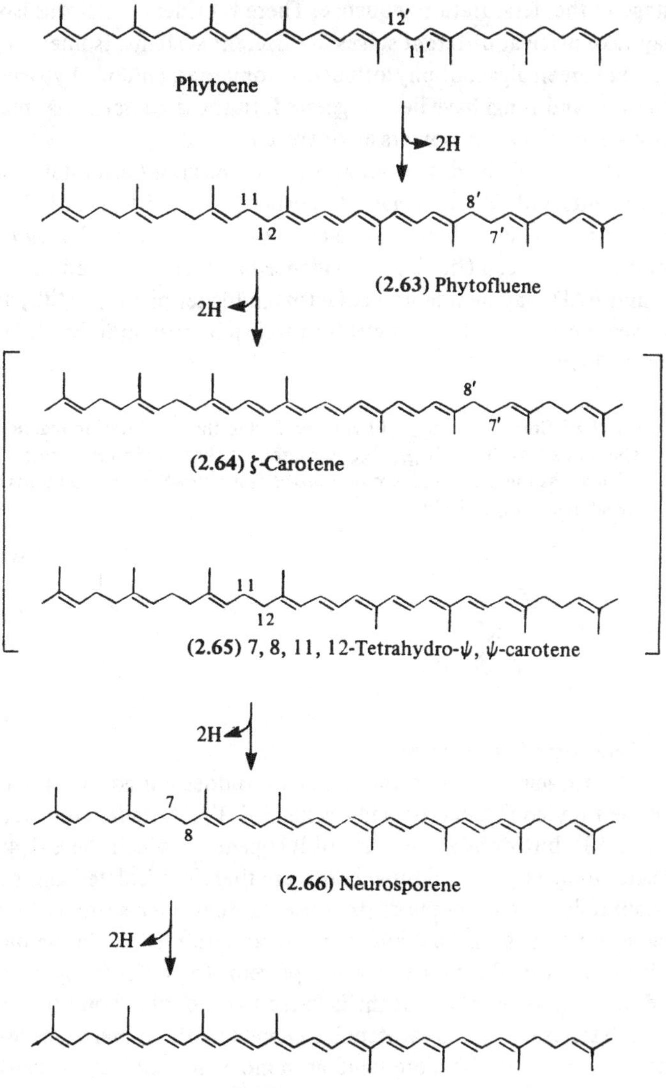

Phytoene

2H

(**2.63**) Phytofluene

2H

(**2.64**) ζ-Carotene

(**2.65**) 7, 8, 11, 12-Tetrahydro-ψ, ψ-carotene

2H

(**2.66**) Neurosporene

2H

(**2.67**) Lycopene

phytofluene and ζ-carotene into lycopene and other carotenes have been achieved with cell-free preparations from tomato plastids and bacteria.

The desaturation process is inhibited in many microorganisms by diphenylamine, and in higher plants by certain herbicides; phytoene accumulates. If the inhibitor is then removed the more desaturated carotenes are formed, apparently at the expense of the accumulated phytoene.

Almost all natural carotenoids have the all-*trans* configuration, so in those tissues which produce 15-*cis*-phytoene an isomerisation must be involved at some stage in the desaturation sequence. There is evidence that this isomerisation may take place at different stages in different systems. Isomerisations of ζ-carotene in green algae, of phytofluene in tomatoes, and of phytoene in some bacteria and fungi have been suggested. In those bacteria that make *trans*-phytoene only *trans* isomers are involved.

The mechanism of the desaturation reactions has not been established, but experiments with MVA labelled stereospecifically with $^3$H at C-2 or C-5 have shown that the introduction of each double bond occurs by *trans* elimination of hydrogen (fig. 2.17). Evidence has been presented that NADP$^+$ and FAD may be required cofactors in higher plant plastids, and the involvement of some form of electron transport system in bacteria is considered likely.

Fig. 2.17. Stereochemistry of hydrogen loss in the desaturation reaction. The terms $H_{2R}$, $H_{2S}$, $H_{5R}$, $H_{5S}$ indicate that these hydrogen atoms originate as the 2-*pro-R*, 2-*pro-S*, 5-*pro-R* and 5-*pro-S* hydrogen atoms of mevalonate, respectively.

### 2.6.5    *Later reactions – general*

The sequence of desaturation reactions does not go to completion to produce the fully conjugated pentadecaene, 3,4,3′,4′-tetradehydro-ψ,ψ-carotene (2.68), but stops at the stage of lycopene, in which the C-3,4 bonds remain saturated. The C-1,2 double bonds are therefore isolated and not part of the main polyene chromophore. In most carotenogenic systems, however, lycopene is not an end product but is merely an intermediate in the biosynthesis of the normal main carotenoids present. In particular, lycopene may undergo various reactions at the isolated C-1,2 double bond to give rise either to the series of acyclic carotenoids characteristic of many photosynthetic bacteria or to the more familiar monocyclic and bicyclic carotenoids that are typical of plants.

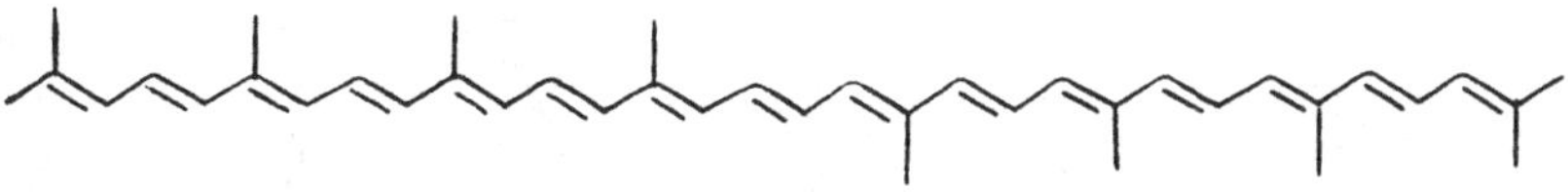

(2.68) 3, 4, 3', 4'-Tetradehydrolycopene

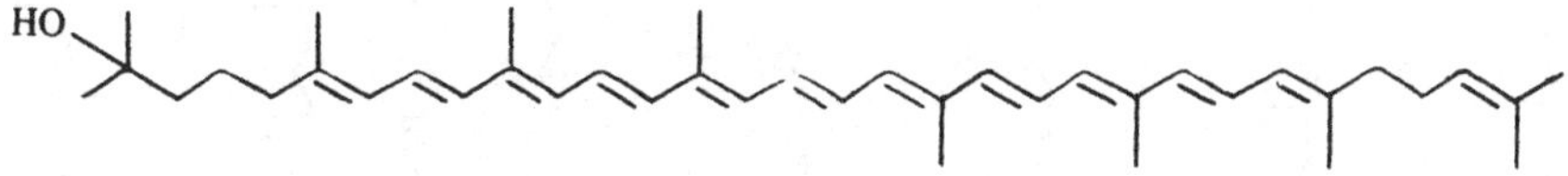

(2.69) Rhodopin

## 2.6.6 *Later reactions – acyclic carotenoid biosynthesis*

Of the several addition reactions which can occur at the C-1,2 double bond, the simplest are seen in the acyclic carotenoid series. The most obvious example is the addition of water to give the 1-hydroxy- and 1-methoxycarotenoids characteristic of photosynthetic and some other bacteria. The simplest case is the hydration of the C-1,2 double bond of lycopene to give rhodopin (1,2-dihydro-$\psi$,$\psi$-caroten-1-ol, **2.69**) the main carotenoid of the budding purple bacterium *Rhodomicrobium vannielii*. The probable mechanism of this reaction is given in fig. 2.18.

Fig. 2.18. A mechanism for formation of 1-hydroxy-1,2-dihydrocarotenoids.

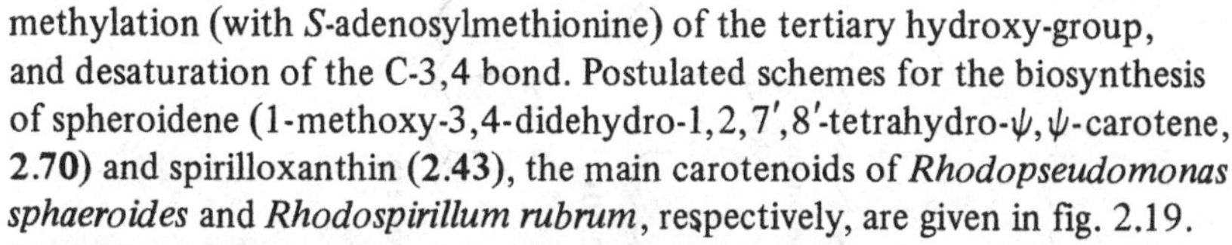

In photosynthetic bacteria, this hydration is commonly followed by methylation (with *S*-adenosylmethionine) of the tertiary hydroxy-group, and desaturation of the C-3,4 bond. Postulated schemes for the biosynthesis of spheroidene (1-methoxy-3,4-didehydro-1,2,7',8'-tetrahydro-$\psi$,$\psi$-carotene, **2.70**) and spirilloxanthin (**2.43**), the main carotenoids of *Rhodopseudomonas sphaeroides* and *Rhodospirillum rubrum*, respectively, are given in fig. 2.19. C-1 hydroxylation and *O*-methylation can, however, occur at earlier desaturation levels; hydroxy- and methoxy-derivatives of phytoene, phytofluene and 1,2,7,8-tetrahydro-$\psi$,$\psi$-carotene have been detected in these bacteria under certain culture conditions.

These pathways operate under anaerobic conditions. Admission of $O_2$ to cultures of *R. sphaeroides* causes the rapid conversion of the yellow spheroidene into the red keto-derivative spheroidenone (1-methoxy-3,4-didehydro-1,2,7',8'-tetrahydro-$\psi$,$\psi$-caroten-2-one, **2.71**) (fig. 2.20).

Fig. 2.19. Postulated schemes for the biosynthesis of spheroidene and spirilloxanthin in photosynthetic bacteria.

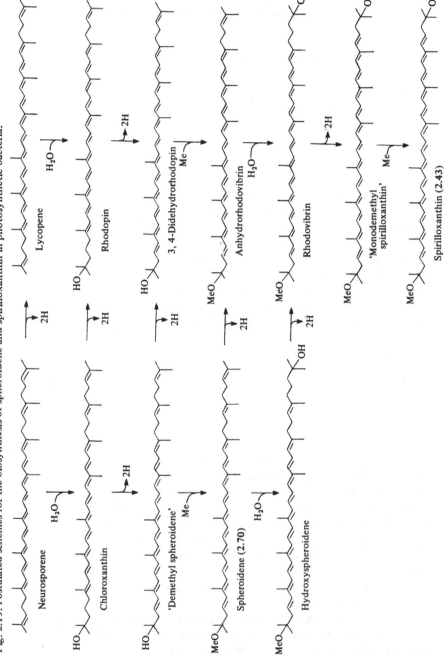

Fig. 2.20. Aerobic conversion of spheroidene into spheroidenone.

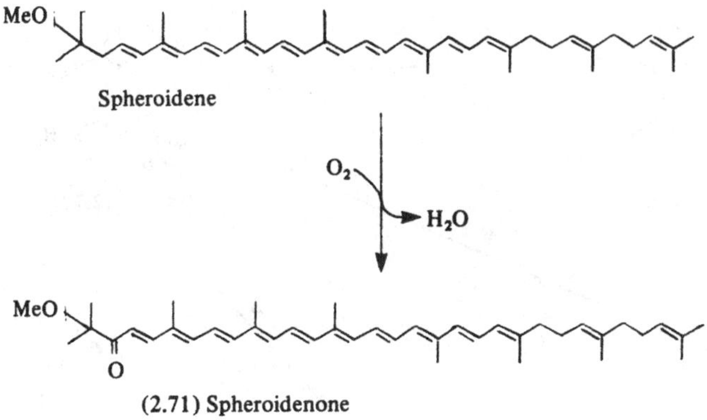

Spheroidene

$O_2$

$H_2O$

(2.71) Spheroidenone

Fig. 2.21. Biosynthesis of acyclic $C_{45}$ and $C_{50}$ carotenoids.

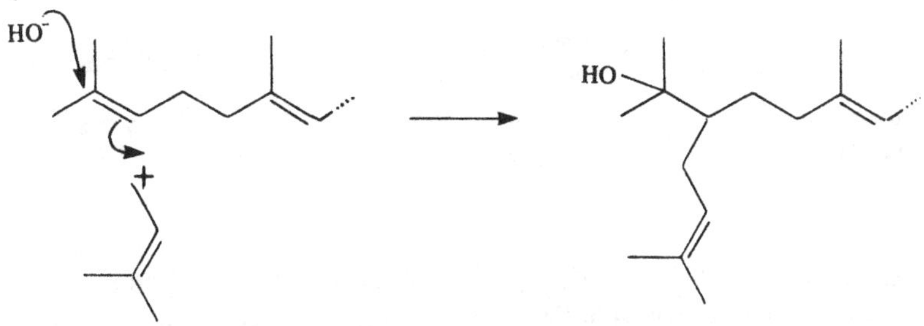

The Halobacteria, which are specially adapted to life in conditions of high salinity, contain acyclic $C_{45}$ and $C_{50}$ carotenoids such as bacterioruberin (2.12). These are believed to be produced by a reaction analogous to the C-1,2 hydration (fig. 2.18) but with a $C_5$ species as the initiating electrophile (fig. 2.21).

### 2.6.7    Cyclisation

The rigidity of the conjugated polyene system of carotenoids precludes extensive cyclisations of the kind that occur in di- and triterpenoids. Cyclisation in carotenoids is limited to the formation of a single six-membered ring at one end or both ends of the acyclic precursor molecule. Cyclisation of carotenoid intermediates may be considered as an addition process initiated by proton attack at C-2 of the terminal C-1,2 double bond. Cyclisation then occurs as illustrated in fig. 2.22 to give a 'carbonium ion' (2.72) which can be stabilised by proton loss from either C-6, C-4 or C-18 to give, respectively, the β-ring (2.73), ε-ring (2.74) or, in rare cases, the γ-ring (2.75). The different ring types are not interconverted.

Fig. 2.22. General mechanism of carotenoid cyclisation to give the three ring types, $\beta$, $\gamma$ and $\epsilon$.

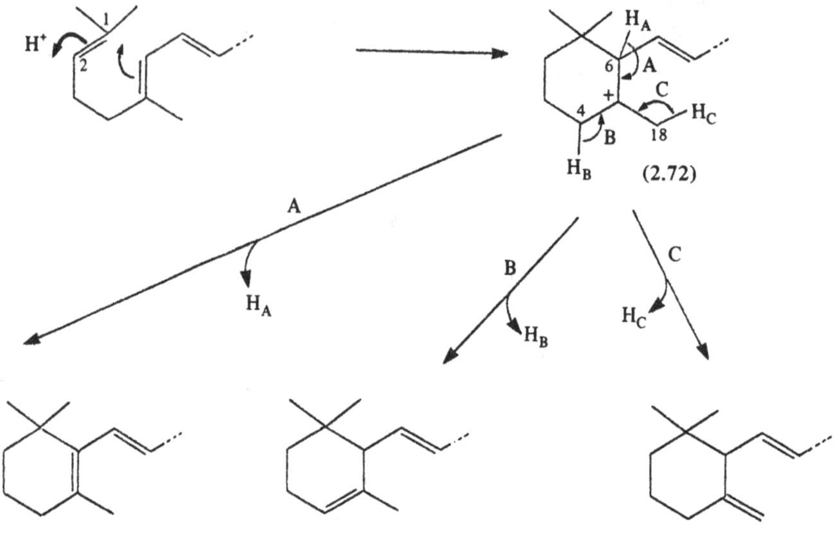

$\beta$-ring
(2.73)

$\epsilon$-ring
(2.74)

$\gamma$-ring
(2.75)

A general overall scheme for the biosynthesis of the common $\beta$- and $\epsilon$-ring cyclic carotenes is given in fig. 2.23. According to this scheme there are two major points at which cyclisation may occur. If normal desaturation goes to completion before cyclisation then lycopene is the immediate precursor of the monocyclic $\gamma$-carotene (2.78) and $\delta$-carotene ($\epsilon,\psi$-carotene, 2.79) and thence of the bicyclic $\beta$-carotene (2.80), $\alpha$-carotene (2.81) and $\epsilon$-carotene (2.82). Alternatively if cyclisation occurs before desaturation is complete, neurosporene and $\beta$- and $\alpha$-zeacarotene ($7',8'$-dihydro-$\beta,\psi$-carotene, 2.76, and $7',8'$-dihydro-$\epsilon,\psi$-carotene, 2.77) are key intermediates. In all cases, however, cyclisation always occurs in a carotenoid 'half-molecule' that has reached the lycopene level of desaturation; end-groups with a C-7,8 single bond cannot cyclise.

Many of the transformations illustrated in fig. 2.23 have been demonstrated in experiments with microbial or tomato plastid enzyme systems, or in bacteria after removal of the cyclisation inhibitor, nicotine.

*Stereochemistry.* Experiments involving labelling with stable isotopes have revealed the stereochemistry of folding, $H^+$ attack and cyclisation in formation of the $\beta$-ring (fig. 2.24). The stereochemistry of formation of the $\epsilon$-ring is still not certain, though the chirality at C-6 and the stereochemistry of

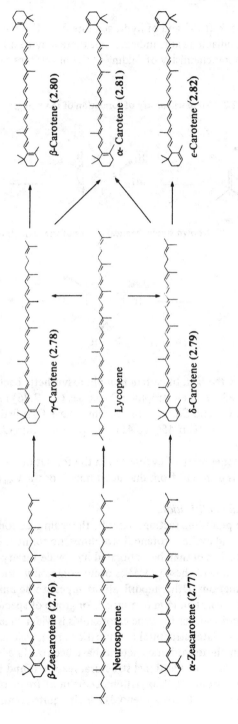

Fig. 2.23. Overall scheme for the biosynthesis of the common β- and ε-ring carotenoids.

hydrogen attack at C-2 and of hydrogen loss from C-4 have been determined (fig. 2.25). Available results indicate that there may be a fundamental difference in the stereochemistry of folding in the biosynthesis of the two ring types.

Fig. 2.24. Stereochemistry of formation of the β-ring.

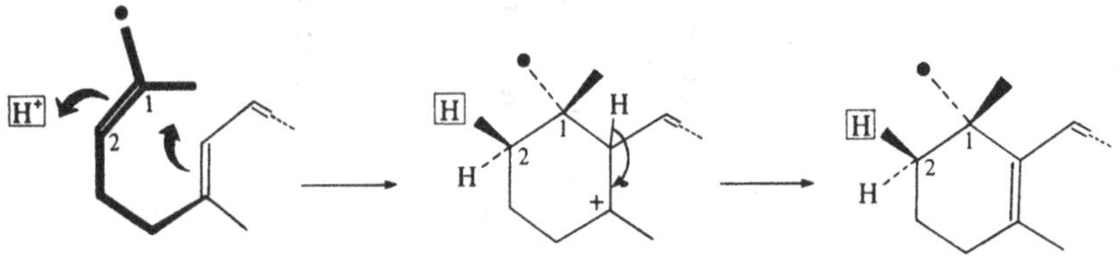

Fig. 2.25. Known stereochemical details of the formation of the ε-ring.

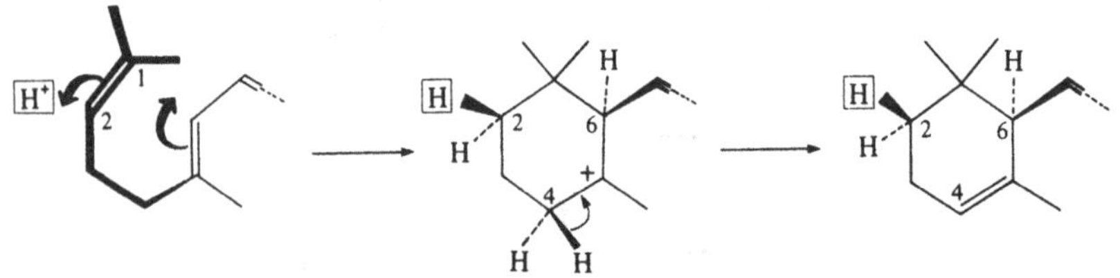

*Cyclic $C_{50}$ carotenoids.* In certain non-photosynthetic bacteria, the cyclisation can be initiated by an electrophilic $C_5$ species (fig. 2.26) resulting in the formation of $C_{45}$ and $C_{50}$ carotenoids which have $C_5$ substituents at C-2 of the β-, ε- or γ-ring, *e.g.* 'C.p.450' (**2.41**), decaprenoxanthin (**2.11**) and sarcinaxanthin (**2.40**).

The stereochemistry of cyclisation in the formation of these 'higher carotenoids' is different from that determined for the $C_{40}$ compounds.

### 2.6.8    Final modifications

The preceding sections describe the main reactions by which the basic acyclic and cyclic carotenoid structures are formed. Individual carotenoid structures can then be elaborated by a wide variety of further transformations. Some of these processes occur commonly, notably introduction of oxygen functions. Other modifications appear to be unique to the formation of a single carotenoid in one species or group of species. The range of structural modifications in cyclic carotenoids is wider than in the acyclic series. The main later structural modifications to acyclic carotenoid structures, particularly in photosynthetic bacteria, have been outlined earlier (see §2.6.6).

The most common additional structural feature found in cyclic carotenoids is the hydroxy-group. Hydroxylation occurs most frequently at C-3, but 2-hydroxy- and 4-hydroxycarotenoids are also quite common. The latter are

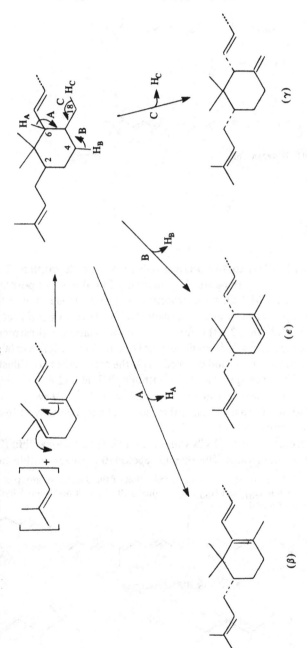

Fig. 2.26. Formation of the C$_5$-substituted rings of the cyclic C$_{50}$ carotenoids.

Fig. 2.27. Formation of the oxocarotenoid canthaxanthin.

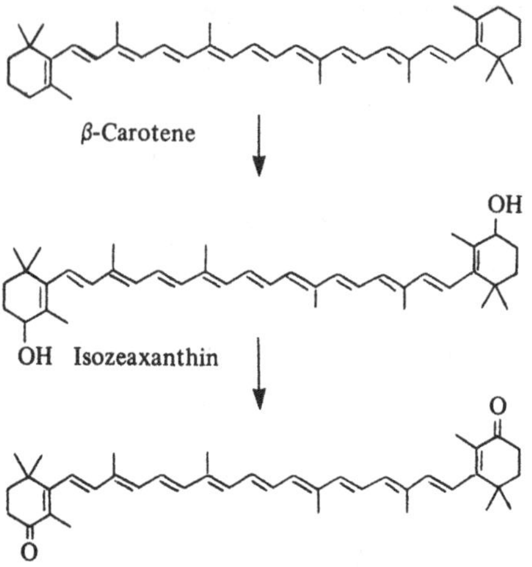

β-Carotene

OH    Isozeaxanthin

O

(2.27) Canthaxanthin

often oxidised further to 4-oxocarotenoids such as canthaxanthin (**2.27**) (fig. 2.27). Hydroxy-groups are also sometimes found in other positions in the molecule, *e.g.* C-19. Of these processes, only the introduction of the C-3 hydroxy-group has been studied. In plants and bacteria, ($3R,3'R$)-zeaxanthin (β,β-carotene-3,3'-diol, **2.83**) is formed by hydroxylation of β-carotene. The hydroxy-group arises from molecular $O_2$ and the reaction is catalysed by a mixed-function oxidase and proceeds with the stereochemistry illustrated in fig. 2.28. The most abundant leaf xanthophyll, lutein (**2.6**), is probably formed similarly from α-carotene. The chirality at C-3', however, is opposite to that at C-3 and to that in zeaxanthin, so the stereochemistry of hydroxylation is obviously different.

The chloroplast xanthophylls violaxanthin (**2.8**) and neoxanthin (**2.32**) contain 5,6-epoxy-groups. The enzymic epoxidation of zeaxanthin to viola-

Fig. 2.28. Stereochemistry of introduction of the 3-hydroxy-group of zeaxanthin. $H_{5R}$ and $H_{5S}$ were originally the 5-*pro-R* and 5-*pro-S* hydrogen atoms of mevalonate.

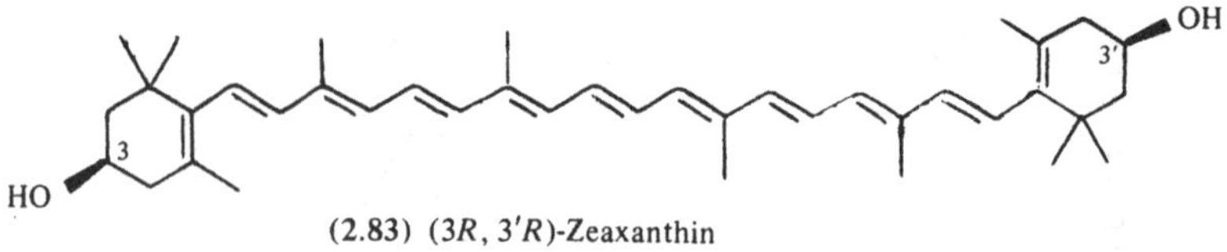

OH

HO

(2.83)  ($3R, 3'R$)-Zeaxanthin

$H_{5R}$  3
$H_{5S}$

HO  3
$H_{5S}$

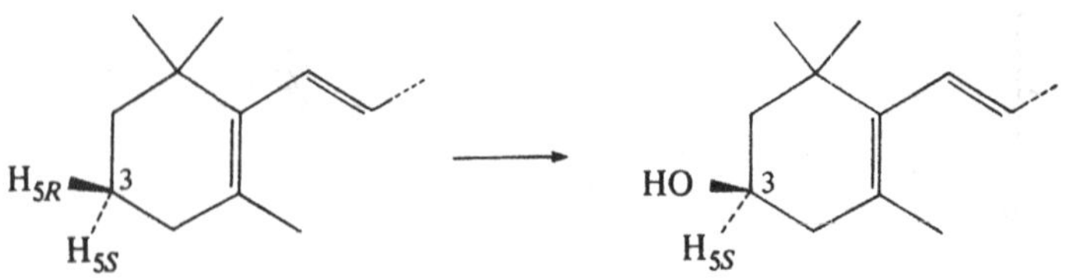

xanthin has been described. Such epoxides (or the related peroxide structure **2.84**) may be important intermediates in several interesting carotenoid modifications. Fig. 2.29 illustrates possible mechanisms for the formation of (*a*) the allenic end-group of neoxanthin, (*b*) acetylenic end-groups as in the algal carotenoid alloxanthin (7,8,7′,8′-tetradehydro-β,β-carotene-3,3′-diol, **2.85**), (*c*) the cyclopentane ring of capsanthin (**2.35**) and capsorubin (**2.36**), and (*d*) the *retro*-carotenoid eschscholtzxanthin (**2.34**). Although these schemes seem very reasonable, there is little or no biochemical evidence to support them.

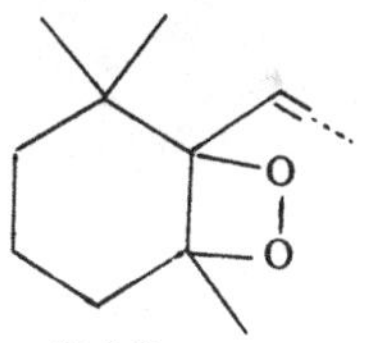

(**2.84**)

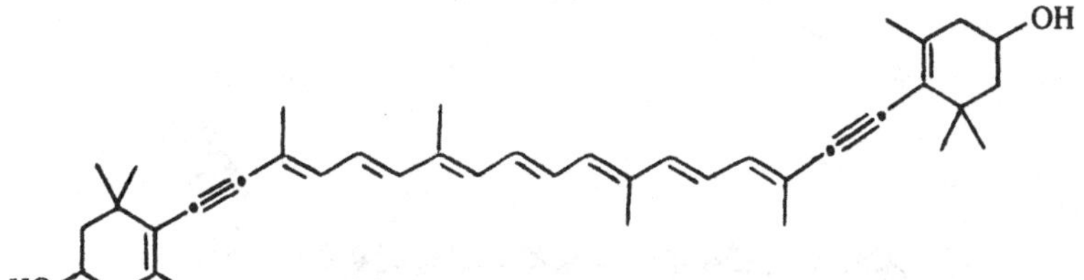

(**2.85**) Alloxanthin

Fig. 2.29. Possible mechanisms for the formation of various xanthophyll structural features from a 5,6-epoxide or 5,6-peroxide.

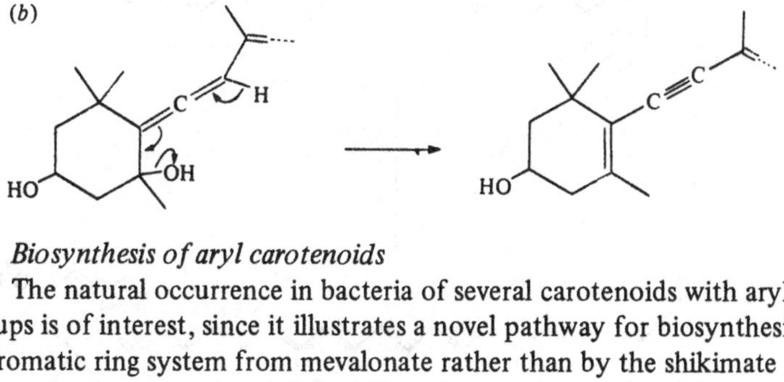

(*a*)

(*b*)

### 2.6.9     Biosynthesis of aryl carotenoids

The natural occurrence in bacteria of several carotenoids with aryl end-groups is of interest, since it illustrates a novel pathway for biosynthesis of the aromatic ring system from mevalonate rather than by the shikimate pathway or from acetate *via* a polyketide mechanism (see chapter **3**).

Fig. 2.29 – *continued*

(*c*)

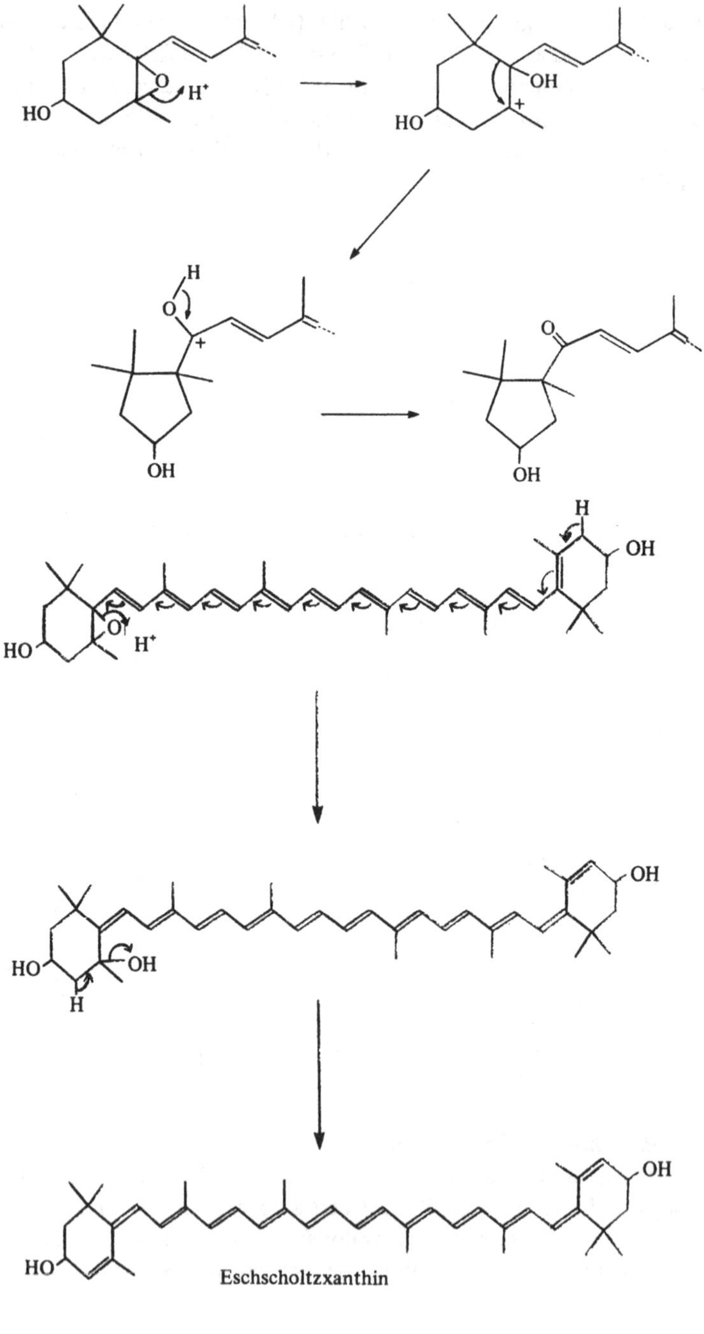

Eschscholtzxanthin

Formation of the 1,2,5-trimethylphenyl end-group as in chlorobactene (**2.45**) involves simply the migration of one of the C-1 methyl groups to C-2. A much more complex rearrangement is required for the biosynthesis of the 1,2,3-trimethylphenyl end-group present in okenone (**2.46**). Although no supporting evidence is yet available, it has been suggested that an intermediate Ladenburg prism structure may be involved (fig. 2.30).

Fig. 2.30. Postulated mechanism for biosynthesis of the 1,2,3-trimethylphenyl end-group.

Ladenburg prism

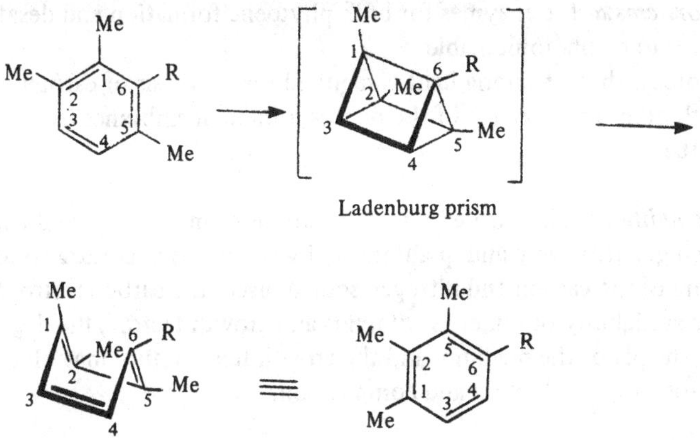

## 2.7 Regulation and control of biosynthesis

### 2.7.1 Fungi and bacteria

*Genetic control.* Mutants of the mould *Phycomyces blakesleeanus* have been used extensively in studies of carotenoid biosynthesis. Three main groups of mutants are available, those that accumulate lycopene, those accumulating phytoene, and those that are unable to synthesise any carotenoids. Extensive complementation studies have revealed that only three genes, termed *car R*, *car B* and *car A*, are involved in normal carotenogenesis, and that the three mutant groups each correspond to mutation in a single gene. Cyclisation (of lycopene?) is carried out by the product of gene *car R*; two copies of its product, *i.e.* two cyclase enzymes, in an enzyme complex, are considered to be concerned in β-carotene formation. Similarly four copies of the product of gene *car B* are considered to act in a dehydrogenase complex which carries out the four successive desaturations required to convert phytoene into lycopene. The whole of the biosynthesis is suggested to take place on a multi-enzyme aggregate containing the desaturases and the cyclases.

*Light.* Light stimulates additional carotenoid synthesis in many fungi and bacteria that normally form reasonable amounts in the dark. In many other fungi and bacteria, however, carotenoid synthesis occurs to a very limited

extent or not at all in the dark, but can be initiated in response to a short
simultaneous exposure to light and oxygen. After photoinduction, there is
usually a time lag (about 4 h) for enzyme synthesis before carotenogenesis
begins. This photoinduction mechanism ensures that carotenoid is available
only when it is needed to protect the organism against the harmful effects of
excessive light and oxygen.

It is now becoming clear which of the steps of carotene formation are
influenced by light. In a *Mycobacterium*, GGPP formation is enhanced by
illumination and PPPP formation is fully photoinduced. In the fungus
*Neurospora crassa*, the enzymes for both phytoene formation and desaturation appear to be photoinducible.

In photosynthetic bacteria light is required for synthesis of carotenoids
along with other components of the photosynthetic membranes (see
chapter **10**).

*Culture conditions*. The carotenoid compositions of many fungi and bacteria
are altered quantitatively and qualitatively by variations in culture conditions.
The nature of the carbon and nitrogen sources used, the carbon:nitrogen
ratio, the availability of minerals, vitamins and growth factors, the degree of
aeration, the pH of the medium, and the growth temperature may all greatly
affect carotenoid production and composition.

*Chemical control*. Many substances are known to stimulate or inhibit carotenoid synthesis or to cause qualitative modifications in the carotenoid
composition of microorganisms. Some, *e.g.* diphenylamine, nicotine, have
been used extensively in biosynthetic studies (see §2.6.4, 2.6.5).

One interesting example of chemical control of carotenogenesis in a
natural biological system is found in certain heterothallic fungi, especially
*Blakeslea trispora*. The (+) and (−) strains do not produce appreciable
amounts of carotenoid when cultured separately, but in mixed (mated)
cultures a large synthesis of β-carotene occurs. This synthesis is induced
by trisporic acid (**2.86**), a hormone produced by a combination of enzymes
from the two strains. Trisporic acid is a metabolite of β-carotene and its main
function is to stimulate sporulation and reproduction. The stimulation of
carotenogenesis is probably part of the mechanism for increased trisporic
acid production.

(**2.86**) Trisporic acid

### 2.7.2 Plants and algae

*Genetic control*. Genetic evidence suggests that in green tissues of higher plants and in algae the desaturation and cyclisation reactions are under direct nuclear control. Thus mutants of maize (*Zea mays*) and the green alga *Scenedesmus obliquus* have been isolated which accumulate phytoene, ζ-carotene or lycopene.

The most important genetic studies have been with tomatoes. Tentative proposals have been made about the sites of action of the genes that control carotenoid biosynthesis in tomato fruit. Several genes act before the phytoene stage and control quantitatively the flow of precursors into the carotenoid pathway. Other genes have been recognised which control, respectively, the desaturation of phytoene, the cyclisation of lycopene to β-carotene, cyclisation to produce ε- rather than β-rings, or the stereochemistry of the final product. Thus the 'ghost' mutant has 'white' fruit containing large amounts of phytoene, the 'high-β' fruit accumulate high concentrations of β-carotene in place of the normal lycopene, the 'delta' strain has δ-carotene as its main pigment, and prolycopene replaces all-*trans*-lycopene as the main carotenoid in the 'tangerine' variety.

*Carotenoid synthesis in chloroplasts*. In photosynthetic tissues, carotenoids are located in the chloroplasts and it is likely that they are synthesised in these organelles. Etiolated seedlings and dark-grown cultures of *Euglena gracilis* produce only small amounts of carotenoids, mainly xanthophylls. In response to a short period of light, the normal chloroplast carotenoids are synthesised as functional chloroplasts are formed. The light effect is thought to be mediated by phytochrome. The carotenoids are an integral part of the chloroplast structure and the regulation of their synthesis is closely inter-related with the synthesis of chlorophyll and other chloroplast constituents (see chapter **10**).

Many algae produce chloroplasts and thus also the normal chloroplast carotenoids even when grown in the dark.

*Carotenoid synthesis in ripening fruit*. In many fruits, ripening is accompanied by massive synthesis of carotenoids, as chloroplasts change into chromo-plasts and the colour changes from green to red. The carotenoids produced during ripening are often ones that were not present in the original chloro-plasts, *e.g.* lycopene in tomatoes, capsanthin in peppers. These changes can take place in fruit that have been removed from the plant. Light usually has no significant effect on the ripening process or the carotenoid accumulation but temperature is important.

*Chemical control*. Many substances are known that stimulate or inhibit caro-tenoid synthesis. Included among these are several herbicides, *e.g.* Sandoz

6706 (4-chloro-5-(dimethylamino)-2-($\alpha,\alpha,\alpha$-trifluoro-*m*-tolyl)-3(2H)-pyridazinone) which may well act by inhibiting carotenoid synthesis (blocking desaturation of phytoene) and hence preventing proper chloroplast development.

## 2.8    Metabolism of carotenoids by animals

### 2.8.1    *Intact carotenoids*

Although many animals use carotenoids for coloration, it is believed that they cannot synthesise them but have to obtain the required amounts from the diet. For example, flamingos in a zoo lose their characteristic attractive pink colour if they become deficient in the ketocarotenoids which, in the wild, they obtain from small crustaceans in the diet. Many animals (birds, fish, and insects and other invertebrates) can, however, modify the structures of dietary carotenoids particularly by introducing oxo-groups at C-4. The conversion of $\beta$-carotene (**2.3**) and zeaxanthin (**2.7**) into canthaxanthin (**2.27**) and astaxanthin (**2.9**), respectively, has been demonstrated with several animals, especially birds and invertebrates.

An interesting and unusual modification is the formation, in sea anemones, of the *nor*-carotenoid actinioerythrin (**2.14**) from astaxanthin esters by a process involving contraction from a 6-membered to a 5-membered ring. The postulated mechanism (fig. 2.31) includes a benzylic acid rearrangement of a trioxo-intermediate.

Some animals have been found to contain carotenoids that would not be present in the diet and could not be formed from dietary carotenoids by any known metabolic process. For example, the ladybird beetle *Coccinella septempunctata* contains unusual $\gamma$-ring carotenoids which are likely to be obtained from microbial symbionts. The origin of the (6$S$,6'$S$)-$\epsilon$-carotene present in coloured oil droplets in the retinas of some birds (chapter **9**) is difficult to explain.

Fig. 2.31. Formation of the nor-carotenoid actinioerythrin.

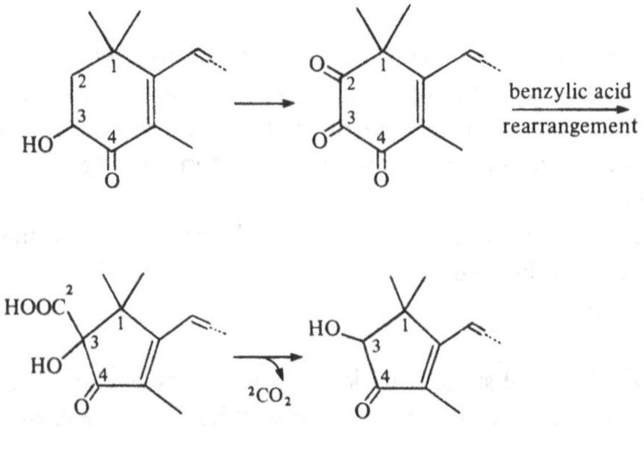

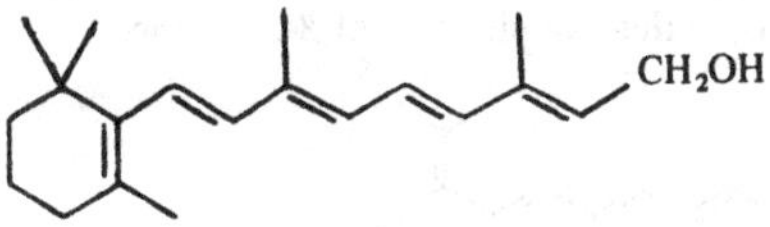

(2.87) Retinol

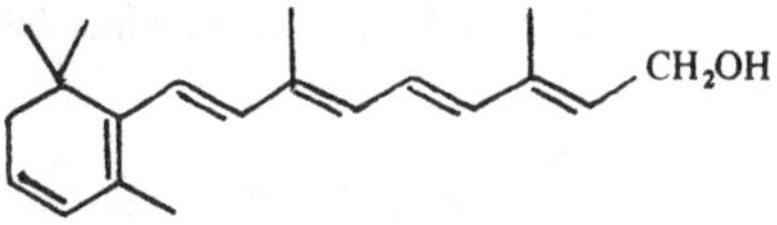

(2.88) 3, 4-Didehydroretinol

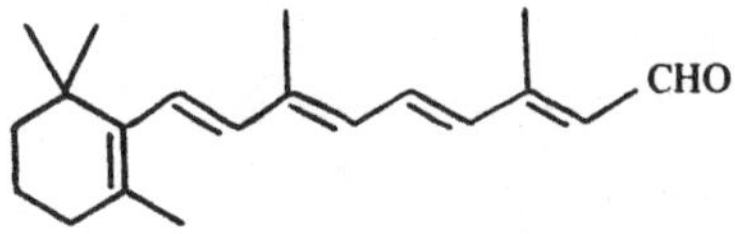

(2.89) Retinaldehyde

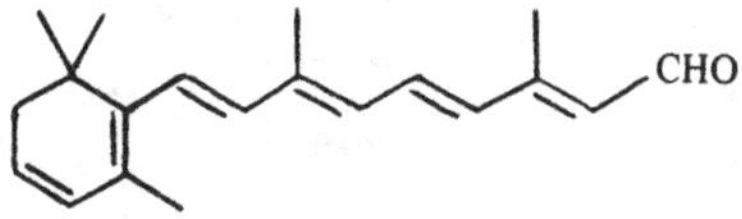

(2.90) 3, 4-Didehydroretinaldehyde

## 2.8.2 Formation of retinaldehyde and visual pigments

In many animals, including man, the most important products of metabolism of carotenoids are the vitamins A (retinol, **2.87**, and 3,4-didehydroretinol, **2.88**) and the corresponding aldehydes, retinaldehyde (**2.89**) and 3,4-didehydroretinaldehyde (**2.90**). Retinaldehyde ($\equiv$ retinal, retinene) is formed in the intestinal mucosa by oxidative cleavage of $\beta$-carotene (fig. 2.32). This process is catalysed by a $\beta$-carotene 15,15′-dioxygenase enzyme and proceeds *via* a transient peroxide intermediate (**2.91**). Retinaldehyde and retinol are readily interconverted in the presence of NAD(H) or NADP(H) by alcohol dehydrogenase enzymes present in various tissues, notably (in mammals) the liver and retina. For its transport in the blood, vitamin A is complexed with a lipoprotein, retinol binding protein, and stores of retinyl esters are maintained in the liver.

The processes of vision depend upon a group of photosensitive pigments located in the retina of the eye. These visual pigments are complexes between opsins (glycolipoproteins) and 11-*cis*-retinaldehyde or 11-*cis*-dehydroretinaldehyde. More details of these complexes and their function in vision are discussed in chapter **9**.

## 2.9 Functions of carotenoids

The main functions of carotenoids in biological tissues are largely a consequence of their light absorption properties. The role of carotenoids in photosynthesis, photoprotection, phototropism and photoreception (vision) and their contribution to the colour of tissues that contain them are discussed in section II.

No other functions of carotenoids have been well documented. The frequent occurrence of high concentrations of carotenoid in reproductive tissues of fungi, algae, plants and animals suggests a possible involvement in reproduction, but no definite role has been established.

Fig. 2.32. Formation of retinaldehyde by oxidative cleavage of β-carotene.

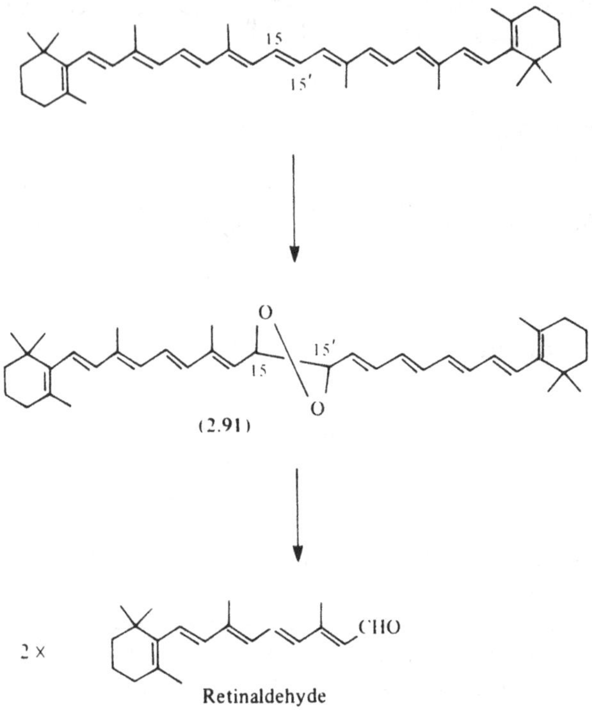

(2.91)

Retinaldehyde

2 x

## 2.10     Carotenoids as food colorants

Carotenoids have been used for many years in the food industry as colouring materials. Their natural occurrence in so many foods makes them ideally suited for this purpose. β-Carotene is used in fatty foods, notably margarine, and has the added advantage of contributing towards the body's vitamin A requirement. Water-soluble, or at least water-dispersible, formulations of β-carotene, canthaxanthin and apocarotenoids are also available and are used in colouring drinks and other food preparations. If the range of carotenoids that can be used, and especially the range of colours that can be produced, is extended, then the use of carotenoids in the food industry will become even wider.

## 2.11     Medical uses of carotenoids

In man, the primary use of carotenoids, whether taken as carotenoid-containing food or as dietary supplements, is the prevention or correction of vitamin A deficiency. In recent years, however, carotenoids have found use as protective agents against certain skin diseases which are aggravated by light.

In particular, the symptoms of erythropoietic porphyria – a condition in which porphyrin metabolism is abnormal and accumulated porphyrins serve as photosensitisers so that when the patient is exposed to sunlight symptoms such as itching, burning and swelling of the skin are seen – are greatly alleviated by administration of high doses of β-carotene. The possible use of carotenoids, especially β-carotene and canthaxanthin, as well as retinoids (vitamin A derivatives), to protect against the incidence of certain kinds of skin cancer produced by the effects of u.v. radiation and chemicals is also now under consideration.

## 2.12    Other polyene pigments

The light-absorption properties of carotenoids are due, in the main, to the conjugated polyene system. It is not surprising therefore that the absorption spectra of non-carotenoid conjugated polyenes are very similar to those of the carotenoids themselves, so much so in fact that the few known examples of other polyene pigments were at first thought to be carotenoids.

These pigments are all of fungal or bacterial origin. The fruiting bodies of the fungus *Corticium salicinum* contain the red pigment cortisalin **(2.92)** and another fungus, *Wallemia sebi*, usually found inside beehives, surprisingly contains pyrrole polyenes such as wallemia A **(2.93)**. Aryl polyenes have been obtained from some bacteria, where they occur along with carotenoids. The flexirubins, esters such as **(2.94)** and its chloro-derivative, are present in gliding bacteria of the *Cytophaga–Flexibacterium* type. Some marine *Xanthomonas* species contain brominated aryl polyene esters, *e.g.* xanthomonadin I **(2.95)**.

The biosynthesis of these pigments has not been studied. They are not likely to be isoprenoid in origin, but are probably formed by the polyketide pathway (see chapter 3). No information is available about the distribution, location within the cell, or functions of any of these pigments.

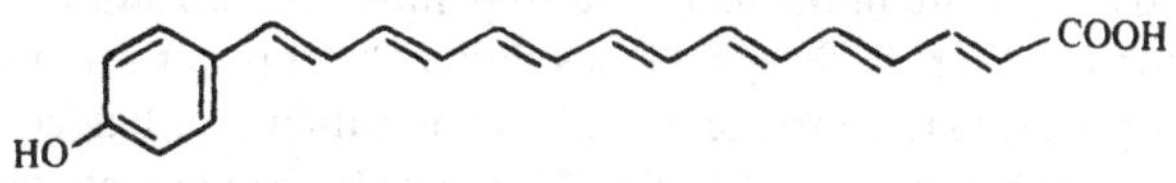

**(2.92)** Cortisalin

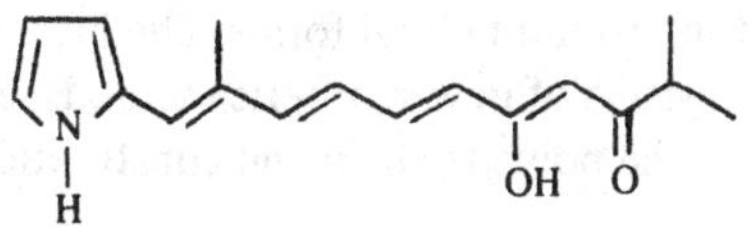

**(2.93)** Wallemia A

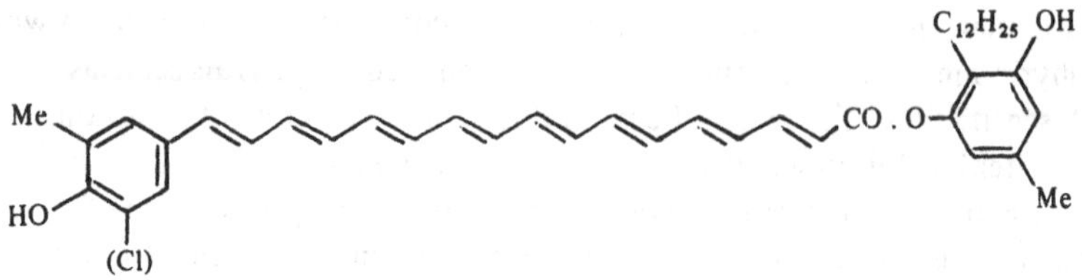

(2.94) Flexirubin

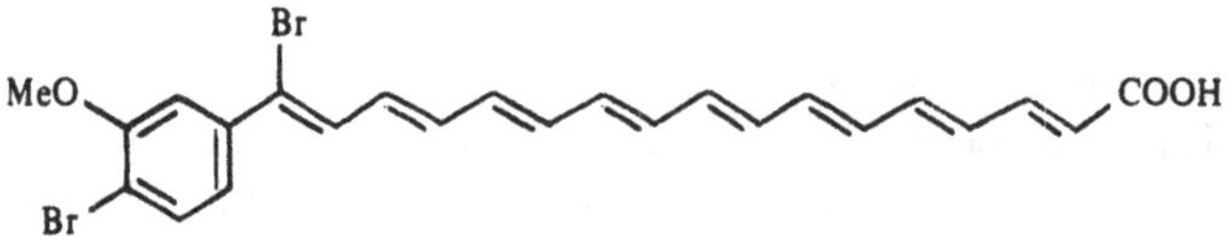

(2.95) Xanthomonadin I

## 2.13   Conclusions and comments

The carotenoids constitute a classic example of a group of natural pigments. All members of the group are closely related structures based on the conjugated polyene chromophore which is responsible for the light-absorption properties. They illustrate perfectly the correlation between absorption maxima and chromophore length. Carotenoids have a particularly wide distribution in living organisms and are involved in all of the photo-functions usually associated with natural pigments. For these reasons (not to mention the author's greater familiarity with them) the carotenoids have been dealt with in this book in rather more detail than some other groups will be.

Carotenoids continue to provide the organic chemist with opportunities to exercise his skills in structure elucidation; new structures turn up with considerable frequency, especially in microorganisms and marine animals. These new structures are in the main relatively minor modifications of the basic structure but there is always the possibility that a major variant, such as a novel ring type, may be found or a nitrogen-, sulphur- or halogen-containing carotenoid may come to light. It must also be remembered that full characterisation of a carotenoid must include determination of its stereochemistry, especially chirality, since it is now known that the 'same' carotenoid may exist in different chiral forms. The ingenuity of the chemist is also tested by the devising of new synthetic methods, especially stereo-controlled synthesis of the polyene chain and construction of chiral end-groups.

Perhaps the greatest problems in the carotenoid field are biochemical. The main biosynthetic pathways and reaction sequences are well established, and progress is being made in defining the stereochemistry and mechanism of some of these reactions. The real challenge, however, is at the enzyme level; the enzymes may well be organised as membrane-bound complexes, and thus present considerable technical difficulties. Until progress is made in this area, details of the mechanisms of regulation and control, especially photoregulation, cannot be worked out. The question of the origin of animal carotenoids may hold some surprises; the old idea that carotenoids can only be obtained from the diet now seems untenable. Animals can modify dietary carotenoids in many interesting ways, including stereochemical transformations. This promises to be a rewarding field of study, although the metabolic transformations generally take place very slowly, and are thus difficult to elucidate.

The association of carotenoids with protein, especially to produce the blue, *etc.*, complexes of marine invertebrates, is an area where progress can be expected in the near future. As well as their intrinsic interest, these complexes are very useful models of protein – small lipid molecule interactions, and results obtained should be of value in a wide area of biochemistry. The microenvironment and protein association of carotenoids in other tissues, particularly photosynthetic membranes, is also an important area for study, and should contribute to a better understanding of the mechanism of photosynthesis. There is a great deal of opportunity for original ideas and work on the functions and functioning of carotenoids in all kinds of living organisms and tissues.

The use of carotenoids as additives to food, as colouring agents and provitamins A, is already extensive, and seems likely to increase as the demand for natural as opposed to purely synthetic colouring materials increases, and as the industrial preparation of a wider variety of carotenoids becomes feasible. We can perhaps also look forward to the discovery of new roles for carotenoids in animals, including possibly the human body, and to further medicinal use of carotenoids in their own right, not merely as precursors of vitamin A.

## 2.14    Suggested further reading

An extensive monograph edited by Isler (1971) is used as the standard reference work, as it gives a very detailed survey of the chemistry, spectroscopy, biosynthesis, functions and industrial uses of carotenoids. This book is nicely complemented by the more biological and biochemical approach of Goodwin's *Comparative biochemistry of the carotenoids* (1980) in two volumes dealing with the plant (and microbial) and animal kingdoms, respectively. *Chemistry and biochemistry of plant pigments* (edited by Goodwin, 1976) contains rather less broad but extremely useful articles on the chemistry (Moss and Weedon), distribution (Goodwin) and biosynthesis

(Britton) of carotenoids. Older works such as Fox (1953), Zechmeister (1962), Karrer and Jucker (1950) contain information that can still be useful.

The whole field of carotenoid chemistry and biochemistry is brought up to date each year by exhaustive literature surveys in the *Terpenoids and steroids* series of the Chemical Society's Specialist Periodical Reports (*e.g.* Britton, 1979), and by the publication of the proceedings of the triennial series of International Symposia on Carotenoids (1966, 1969, 1972, 1975, 1978, 1981) (see Plenary and Session Lectures, 1967, 1969, 1973, 1976, 1979, 1982).

The rather specialised topic of carotenoproteins has been reviewed by Cheesman, Lee and Zagalsky (1967), Zagalsky (1976), and Britton (1981 Carotenoid Symposium), and many papers on the subject have been collected into a book by Lee (1977).

Medical uses of carotenoids have been reviewed by Mathews-Roth in the 1981 Carotenoid Symposium proceedings, and possible correlation between dietary carotenoid and cancer protection has been evaluated by Peto, Doll, Buckley and Sporn (1981).

Reviews or descriptions of methods useful for carotenoid work have been compiled by Britton and Goodwin (1971), Davies (1976) and Liaaen-Jensen and Jensen (1965) and Liaaen-Jensen (1971).

## 2.15   Selected bibliography

Britton, G. (1976) Biosynthesis of carotenoids, in *Chemistry and biochemistry of plant pigments*, 2nd edition, vol. 1, ed. T. W. Goodwin, p. 262. London, New York and San Francisco: Academic Press.

Britton, G. (1979) Carotenoids and polyterpenoids, in *Specialist Periodical Reports: Terpenoids and steroids*, Vol. 9, ed. J. R. Hanson, p. 218. London: The Chemical Society.

Britton, G. and Goodwin, T. W. (1971) Biosynthesis of carotenoids, *Methods Enzymol.*, 18C, 654.

Cheesman, D. F., Lee, W. L. and Zagalsky, P. F. (1967) Carotenoproteins in invertebrates, *Biol. Rev.*, 42, 131.

Davies, B. H. (1976) Carotenoids, in *Chemistry and biochemistry of plant pigments*, 2nd edition, vol. 2, ed. T. W. Goodwin, p. 38. London, New York and San Francisco: Academic Press.

Fox, D. L. (1953) *Animal biochromes and structural colours*. Cambridge University Press.

Goodwin, T. W. (1952) *The comparative biochemistry of carotenoids*. London: Chapman and Hall.

Goodwin, T. W. (1976) Distribution of carotenoids, in *Chemistry and biochemistry of plant pigments*, 2nd edition, vol. 1, ed. T. W. Goodwin, p. 225. London, New York and San Francisco: Academic Press.

Goodwin, T. W. (1980) *The comparative biochemistry of the carotenoids*, 2nd Edition, vol. 1. London: Chapman and Hall. (Vol. 2 in press.)

Isler, O. (ed.) (1971) *Carotenoids*. Basel and Stuttgart: Birkhäuser-Verlag.

Karrer, P. and Jucker, E. (1950) *Carotenoids*. (trans. E. A. Braude) Amsterdam: Elsevier.

Lee, W. L. (1977) *Carotenoproteins in animal coloration*. Stroudsberg, USA: Dowden, Hutchinson and Ross.

Liaaen-Jensen, S. (1971) Isolation, reactions, in *Carotenoids*, ed. O. Isler, p. 61. Basel and Stuttgart: Birkhäuser-Verlag.

Liaaen-Jensen, S. and Jensen, A. (1965) Recent progress in carotenoid chemistry, *Prog. Chem. Fats other Lipids*, 8, 129.

Moss, G. P. and Weedon, B. C. L. (1976) Chemistry of the carotenoids, in *Chemistry and biochemistry of plant pigments*, 2nd edition, vol. 1, ed. T. W. Goodwin, p. 149. London, New York and San Francisco: Academic Press.

Peto, R., Doll, R., Buckley, J. E. and Sporn, M. B. (1981) Can dietary beta-carotene materially reduce human cancer rates?, *Nature*, **290**, 201.

Plenary and Session Lectures, First International Symposium on Carotenoids other than vitamin A, 1966 (1967) *Pure Applied Chem.*, **14**, 227. Also published as *Carotenoids other than vitamin A-I*. London: Butterworth.

Plenary and Session Lectures, Second International Symposium on Carotenoids other than vitamin A (1969) *Pure Applied Chem.*, **20**, 365. Also published as *Carotenoids other than vitamin A-II*. London: Butterworth.

Plenary and Session Lectures, Third International Symposium on Carotenoids other than vitamin A, 1972 (1973) *Pure Applied Chem.*, **35**, 1. Also published as *Carotenoids other than vitamin A-III*. London: Butterworth.

Plenary and Session Lectures, Fourth International Symposium on Carotenoids, 1975 (1976) *Pure Applied Chem.*, **47**, 97. Also published as *Carotenoids-4 (Berne, 1975)*, ed. B. C. L. Weedon. Oxford: Pergamon.

Plenary and Session Lectures, Fifth International Symposium on Carotenoids, 1978 (1979) *Pure Applied Chem.*, **51**, 436–675, 857–886. Also published as *Carotenoids-5 (Madison, 1978)*, ed. T. W. Goodwin. Oxford: Pergamon.

Plenary and Session Lectures, Sixth International Symposium on Carotenoids 1981. Published (1982) as *Carotenoid chemistry and biochemistry*, eds G. Britton and T. W. Goodwin. Oxford: Pergamon.

Zagalsky, P. F. (1976) Carotenoid–protein complexes, *Pure Applied Chem.*, **47**, 103.

Zechmeister, L. (1962) Cis-trans *isomeric carotenoids, vitamin A and arylpolyenes*. Vienna: Springer.

# 3 Quinones

## 3.1 Introduction

The quinones are a large and rather heterogeneous collection of compounds. They range in colour from pale yellow, through orange, red, purple and brown to almost black, and are important pigments in certain fungi and lichens and in some groups of invertebrate animals. Quinones are also widely distributed in higher plants, but are usually present in tissues that are not normally seen, *e.g.* bark, heartwood and roots. They rarely make any contribution to the external colour of higher plant tissues. Some, however, formed the basis of important dyestuffs in the ancient world, *e.g.* henna, madder.

Some isoprenylated quinones, *e.g.* ubiquinone, menaquinone, phylloquinone and plastoquinone, are extremely important biological molecules. One or more of these molecules may be found in all living tissues, but they do not serve as pigments and hence will not be considered in detail.

## 3.2 Structures

The basic quinone structure is that of an unsaturated cyclic diketone derived from a monocyclic or polycyclic aromatic hydrocarbon. The quinone structures may be derived formally by oxidation of appropriate dihydroxyphenols. Thus the simplest examples are *ortho*- or 1,2-benzoquinone (**3.1**) and *para*- or 1,4-benzoquinone (**3.2**), benzene derivatives which may be regarded as oxidation products of the dihydroxyphenols catechol (**3.3**) and quinol (hydroquinone, **3.4**), respectively.

Other quinones may be obtained from other aromatic hydrocarbons, and normally take their names and numbering scheme from the parent hydrocarbon. Thus the two naphthalene derivatives (**3.5**) and (**3.6**) are, respectively, 1,2- and 1,4-naphthaquinone (or naphthoquinone). Of the several possible anthraquinones (or anthroquinones), only the 9,10-quinone structure (**3.7**) occurs at all frequently in natural pigments, though the 1,2-quinone structure is occasionally encountered.

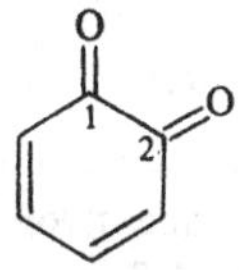

(3.1) 1, 2-Benzoquinone

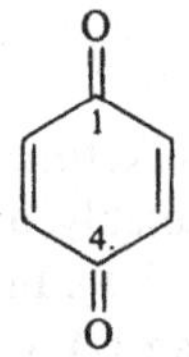

(3.2) 1, 4-Benzoquinone

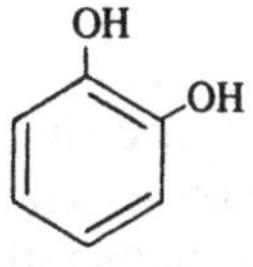

(3.3) Catechol

(3.4) Quinol

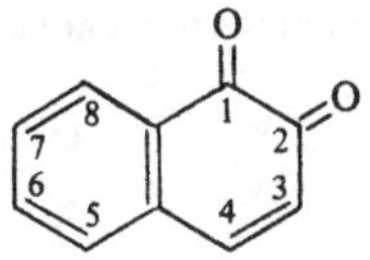

(3.5) 1, 2-Naphthaquinone

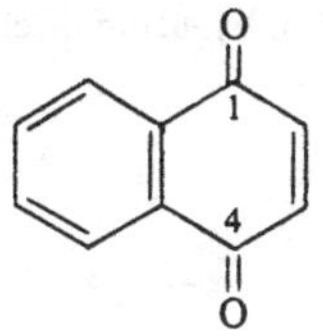

(3.6) 1, 4-Naphthaquinone

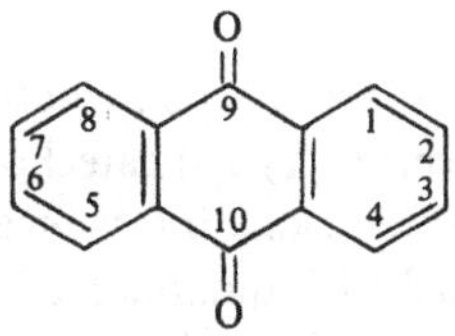

(3.7) 9, 10-Anthraquinone

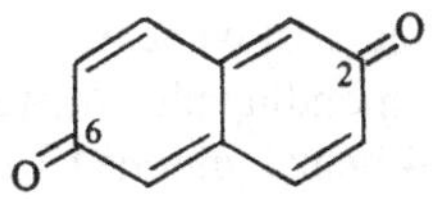

(3.8) 2, 6-Naphthaquinone

The quinone system need not be confined to one ring. The two keto-groups may be in different rings but conjugated with a suitable $\pi$-electron system to give extended quinone structures. 2,6-Naphthaquinone (3.8) is one of the simplest examples. Examples of complex structures containing an extended quinone system are seen in the aphid pigments (aphins) (see §*3.4.4*).

The quinones that are most widespread and important as pigments are the 1,4-naphthaquinones and 9,10-anthraquinones, but a few examples of other quinone structures will be encountered in the subsequent survey.

In the natural quinone pigments, the basic quinone skeleton usually has other substituent groups, methyl, hydroxy- and methoxy-groups being most frequent. The phenolic hydroxy-groups may be glycosylated (especially in higher plants). Some of the more complex quinones have longer sidechains, which may in some cases be folded into additional rings. Other complex quinones may be regarded as dimers of the common simple naphthaquinone and anthraquinone structures.

Structures of some of the more important naturally-occurring quinones are to be found in later sections of this chapter.

## 3.3    Properties
### 3.3.1    *Chemical properties*

The quinones are conjugated cyclic diketones rather than aromatic systems. They are, however, stabilised considerably by resonance involving charged aromatic contributing structures (fig. 3.1). In general, 1,2-quinones are less stable and therefore more reactive than 1,4- or extended quinone systems.

Fig. 3.1. Contribution of charged aromatic structures to quinone stabilisation.

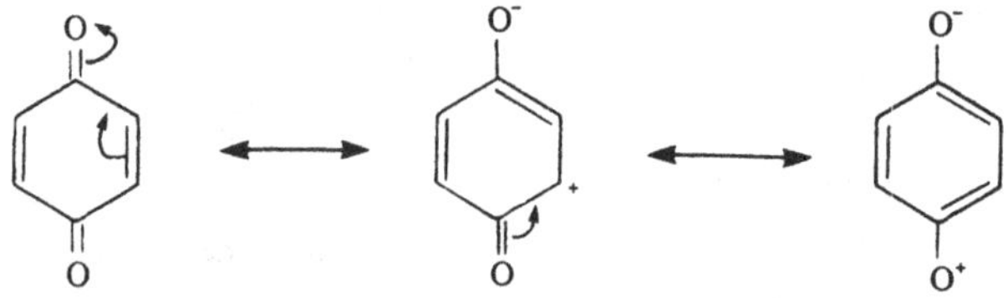

The characteristic chemical reaction of quinones is reversible reduction to the corresponding phenol. Thus 1,2-benzoquinone (**3.1**) and catechol (**3.3**), or 1,4-benzoquinone (**3.2**) and quinol (**3.4**) are readily interconverted. Quinones with relatively little substitution, especially 1,2-quinones, are also particularly susceptible to polymerisation by free-radical coupling processes.

Many natural quinones have additional phenolic hydroxy-substituents and are thus slightly acidic, consequently ionising and forming salts in alkaline solution. This property is especially pronounced in the case of hydroxy-benzoquinone (**3.9**) and 2-hydroxy-1.4-naphthaquinone (**3.10**) which are vinylogous carboxylic acids.

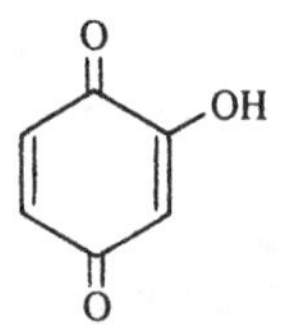

(**3.9**) Hydroxybenzoquinone 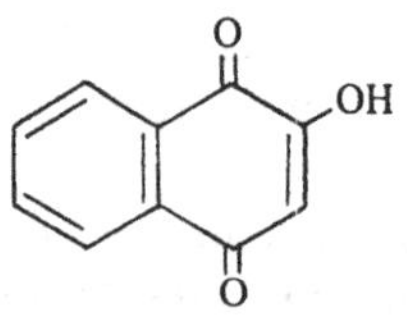          (**3.10**) 2-Hydroxy-1, 4-naphthaquinone

### 3.3.2    *Physical properties*

Almost all natural quinones are solids and crystallise readily. Most are readily soluble in organic solvents, though glycosides and some carboxylic acids will dissolve in water. Those quinones that are also phenols or carboxylic acids are soluble in alkaline aqueous solution.

Higher polycyclic quinones behave rather like polymers and cannot readily be dissolved in any aqueous or organic medium.

### 3.3.3    Light absorption

A detailed theoretical explanation of the light-absorption properties of quinones is beyond the scope of this book. A brief descriptive account of the absorption spectra of quinones and the factors that affect the wavelengths of maximal absorption is, however, essential to any assessment of their value as pigments.

The absorption spectra of the parent quinones are illustrated in fig. 3.2. The simplest spectrum, that of 1,4-benzoquinone, has an intense absorption band (I) near 240 nm, a medium-intensity 'electron-transfer transition' band (II) at *ca* 285 nm and only weak quinonoid n → $\pi^*$ absorption in the visible region, *ca* 434 nm (band III). For the unstable 1,2-benzoquinone, the absorption maxima are at considerably longer wavelengths, bands II and III appearing at 375 and 568 nm, respectively. Quinonoid n → $\pi^*$ transitions are 'forbidden', consequently band III absorption is of low intensity. Although this absorption lies in the visible region it is nevertheless too weak to make the compounds coloured except in very high concentrations.

The spectra of naphthaquinones, anthraquinones and higher quinones are considerably more complex, because of the presence of absorption bands due to benzenoid transitions, in addition to quinonoid absorptions. The main absorption bands of 1,4-naphthaquinone are at 245, 257 and 335 nm, and those of anthraquinone at 243, 263, 332 and 405 nm (fig. 3.2).

Fig. 3.2. The main absorption bands in the electronic spectra of 1,2-benzoquinone (++++), 1,4-benzoquinone (-----), 1,4-naphthaquinone (- - - -) and anthraquinone (———). The intensities (logarithmic scale) of the bands for each compound are only approximate, and no comparison between the relative intensities of absorption bands of different compounds is implied.

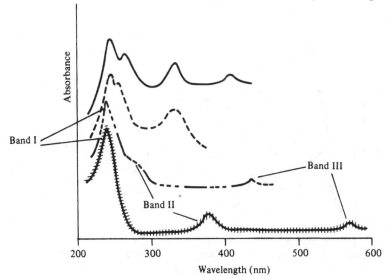

Substituent groups influence quinone spectra very greatly. Introduction of a substituent into 1,4-benzoquinone produces only small effects on bands I and III, but band II undergoes a significant red shift, in the order —Me (27 nm), —OMe (69 nm), —OH (81 nm). The effect of a second substituent is smaller. The result is that in substituted benzoquinones, band II absorption may be shifted into the visible region, so that the compounds appear coloured.

In the case of naphthaquinones (fig. 3.3), substitution in the quinone ring has little effect on the absorption maxima, but substitution in the aromatic ring, especially with hydroxy-groups, causes bathochromic effects which may shift certain u.v. absorption bands into the visible region. *Peri*-hydroxylation (*e.g.* positions 5 and 8 of 1,4-naphthaquinone) causes the most marked effect; the benzenoid band may shift by almost 100 nm (*e.g.* 5-hydroxy-1,4-naphthaquinone (**3.11a**) from 335 to 429 nm). Naphthazarins (**3.11b**) with two *peri*-hydroxy groups, exhibit strong multibanded benzenoid absorption centred around 525 nm. This structure forms the basis of the spinochromes, strongly coloured pigments found in echinoid animals (see §*3.4.2*).

Similarly with anthraquinones, complex benzenoid absorption is observed and the dominating influence on $\lambda_{max}$ is the presence of hydroxy- and alkoxy-groups, especially multiple substituents. The influence of α-hydroxy-

Fig. 3.3. Absorption spectra of 1,4-naphthaquinone (———), 5-hydroxy-1,4-naphthaquinone (– – – –) and 5,8-dihydroxy-1,4-naphthaquinone (—— —— ——). Intensities (logarithmic scale) are only approximate, and those for different compounds should not be compared directly.

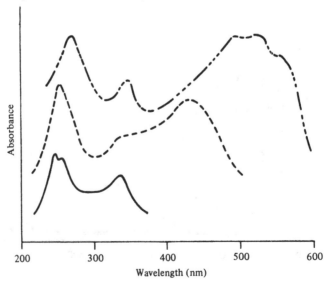

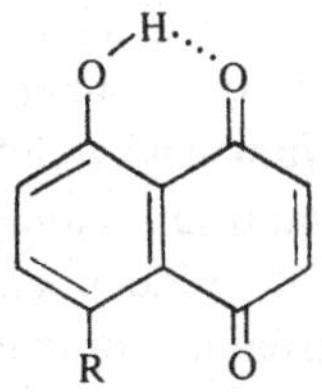

(3.11 a)  R = H; 5-hydroxy-1, 4-naphthaquinone
(3.11 b)  R = OH; 5, 8-dihydroxy-1, 4-naphthaquinone

groups is greater than that of $\beta$-hydroxyls (except when the $\beta$-hydroxyl is adjacent to an $\alpha$-hydroxyl).

Hydroxy-derivatives of quinones are capable of ionising under basic conditions, and this results in considerable bathochromic shifts. These alkali shifts are of great diagnostic value, especially in the naphthaquinone and anthraquinone series. Thus, for example, the anion of 2-hydroxy-1,4-naphthaquinone (3.12) is orange ($\lambda_{max}$ at 459 nm). Similarly, the anion of 5-hydroxy-1,4-naphthaquinone is violet ($\lambda_{max}$ 538 nm) and those of 2,3-, 5,6- and 5,8-dihydroxy-1,4-naphthaquinone blue ($\lambda_{max}$ 650, 571 and 655 nm, respectively). Another useful diagnostic feature is that reduction (with NaBH$_4$) to the quinol produces an aromatic system and a benzenoid type of spectrum.

Quinone absorption spectra are usually determined in ethanol; when other solvents are used, *e.g.* CHCl$_3$, somewhat different $\lambda_{max}$ values are obtained. Tables of absorption maxima of natural and model quinones are presented in the monograph by Thomson (1971), which also includes discussion of infrared, n.m.r. and mass spectra.

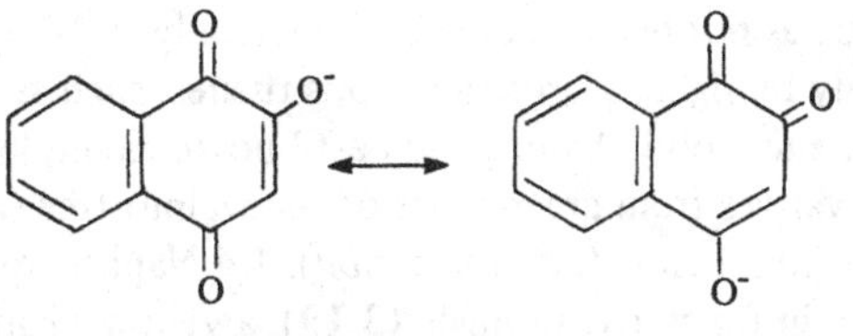

(3.12)

## 3.4    Occurrence and distribution

Coloured quinones are found chiefly in higher plants, fungi and bacteria, and, in the animal kingdom, in arthropods and echinoderms. Quinones are very readily formed by oxidation of the corresponding quinols. In many cases the quinones that are isolated may be artefacts of the extraction procedure or may be produced during air-drying of plants or tissues.

### 3.4.1    Benzoquinones

About 100 benzoquinones, almost all of them 1,4-benzoquinones, have been isolated from natural sources, mainly fungi and various tissues of higher plants. The benzoquinones include such important biological molecules as plastoquinone (**3.13**) and ubiquinone (**3.14**). Ubiquinone samples are usually isolated as mixtures of molecules with different isoprenoid chain lengths (isoprenylogues). In any tissue or species one isoprenylogue usually predominates. Ubiquinones-9 and -10, with nine and ten isoprene residues, respectively, are the most common. In general, benzoquinones are not strongly coloured and make no appreciable contribution to natural colour.

(**3.13**) Plastoquinone                    (**3.14**) Ubiquinone-*n*

### 3.4.2    Naphthaquinones

The natural naphthaquinones include the vitamins K, phylloquinone (**3.15**) and menaquinone (**3.16**), widely distributed, biologically important molecules that do not serve as pigments. These substances have the same naphthaquinone nucleus but different isoprenoid sidechains. Phylloquinone (vitamin $K_1$), from higher plants, has a phytyl sidechain, whereas the bacterial menaquinones (vitamin $K_2$), like ubiquinone, are variable in chain length. Six to nine isoprene units are most frequent. The distribution of other naphthaquinones in plants has not been studied systematically. 1,4-Naphthaquinones occur sporadically in higher plants in various tissues, such as leaves, flowers, fruits, roots, bark and wood. Among the best-known examples are juglone (**3.17**) and its derivatives from green parts of the walnut tree (*Juglans regia*), and lawsone (**3.18**) from henna (*Lawsonia alba*). 1,4-Naphthaquinones are also occasionally found in fungi, *e.g.* mollisin (**3.19**), a yellow pigment from cultures of *Mollisia fallens*. Other naphthaquinone types (1,2-, 1,5-, 2,6-) are rarely encountered in Nature.

*Spinochromes.* In the animal kingdom, some 20 closely related, strongly coloured, red, purple or blue naphthaquinone pigments have been found in echinoderms, mostly in sea urchins, but also in brittle stars and starfish. These compounds, known as spinochromes or echinochromes, were first isolated from the calcareous parts, *i.e.* spines and test (shell) of sea urchins, but are also present in the perivisceral fluid, eggs and internal organs. Most

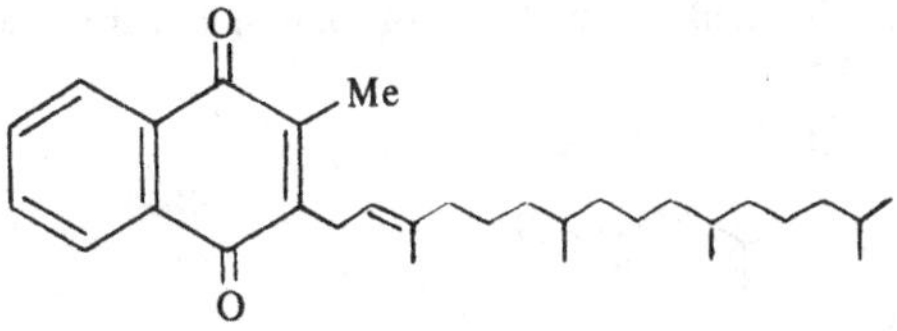

(3.15) Phylloquinone

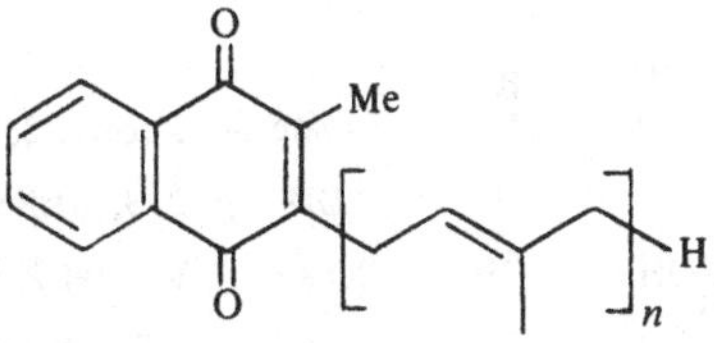

(3.16) Menaquinone-*n*

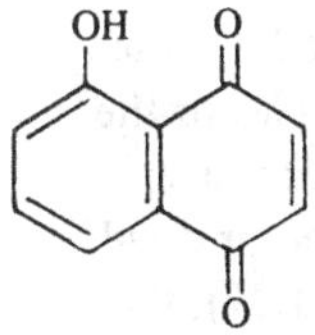

(3.17) Juglone 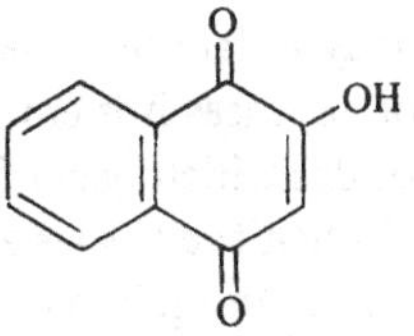                (3.18) Lawsone 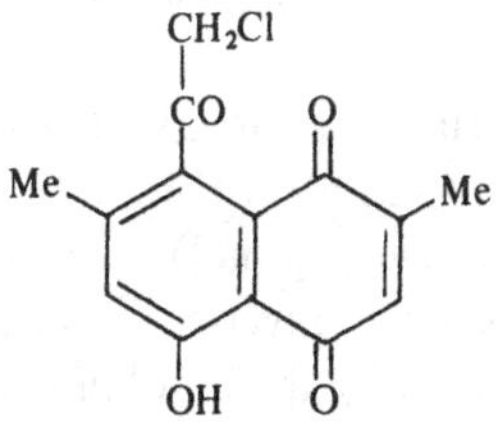                (3.19) Mollisin

species yield a mixture of six or more pigments, which are present in the spines and test chiefly as calcium and magnesium salts, and in other tissues probably as protein complexes. Structurally these echinoderm pigments are highly substituted and highly oxygenated derivatives of juglone or naphthazarins.

Considerable confusion has arisen through the extensive use of trivial names. All the main compounds have been given several synonyms. The names spinochromes A–E and echinochrome A are now accepted for the six major members of the group, but the use of a semi-systematic nomenclature whereby the pigments are named as substituted juglones or naphthazarins is recommended. The structures (3.20–3.25) and semi-systematic names of these main compounds are given in fig. 3.4.

### 3.4.3     Anthraquinones

The anthraquinones are the largest group of quinones in Nature. Almost 200 examples have been found in plants, fungi and lichens. The anthraquinone which probably has the widest apparent distribution is emodin (3.26) which has been isolated from moulds, higher fungi, lichens, flowering

Fig. 3.4. Structures and nomenclature of some spinochrome and echinochrome pigments.

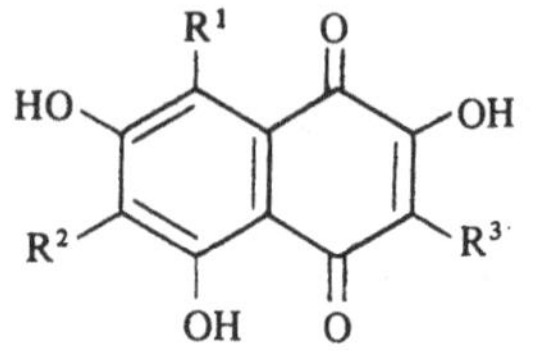

|            |                | R¹     | R²        | R³   |                                      |
|------------|----------------|--------|-----------|------|--------------------------------------|
| (3.20)     | Spinochrome A  | OH     | _H        | COMe | 3-acetyl-2, 7-dihydroxynaphthazarin  |
| (3.21)     | Spinochrome B  | H      | H         | OH   | 2, 3, 7-trihydroxyjuglone            |
| (3.22)     | Spinochrome C  | OH     | OH        | COMe | 3-acetyl-2, 6, 7-trihydroxynaphthazarin |
| (3.23)     | Spinochrome D  | OH     | H         | OH   | 2, 3, 6-trihydroxynaphthazarin       |
| (3.24)     | Spinochrome E  | OH     | OH        | OH   | 2, 3, 6, 7-tetrahydroxynaphthazarin  |
| (3.25)     | Echinochrome A | OH     | CH₂CH₃    | OH   | 6-ethyl-2, 3, 7-trihydroxynaphthazarin |

plants and insects. However, there may in fact be little or no emodin present
in fresh plant tissues. In most, if not all, cases it is emodin glycosides or the
reduced (quinol) form and its glycosides which occur *in vivo* and these are
readily converted into emodin itself when the tissues are disrupted or dried.
Probably the best-known higher plant anthraquinone is alizarin (3.27), the
principal pigment of madder, the ground root of *Rubia tinctorum*, which was
perhaps the most important of ancient dyestuffs.

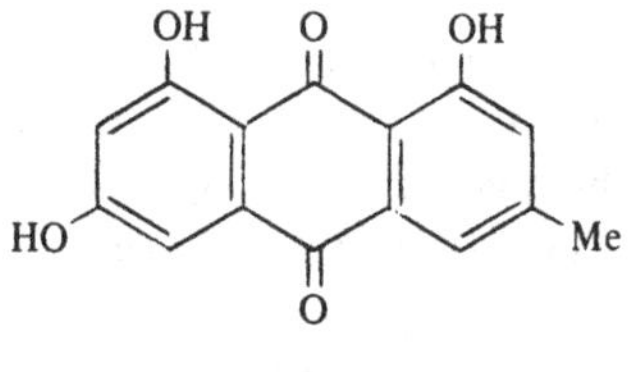

(3.26) Emodin 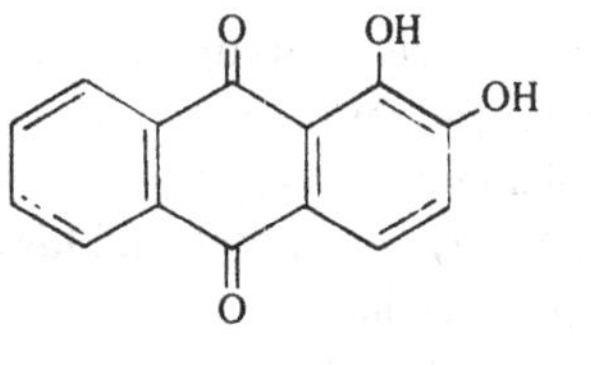                    (3.27) Alizarin

*Animal anthraquinones*. Some insects of the Coccidae are coloured by anthraquinones. Most of these pigments are carboxylic acids, and some, notably
the C-glucoside carminic acid (3.28), the carmine-red colouring matter of
cochineal, and the red kermesic acid (3.29), were used for many centuries
as dyes.

It has recently been discovered that a group of Australian crinoids (sea
lilies) are coloured by red or purple anthraquinones such as rhodoptilometrin
(3.30) and similar pigments which have the extended sidechain at position 4.
These compounds are very similar to some anthraquinone fungal metabolites.
Hallachrome, a pigment of certain polychaete worms is interesting as an
example of the rare 1,2-anthraquinone structure (3.31).

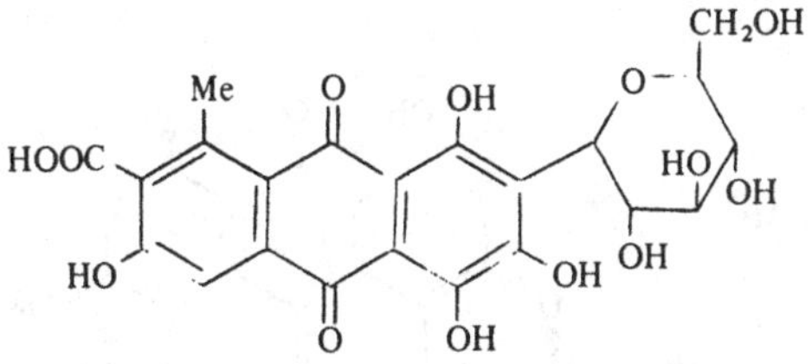

(3.28) Carminic acid

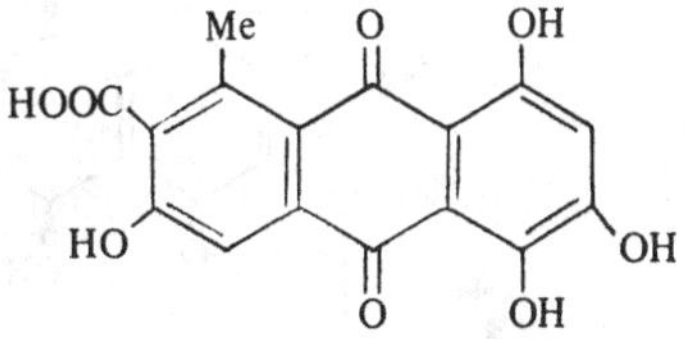

(3.29) Kermesic acid

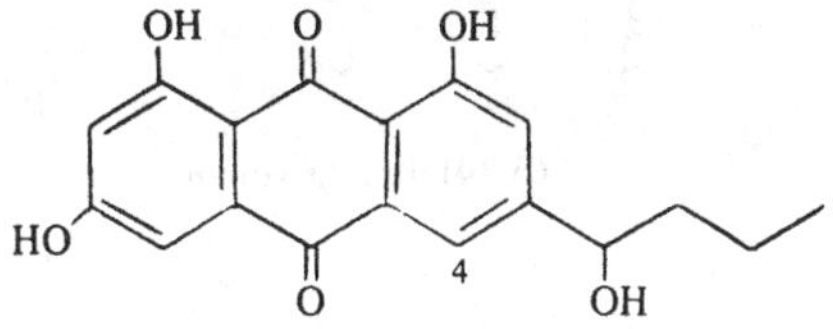

(3.30) Rhodoptilometrin

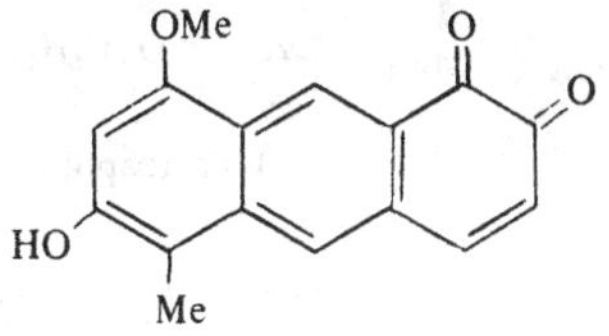

(3.31) Hallachrome

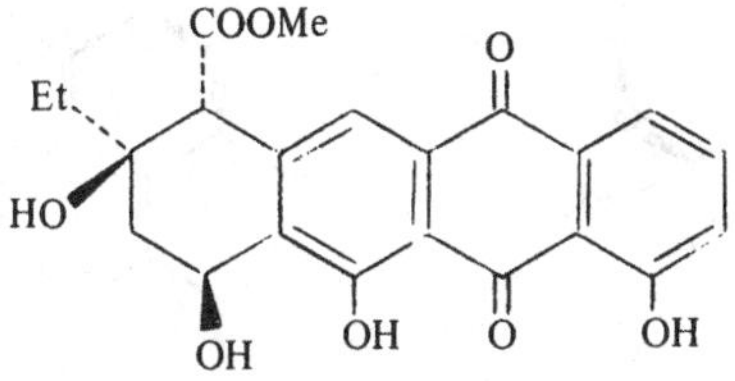

(3.32) Aklavinone

*Anthracyclinones*. The anthracyclinones from cultures of some streptomycetes are obviously related to the tetracycline antibiotics. Structurally, however, they may be regarded as substituted anthraquinones, *e.g.* aklavinone (3.32).

## 3.4.4 *Extended quinones*

Amongst the higher, or extended quinones, the aphins are particularly interesting and have been studied most intensively. They are derived from protoaphins which are present in the haemolymph of various aphids. The protoaphins, *e.g.* (3.33), are clearly naphthaquinone derivatives, which are converted, after the death of the insect, into a stable, red end product erythroaphin (3.34) *via* less stable intermediates, the yellow xanthoaphin (3.35) and the orange chrysoaphin (3.36). Different aphid species elaborate different proto-, xantho-, chryso- and erythroaphin structures with different stereochemistry. The ones illustrated are those of pigments first isolated from *Aphis fabae* (blackfly). All these compounds are extended quinones, in which the two carbonyl groups are separated by a considerable distance in a complex polycyclic conjugated structure.

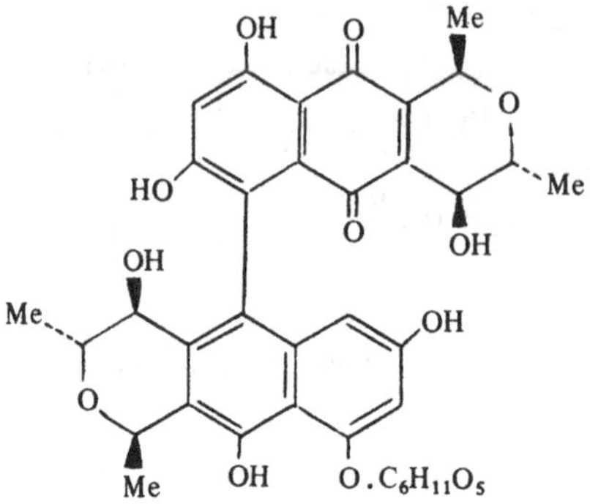

(3.33) Protoaphin

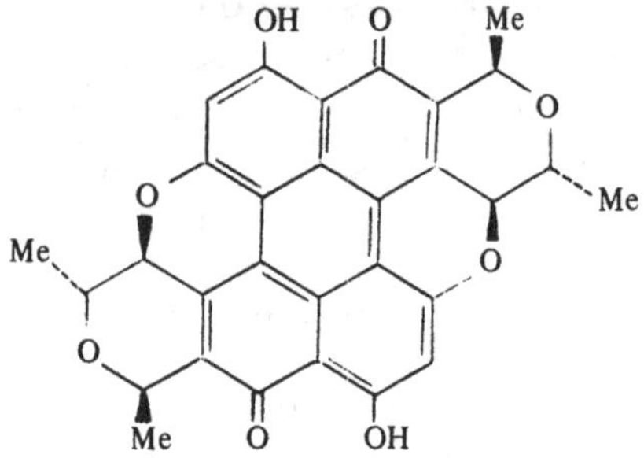

(3.34) Erythroaphin

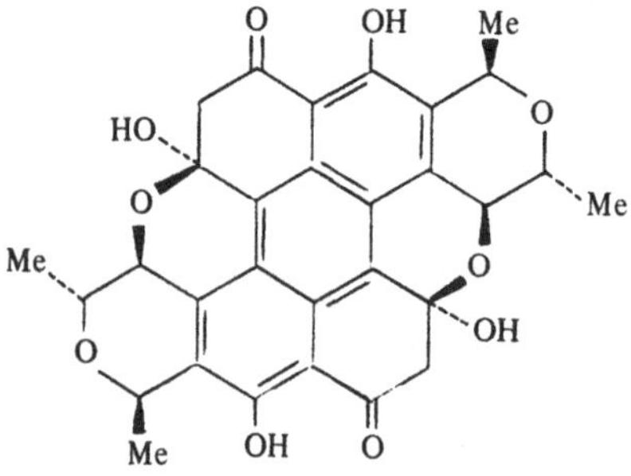

(3.35) Xanthoaphin

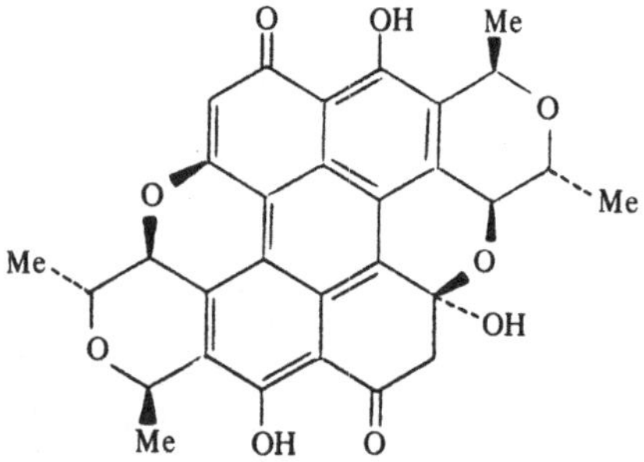

(3.36) Chrysoaphin

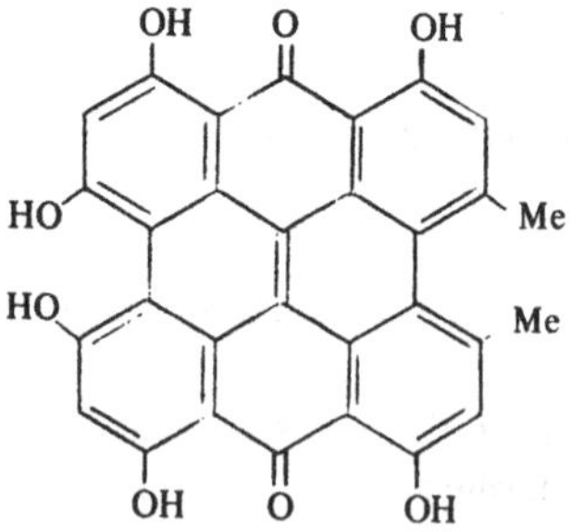

(3.37) Hypericin

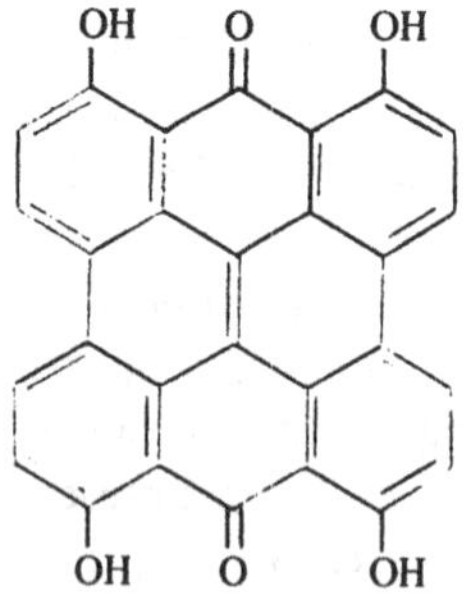

(3.38) Fringelite

Other interesting complex quinones include hypericin (3.37), present in plants of the genus *Hypericum* (see §3.7.2) and several compounds known as fringelites, *e.g.* (3.38) which have been found in fossil crinoids of Jurassic age. Hypericin and fringelites are essentially anthraquinone dimers.

## 3.5     Contribution to colour

The isoprenylated quinones, ubiquinone, plastoquinone, *etc*., which are found universally in higher plants, make no contribution to colour. Other quinones, especially naphthaquinones and anthraquinones, are also distributed quite widely in various tissues of higher plants, but again they rarely make much contribution to external colour. They are, however, often responsible for the colours, usually yellow, orange or brown, of moulds, higher fungi and lichens.

It is in the animal kingdom that the best examples of pigmentation by quinones are found. Thus the spinochromes and echinochromes provide the purple, blue or green colours of most sea urchins; the beautifully coloured, highly decorative tests of some species are frequently displayed in the home. The anthraquinone acids are responsible for the red colours of some coccid insects and thence of the food colorant cochineal. The protoaphins present in the haemolymph of many aphid species are ionised at biological pH to form violet-red anions which are chiefly responsible for the dark colours of the haemolymph. After death of the insects, haemolymph enzymes convert the protoaphins into stable, coloured products, the aphins, which give colour to the insects *post mortem*. Aphins are characteristic of dark-coloured aphid species but related green, water-soluble glycosides, the aphanins, may be present in some species, such as the well-known greenfly, which do not have aphins.

## 3.6     Biosynthesis
### 3.6.1     Introduction

In the biosynthesis of every other main class of pigments one central pathway is employed to give the basic structure. The various individual compounds are then elaborated by later modification of this basic structure. The quinones do not follow this pattern. They are biosynthetically heterogeneous. In extreme cases different organisms may elaborate the same quinone by different pathways.

Most quinones are formed by one of two alternative major pathways. In some cases a third major pathway contributes some of the quinone skeleton. The same basic pathways, however, give rise to all the various groups of quinones, *i.e.* benzoquinones, naphthaquinones, anthraquinones and extended quinones.

The number of individual quinones whose biosynthesis has been investigated is very small. In many cases the biosynthetic pathway has merely been inferred from structural considerations. As Thomson (1971) puts it, 'A molehill of fact is supporting a mountain of speculation'.

### 3.6.2    General considerations

Quinones are oxidised aromatic compounds, so it is not surprising that they may be biosynthesised by an extension of the major pathways by which aromatic compounds in general are biosynthesised. The two pathways used are the polyketide route from acetate and malonate, and the shikimate pathway. In the case of some naphthaquinones and anthraquinones a third major biosynthetic route, the isoprenoid pathway, is also involved. In general, knowledge of the biosynthesis of quinones is still too fragmentary for even the broadest regulating factors to be recognised.

In this account, only a brief outline of the essential details of the basic pathways will be presented, but some of the more important and more extensively documented examples will be given.

### 3.6.3    The polyketide pathway

This is probably the route used most frequently for the biosynthesis of quinone pigments. It is a process rather similar to the chain-lengthening steps of fatty acid biosynthesis, and is illustrated in fig. 3.5. In the first stage,

Fig. 3.5. The basic polyketide pathway for the biosynthesis of phenols and quinones.

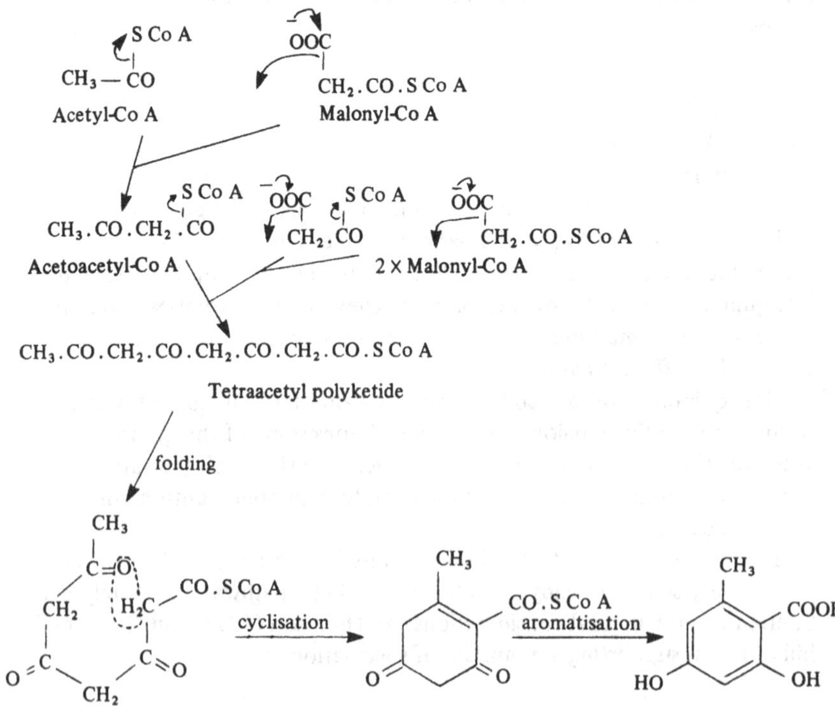

condensation of acetyl-CoA with malonyl-CoA gives acetoacetyl-CoA. In contrast to the situation in fatty acid biosynthesis, the keto-group formed is not reduced at this stage. Instead, the chain-lengthening process continues by addition of further $C_2$ units from malonyl-CoA until a chain of the required length has been built up. This pattern of alternating CO and $CH_2$ groups is the polyketide system.

When a polyketide is folded in a suitable way, cyclisation can take place by elimination of water from appropriately positioned CO and $CH_2$ groups, leading to an aromatic or quinone molecule.

The pattern of oxygen functions on alternate carbon atoms can frequently be discerned in molecules biosynthesised by this route, although addition or removal of substituent groups often disguise the polyketide pattern. It may be possible to demonstrate that a quinone is formed by the polyketide pathway by incorporation experiments with acetate (or malonate) labelled with radioactive or stable isotopes. The manner of polyketide folding may be revealed by degradation of the labelled molecule.

*Benzoquinones*. A simple example in the benzoquinone series is the fungal metabolite spinulosin (**3.41**) which is formed from a tetraacetate polyketide (**3.39**) *via* orsellinic acid (**3.40**) (fig. 3.6).

Fig. 3.6. Polyketide mechanism for biosynthesis of the benzoquinone, spinulosin.

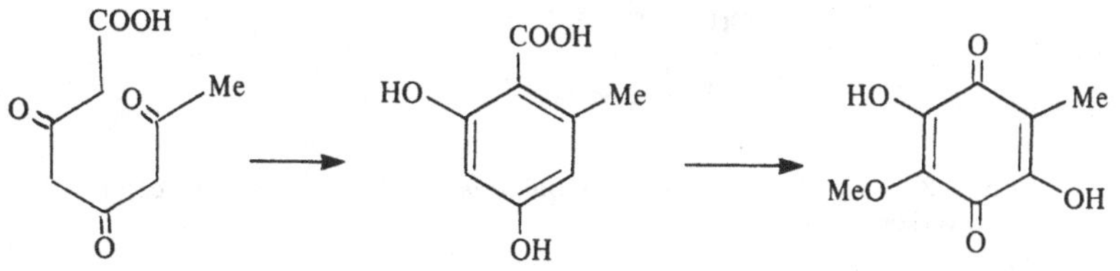

(3.39)          (3.40) Orsellinic acid          (3.41) Spinulosin

*Naphthaquinones*. In the naphthaquinone series, feeding experiments with labelled acetate and malonate in cultures of the mould *Fusarium javanicum* have established the origin of javanicin (**3.43**) from a heptaacetyl polyketide (**3.42**) as illustrated in fig. 3.7. Reduction of the terminal carboxyl group to methyl is unusual. A less simple example is provided by mollisin (**3.19**), another fungal metabolite, produced by cultures of *Mollisia caesia*. Biosynthesis again occurs from acetate and malonate, but in this instance two polyketide chains, of three and four $C_2$ units, respectively, are used rather than one longer chain.

*Anthraquinones*. It is probable that many, perhaps most anthraquinones, especially those of the emodin (**3.26**) type are also produced by polyketide

mechanisms. Emodin itself is thought to arise by suitable folding and
condensation of an octaacetyl polyketide chain as outlined in fig. 3.8. Introduction or removal of substituent groups at later stages can then give rise
to the numerous natural emodin-like quinones. Alternative patterns for
folding of the polyketide chain are required to account for the biosynthesis

Fig. 3.7. Formation of the naphthaquinone javanicin from a heptaacetyl
polyketide.

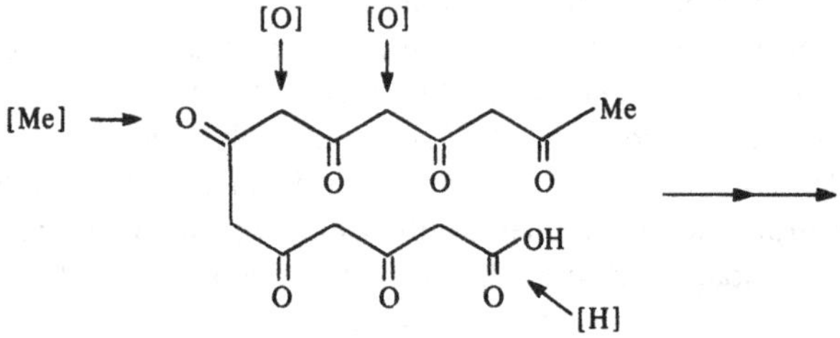

(3.42)

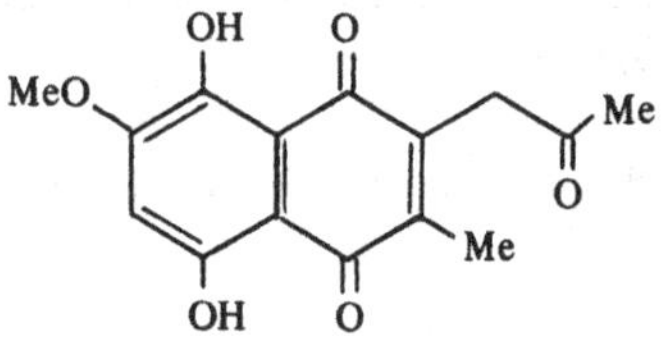

(3.43) Javanicin

Fig. 3.8. Biosynthesis of the anthraquinone emodin from an octaacetyl
polyketide.

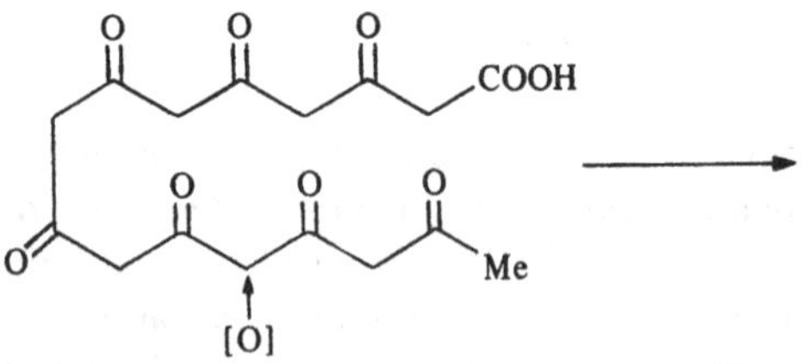

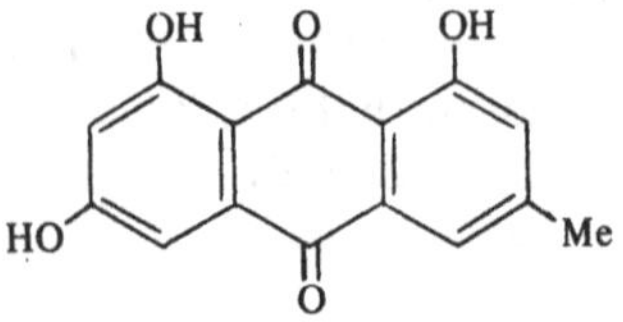

(3.26) Emodin

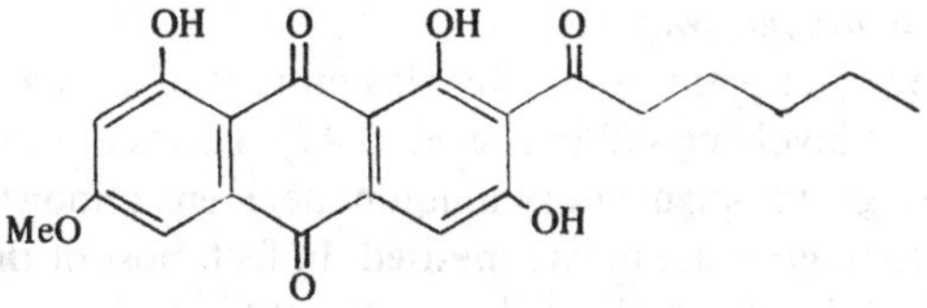

**(3.44)** Solorinic acid

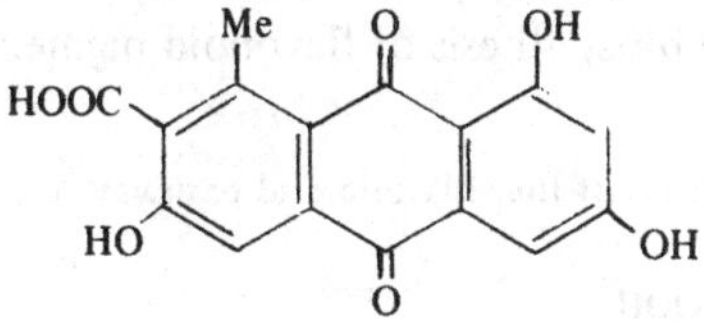

**(3.45)** Laccaic acid D

of other anthraquinones, such as the lichen pigment solorinic acid **(3.44)** and the insect quinone laccaic acid D **(3.45)**.

*Extended quinones*. Many of the larger natural quinones, especially those produced by fungi, also appear to be formed by the acetate–malonate pathway. Tracer studies have shown that the anthraclinone ε-pyrromycinone **(3.46)** arises from nine acetate (malonate) units together with one propionate unit which serves as the starter on which the chain is built up (fig. 3.9).

Fig. 3.9. Formation of the anthraclinone ε-pyrromycinone from a polyketide.

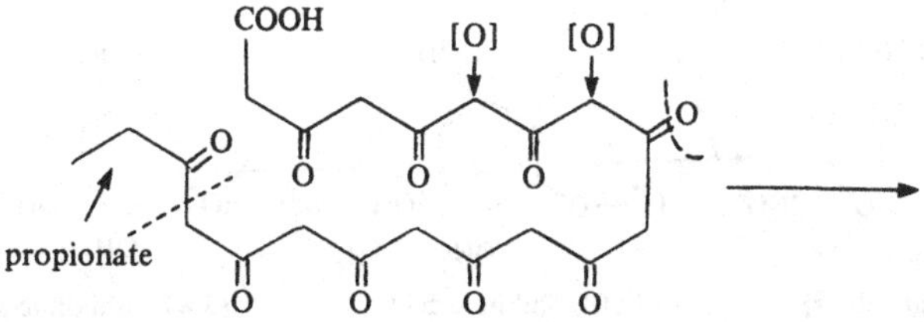

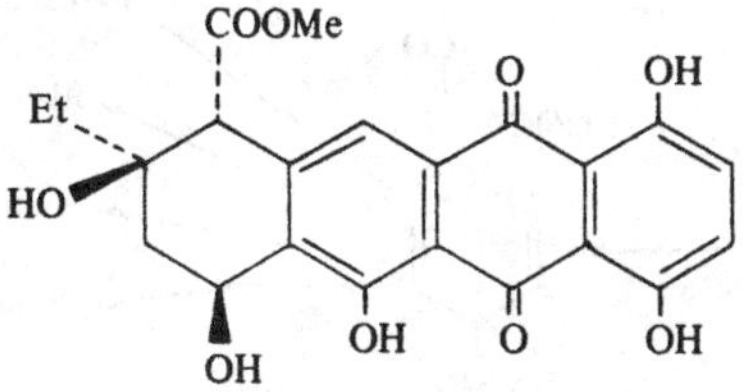

**(3.46)** ε-Pyrromycinone

### 3.6.4 *The shikimate pathway*

The other main pathway for the biosynthesis of aromatic compounds and quinones is that involving shikimic acid **(3.47)**. This is the route by which the important though non-pigmentary ubiquinone, menaquinone, phylloquinone and plastoquinone are biosynthesised. In fact most of the details of quinone biosynthesis by this pathway have been obtained from studies of ubiquinone and menaquinone formation. Many coloured quinones, especially naphthaquinones, also arise by this route. Shikimate will also be encountered as an intermediate in the biosynthesis of flavonoid pigments (see chapter **4**).

Fig. 3.10. Early stages of the shikimic acid pathway of phenol and quinone biosynthesis.

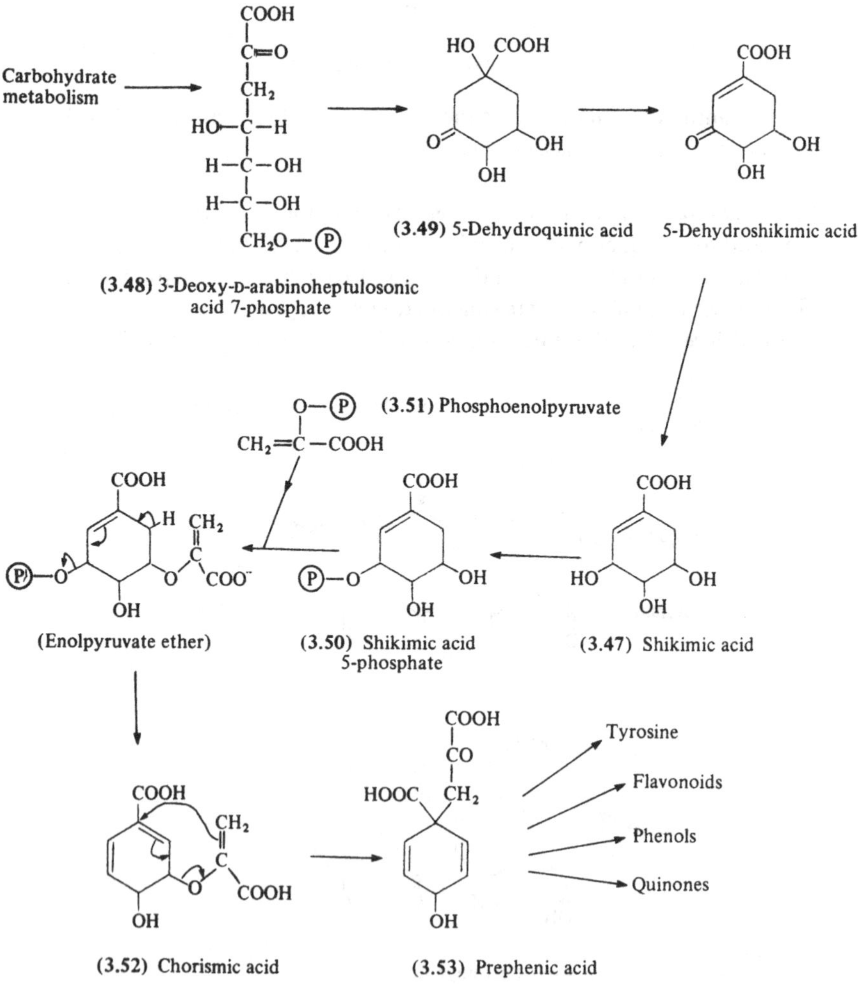
Carbohydrate metabolism
**(3.48)** 3-Deoxy-D-arabinoheptulosonic acid 7-phosphate
**(3.49)** 5-Dehydroquinic acid
5-Dehydroshikimic acid
**(3.51)** Phosphoenolpyruvate
(Enolpyruvate ether)
**(3.50)** Shikimic acid 5-phosphate
**(3.47)** Shikimic acid
**(3.52)** Chorismic acid
**(3.53)** Prephenic acid
Tyrosine
Flavonoids
Phenols
Quinones

Details of the early stages of the shikimate pathway are given in fig. 3.10. Precursors from the general carbohydrate metabolic pool give the $C_7$ sugar 3-deoxy-D-arabinoheptulosonic acid 7-phosphate (DAHP, **3.48**) which undergoes cyclisation to give 5-dehydroquinic acid (**3.49**). This is converted into shikimic acid (**3.47**) and shikimic acid 5-phosphate (**3.50**). Addition of a further $C_3$ unit (as phosphoenolpyruvate, **3.51**) to the latter gives chorismic acid (**3.52**) which rearranges to prephenic acid (**3.53**). These compounds are extremely important biosynthetic intermediates, being precursors to the aromatic amino acids phenylalanine and tyrosine, to the flavonoid pigments and to many aromatic natural products. The natural quinones are formed either *via* prephenic acid or from earlier intermediates such as shikimic or chorismic acids.

*Benzoquinones*. The benzoquinone ring of ubiquinone is formed from shikimate *via* chorismate (**3.52**) and *p*-hydroxybenzoate (**3.54**) (fig. 3.11). It is

Fig. 3.11. Biosynthesis of the benzoquinone ring of ubiquinone. Many alternative sequences are possible for introduction of the prenyl sidechain and methyl and methoxy-substituents. Other benzoquinones may be biosynthesised similarly from shikimate *via* *p*-hydroxybenzoate or homogentisate.

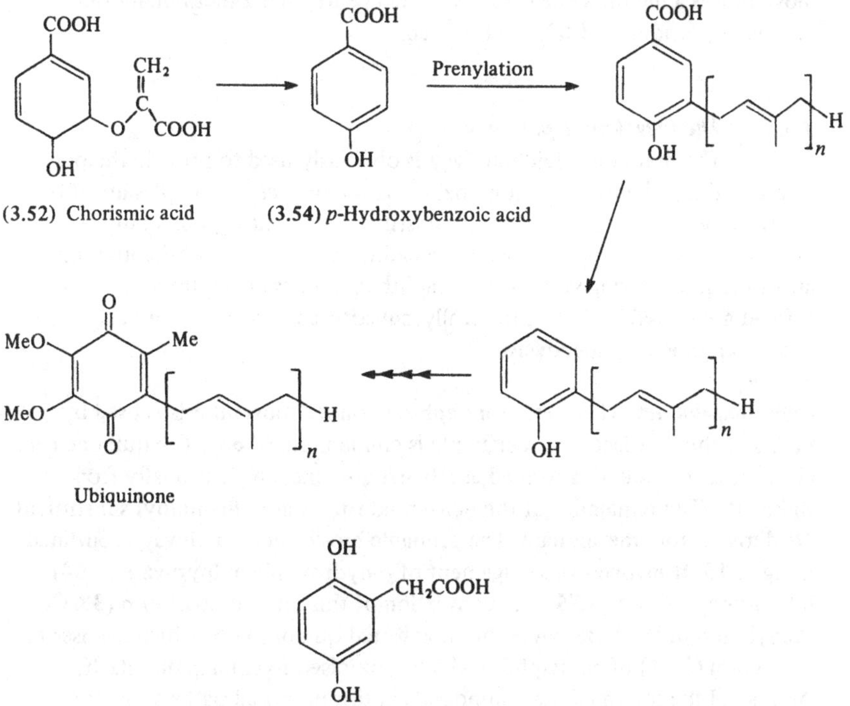

(3.52) Chorismic acid     (3.54) *p*-Hydroxybenzoic acid

Ubiquinone

(3.55) Homogentisic acid

probable that some other natural benzoquinones arise from shikimate either in a similar way from $p$-hydroxybenzoate or perhaps *via* homogentisate (**3.55**) or even the $C_6$-$C_3$ cinnamic acids (see chapter 4) but experimental details are lacking.

*Naphthaquinones.* The role of shikimate in quinone biosynthesis is best understood in the case of naphthaquinones. The main details of the pathway for formation of menaquinone and phylloquinone, and of lawsone (**3.18**) and juglone (**3.17**) have been elucidated (fig. 3.12). Feeding experiments show that the shikimate is incorporated as an intact $C_7$ unit to form the benzenoid (A) ring and one of the quinone carbonyl groups. The three remaining carbon atoms are derived from the non-carboxyl carbons of glutamate (**3.56**) or $\alpha$-oxoglutarate (**3.57**). The key intermediate seems to be $o$-succinylbenzoate (**3.58**) formed by addition to shikimate of the thiamine pyrophosphate derivative of succinic semialdehyde (obtained from glutamate or $\alpha$-oxoglutarate). The final steps by which $o$-succinylbenzoate is converted into lawsone, juglone and related quinones are still not clear. It is thought that 1,4-naphthaquinone itself (**3.59**) and perhaps also 1,4-naphthaquinol (**3.60**) are intermediates in the pathway leading to juglone. In menaquinone biosynthesis a naphthalene carboxylic acid (**3.61**) and 2-methylnaphtha-quinone (menadione, **3.62**) are involved.

### 3.6.5    The mevalonate pathway

The mevalonic acid pathway is obviously used to provide the poly-isoprenoid sidechains of ubiquinone, menaquinone, *etc*. Examples are also known in which a short isoprenoid substituent ($C_5$ and $C_{10}$) in an inter-mediate is used to supply some of the carbon atoms of a naphthaquinone or an anthraquinone ring system. The naphthaquinones and anthraquinones so formed are therefore, biosynthetically, substituted benzoquinones and naphthaquinones, respectively.

*Naphthaquinones.* This route for naphthaquinone biosynthesis is used by plants of the Pyrolaceae. An example is chimaphilin (**3.63**). The quinone ring (B) of this compound is formed as a benzoquinone, conventionally from shikimate. The remainder of the benzenoid ring A and the methyl substituent are derived from mevalonate. The probable biosynthetic pathway is outlined in fig. 3.13. It involves rearrangement of $p$-hydroxyphenylpyruvate (**3.64**) into homogentisate (**3.55**) and conversion of this into homoarbutin (**3.65**). Prenylation of the latter gives the substituted quinol (**3.66**) which cyclises to the quinol (**3.67**) of chimaphilin which is oxidised to chimaphilin itself. Almost all the intermediate compounds in this proposed pathway have been isolated (as glucosides).

Fig. 3.12. Outline of the likely pathway for biosynthesis of naphthaquinones, including menaquinone, phylloquinone, lawsone and juglone, from shikimate. (TPP = thiamine pyrophosphate.)

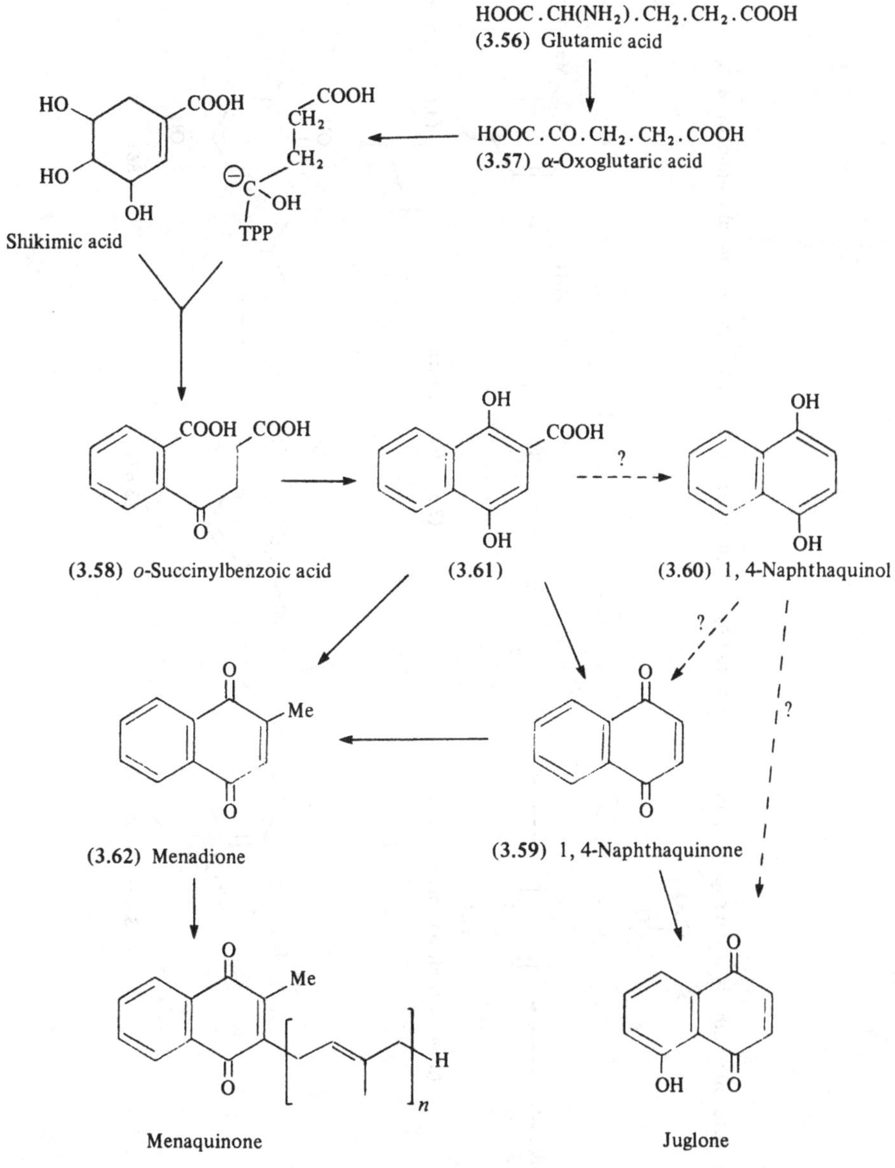

$HOOC.CH(NH_2).CH_2.CH_2.COOH$
(3.56) Glutamic acid

$HOOC.CO.CH_2.CH_2.COOH$
(3.57) α-Oxoglutaric acid

Shikimic acid

TPP

(3.58) o-Succinylbenzoic acid

(3.61)

(3.60) 1, 4-Naphthaquinol

(3.62) Menadione

(3.59) 1, 4-Naphthaquinone

Menaquinone

Juglone

Fig. 3.13. Biosynthesis of the naphthaquinone chimaphilin by a combination of the shikimate and mevalonate pathways.

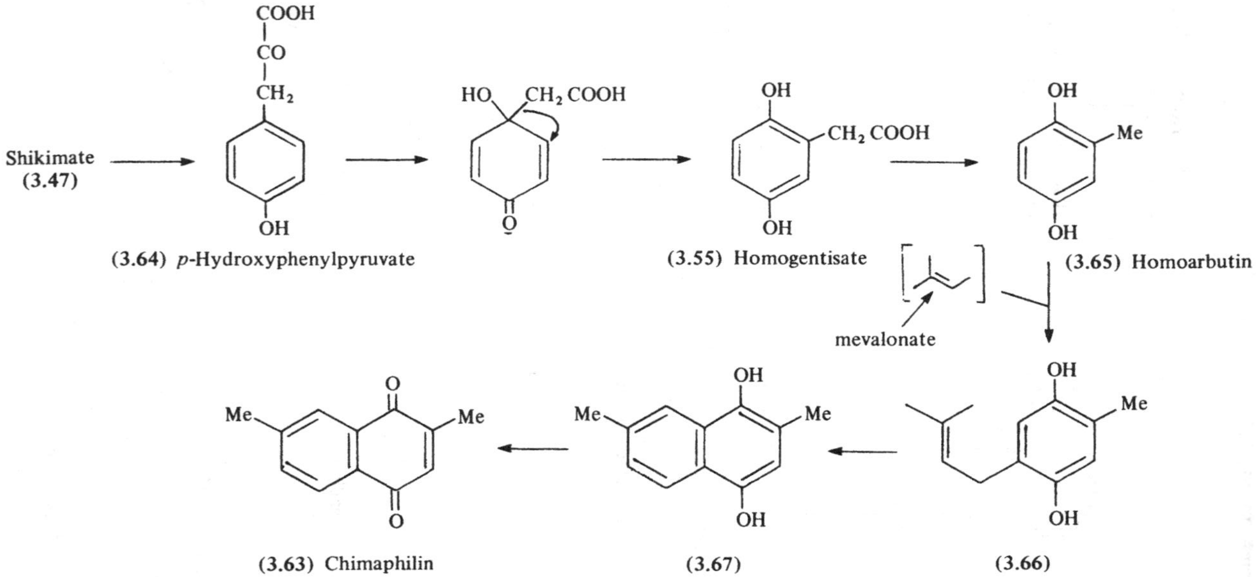

In the case of alkannin (**3.68**) obtained from roots of the Boraginaceae, it is ring A that is derived from shikimate *via* p-hydroxybenzoate, and two molecules of mevalonate then provide the remaining ten carbon atoms, including those of the quinone ring, ring B. The biosynthesis (fig. 3.14) involves essentially prenylation of p-hydroxybenzoate with geranyl pyrophosphate to give (**3.69**) followed by oxidative cyclisation and introduction of oxygen functions.

Fig. 3.14. Formation of the prenylated naphthaquinone alkannin from p-hydroxybenzoate (shikimate pathway) and geranyl pyrophosphate (from mevalonate).

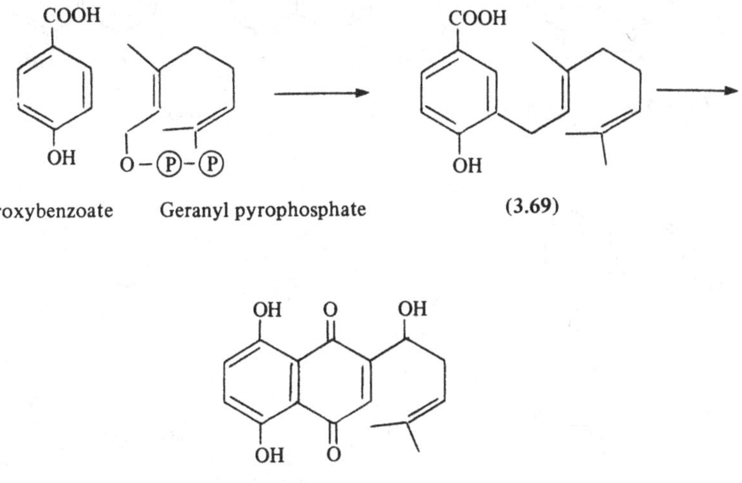

p-Hydroxybenzoate    Geranyl pyrophosphate              (**3.69**)

(**3.68**) Alkannin

*Anthraquinones*. The biosynthesis of anthraquinones in plants of the Rubiaceae, Bignoniaceae and Verbenaceae appears to be rather similar to that of alkannin. Alizarin (**3.27**) biosynthetically is a substituted naphtha-quinone and incorporates label from both shikimate and mevalonate. Rings A and B are formed from shikimate *via* o-succinylbenzoate as in the juglone-lawsone pathway outlined above (fig. 3.12), whereas ring C arises from mevalonate. One methyl group of the prenyl sidechain is lost during the formation of alizarin, but is retained, sometimes in an oxidised form, in other related anthraquinones.

### 3.6.6    *Phenolic coupling*

Phenolic coupling is a biosynthetic process frequently encountered in organisms which elaborate phenols and quinones. In this process, dimerisation occurs by a free-radical reaction in which a bond is formed between very reactive positions on the monomers. Many of the larger, naturally occurring quinones have structures which indicate them to be dimers or other condensation products of naphthaquinones or anthraquinones. Sometimes the monomers may also be present in the same tissues. A convenient and simple illustration is the co-occurrence of 7-methyljuglone (**3.70**) and several dimers (**3.71–3.73**) in *Drosera* and *Diospyros* species, including ebony wood, where they may in part be precursors of the black pigment.

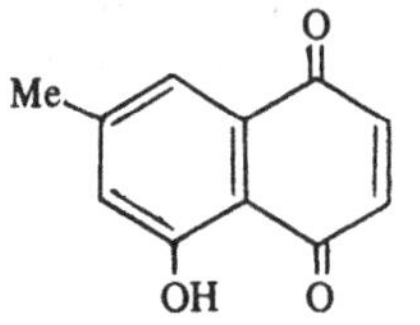

(3.70) 7-Methyljuglone 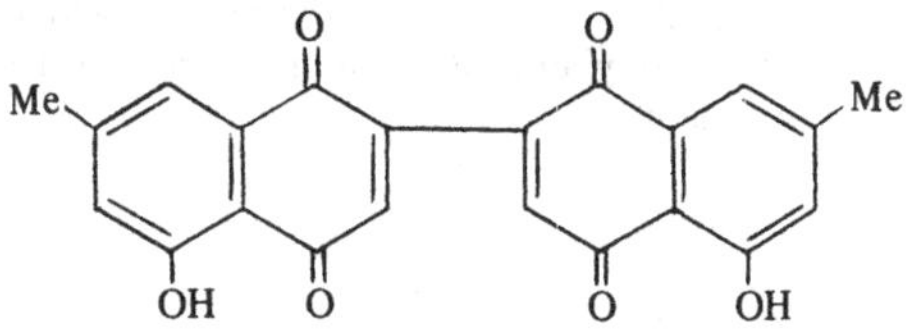                    (3.71)

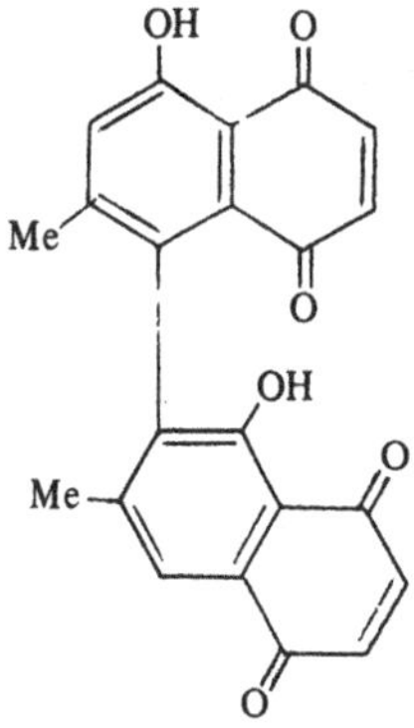

(3.72) 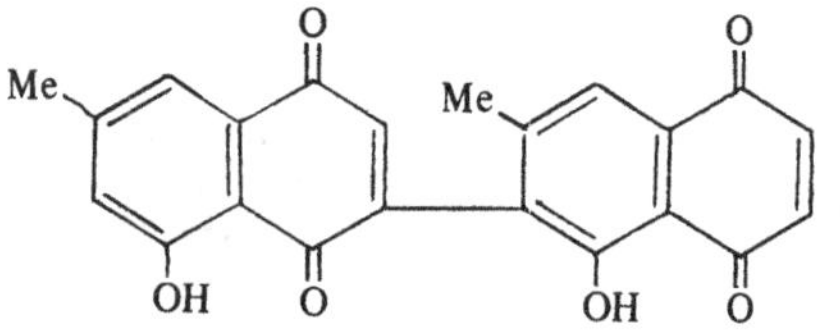                                    (3.73)

In suitable cases several sequential phenolic couplings can occur. Hypericin (**3.37**) is a bisanthraquinone in which the monomers are linked in three places.

Although the chemistry is much more complex, phenolic coupling is also involved in the formation of protoaphins in aphids, and in the subsequent *post-mortem* transformation of protoaphins into xanthoaphins.

The formation of polymers (melanins) by similar phenolic coupling of quinonoid monomers will be outlined in chapter **7**.

### 3.6.7    Overall appraisal

The subject of quinone biosynthesis is complex because completely different alternative pathways may be utilised. Fungi appear to produce their quinones mainly by the polyketide route, whereas in higher plants the several routes involving shikimate are more common, not infrequently in combination with the isoprenoid pathway by which mevalonate supplies part of the naphthaquinone or anthraquinone ring systems.

The animal quinone pigments, *e.g.* the spinochromes and echinochromes of echinoderms and the anthraquinones of insects and crinoids, do not appear to be obtained by the animals from their food. Quinone synthesis takes place within the animals, and proceeds by the polyketide mechanism in those examples that have been studied. The possibility that these quinones may be formed by microbial symbionts has not been disproved; indeed it has been considered very likely in some cases such as the insect pigment carminic acid (3.28).

## 3.7 Functions and biological effects

There is no single general function of quinone pigments. They are not important as pigments in photosynthesis, and probably do not play a role as photoreceptors in other processes. They are rarely of prime importance as external pigments; they usually make little contribution to the external colour of the organisms or tissues that produce and accumulate them.

### 3.7.1 Coloration

Some examples are recognised of the coloration (usually yellow, orange or brown) of fungi and lichens by naphthaquinones and anthra-quinones. The brilliant red, purple and even blue or green colours of certain invertebrate animals, *e.g.* sea urchins and some related forms, some crinoids (feather stars and sea lilies), coccid insects and aphids, are the most obvious illustrations of the pigmentary properties of quinones. Quinones are not generally important as higher plant pigments but do contribute to yellow colours in a small number of examples.

### 3.7.2 Functions not related to light absorption

Many quinones have important biological functions which do not depend upon light-absorption properties. Even though the compounds may absorb some wavelengths of visible light and will therefore be coloured, this property is not used in the functioning of the molecules.

*Electron transport*. Perhaps the most important quinones are the isoprenyl-ated molecules ubiquinone and menaquinone, which serve as respiratory coenzymes in the electron transport systems of animals, plants and micro-organisms, and also the related plastoquinone of the chloroplast photo-synthetic electron transport systems (see chapter 10). The important property here is the easy and reversible reduction to the quinol, *via* a semi-quinone radical.

*Biological activity against other living organisms.* Many quinones have destructive properties towards other tissues or organisms, and may be produced and used for defensive or protective purposes. Simple benzoquinones are irritant, toxic and corrosive, and are used by some insects, *e.g.* the bombardier beetle, as defensive secretions which may be discharged explosively and at high temperature into the 'face' of an attacker. Several fungal naphthaquinones and anthracyclinones have antibacterial or antiviral properties, and hence afford protection to the organism that produces them. The simple naphthaquinone juglone (3.17) produced by the walnut tree is toxic to many other plants.

In at least one case the function of quinones may be offensive rather than defensive. The plant pathogenic mould *Fusarium martii* produces several naphthaquinones, *e.g.* marticin (3.74), which have been shown to induce wilting in the host plant and may therefore be an important part of the pathogen's attacking mechanism.

Several quinones are known to be irritating, corrosive or toxic to man and other mammals. The allergic reaction suffered by many people on contact with the popular houseplant *Primula obconica* is due to a benzoquinone, primin (3.75). Several naphthaquinones present in wood can cause irritation, sneezing and eczema among woodworkers who are exposed to or inhale the wood dust.

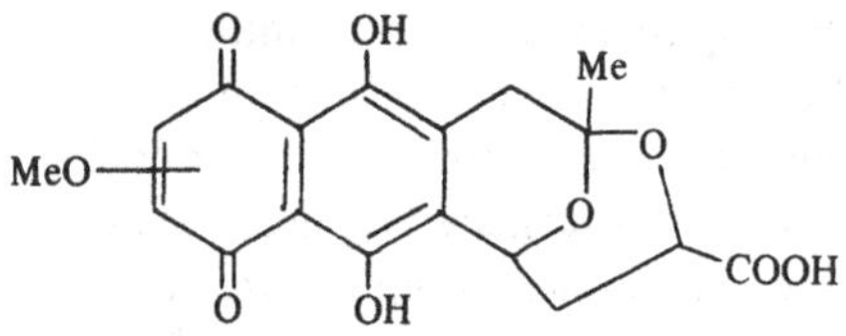

(3.74) Marticin 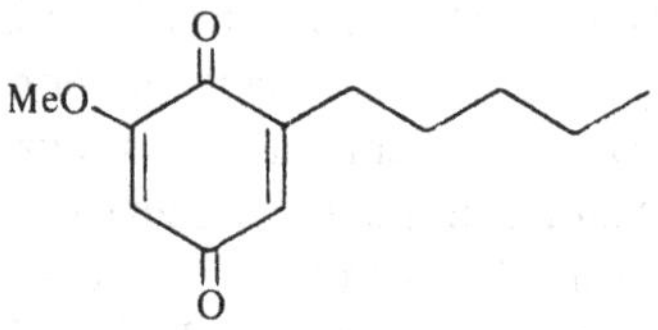                    (3.75) Primin

The complex extended quinone hypericin (3.37) and related compounds are responsible for the toxicity of plants of the *Hypericum* genus (St John's wort) towards animals. Hypericin appears to act in the animal as a photosensitiser which brings about photooxidation of sensitive proteins, *etc.*, thus causing inflammation, oedema and, in extreme cases, death.

### 3.7.3    Spinochromes and echinochromes

Many functions have been suggested for spinochromes and echinochromes in echinoderms. Most of these suggestions have since been discredited or at best not substantiated, but these compounds do have algistatic properties and may be of use to the animals in keeping down populations of parasitic blue-green algae.

## 3.8     Industrial and medicinal uses

### *3.8.1     Dyes*

In the Ancient World some quinone pigments were extremely important as dyestuffs. The plant sources of these pigments were often cultivated in large fields or plantations simply to supply the dyers with raw material. Probably the most widely used of these ancient dyes was madder, the ground root of *Rubia tinctorum*, which was used by the ancient Persians, Egyptians, Greeks and Romans. The dyeing principle of madder is a mixture of anthraquinones, the main one being alizarin (**3.27**), which formed the basis of stable, red mordant dyes. Although alizarin is now virtually obsolete as a dye it still finds use in the study of bone growth.

Henna, a yellow-brown preparation from leaves of *Lawsonia alba*, has long been used in Africa and in the East for dyeing and for cosmetic purposes. Indeed, traces of henna can still be found on the nails of mummies from Ancient Egypt. The dyeing principle is the naphthaquinone lawsone (**3.18**) which is readily extracted from the leaves of *L. alba* with aqueous sodium carbonate.

Amongst several other examples, extracts containing the naphthaquinone alkannin (**3.68**) and its epimer shikonin were used by the Romans and ancient Japanese, and Morinda root, used in India, Java and by the Maoris of New Zealand has anthraquinones such as morindone (**3.76**) as the dyeing principle.

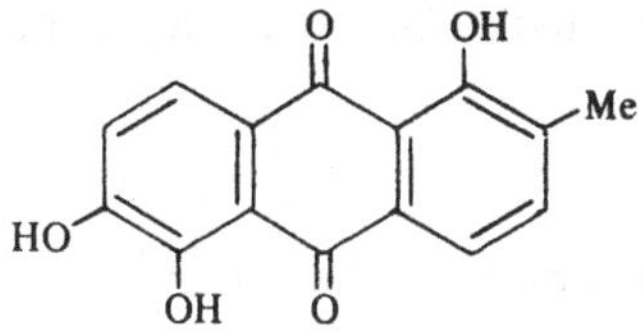

(3.76)  Morindone

Finally mention must be made of the quinone dyes of insect origin. Perhaps the most ancient dyestuff on record is kermes, obtained from vast numbers of the coccid insect *Kermococcus ilicis*. The colouring principle is the anthraquinone kermesic acid (**3.29**). The related compound carminic acid (**3.28**) from another insect *Dactylopius coccus* is the active principle of cochineal, first used perhaps by the Incas of Peru, and still used in small quantities for colouring food and as a biological stain.

With the development of modern synthetic chemistry the large-scale production and use of the natural quinone dyes has largely died out, but some are still used, albeit in small quantities, as colorants for food, wine and cosmetics.

### *3.8.2     Medicinal uses*

From ancient times preparations of quinones have been used medicinally as purgatives. Naphthaquinones and anthraquinones are responsible

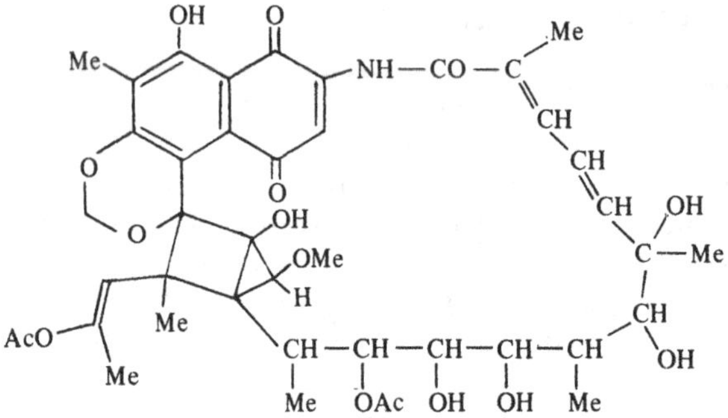

(3.77) Streptovaricin A (stereochemistry not shown)

for the well-known and widely used purgative action of senna, cascara and rhubarb.

Recently, more and more of the naturally occurring quinones have been found to have antiviral, antibacterial or antifungal activity. In some cases it has been possible to use them as antibiotics for treating infections. The streptovaricins (*e.g.* **3.77**), are orange macrolide naphthaquinones which show marked activity, *in vivo*, against *Mycobacterium tuberculosis*. The anthracyclinones also exhibit antibiotic activity, like the tetracyclines which they resemble structurally.

### 3.9    Conclusions and comments

The carotenoids, described in the previous chapter, are a homo-geneous group of pigments, all derived from the same basic carbon skeleton. The quinones present a much different picture. This group contains a miscel-lany of different carbon skeletons and ring systems, but these all have one feature in common, the cyclic enedione or quinone structure. There have been no comprehensive systematic surveys of quinone compositions of natural tissues, so there is surely a substantial number and variety of novel quinones waiting to be discovered and tackled by the organic chemists, and perhaps some complex stereochemical problems to be unravelled.

The variety of opportunities extends also to the biosynthesis of quinones. It is clear that two main pathways (sometimes three) are used, but few individual examples have been studied in any detail. It is an interesting situation in which two completely different major pathways are used for the biosynthesis of very similar, even the same, compounds. It is a useful exercise to devise feasible pathways and mechanisms for the biosynthesis of individual natural quinones. There is much scope for the experimental elucidation of the biosynthetic pathways leading to individual compounds, and the enzyme systems involved, with very few exceptions, remain completely unexplored.

Not surprisingly, knowledge of the regulation of quinone biosynthesis is almost a complete blank. Other areas ripe for biochemical study include the location of quinone pigments within the cell, possible association with protein or other molecules, and the functions of quinones in many of the tissues that possess them, since their contribution to colour is usually not significant.

Greater knowledge of the effects of some quinones on living tissues, and explanations of these effects, may lead to greater interest in possible medicinal uses and other commercial exploitation of quinones, and this in turn may spur the search for novel natural quinone types and structures.

## 3.10    Suggested further reading

For additional reading about natural quinones one need look no further than the comprehensive monograph by Thomson (1971), which gives an exhaustive account of the chemistry of all the natural quinones, including pigments, that had been discovered up to 1970. It also contains lively accounts of the pioneering work on the elucidation of some of the structures by classical methods and entertaining descriptions of the historical use of some important natural quinone dyestuffs. The biosynthesis of quinones is also dealt with in this volume and in a more recent review by the same author (Thomson, 1976a). Details of the biosynthesis of aromatic compounds, including quinones, by the shikimate pathway are included in a review by Haslam (1979). Two volumes on the chemistry of quinonoid compounds (Patai, 1974) include valuable information on many natural quinones. Useful details of earlier work on the chemistry and biochemistry of quinones may be found in an older monograph by Morton (1965).

For general experimental methods the reader is directed to an article compiled by Thomson (1976b). The extensive lists of references in the Thomson (1971) monograph should be consulted to obtain details of methods applicable to individual quinones.

## 3.11    Selected bibliography

Haslam, E. (1979) Shikimic acid metabolites, in *Comprehensive organic chemistry*, vol. 5, ed. E. Haslam, p. 1167. Oxford and London: Pergamon.

Morton, R. A. (ed.) (1965) *Biochemistry of quinones*. London: Academic Press.

Patai, S. (ed.) (1974) *The chemistry of the quinonoid compounds*, vols 1 and 2. London and New York: Wiley.

Thomson, R. H. (1971) *Naturally occurring quinones*, 2nd edition. London and New York: Academic Press.

Thomson, R. H. (1976a) Quinones, nature, distribution and biosynthesis, in *Chemistry and biochemistry of plant pigments*, 2nd edition, vol. 1, ed. T. W. Goodwin, p. 527. London, New York and San Francisco: Academic Press.

Thomson, R. H. (1976b) Isolation and identification of quinones, in *Chemistry and biochemistry of plant pigments*, 2nd edition, vol. 2, ed. T. W. Goodwin, p. 207. London, New York and San Francisco: Academic Press.

# 4    *O*-Heterocyclic pigments – the flavonoids

## 4.1    Introduction

The flavonoids (alternative spelling 'flavanoids') are almost exclusively products of higher plants. They include the anthocyanins which are responsible for perhaps the most striking of all plant colours, those of the brilliant red, purple and blue flowers and fruits which contrast most strongly with the background green of the foliage and thus catch the eye of man and other animals. Other classes of flavonoids may also have a contribution to make to plant colours, though usually this contribution is not immediately obvious.

## 4.2    Structures and nomenclature

### 4.2.1    *Structural classes*

The natural flavonoids all have the same basic structural framework. They are *O*-heterocyclic compounds with structures based on the tricyclic skeleton of flavone, *i.e.* 2-phenylchromone or 2-phenylbenzo-γ-pyrone **(4.1)** and flavan (2-phenylbenzopyran) **(4.2)**. The basic flavonoid structure thus consists of two benzene rings, designated A and B, joined by a $C_3$ unit which, together with an oxygen atom, forms the γ-pyrone or γ-pyran ring. The numbering scheme is illustrated in **(4.1)**. The benzopyrone (-pyran) system is given plain numerals whereas ring B is regarded as a substituent and given primed numerals.

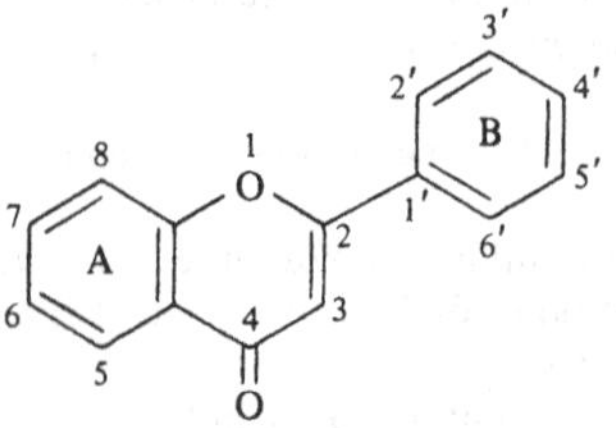

(4.1) Flavone 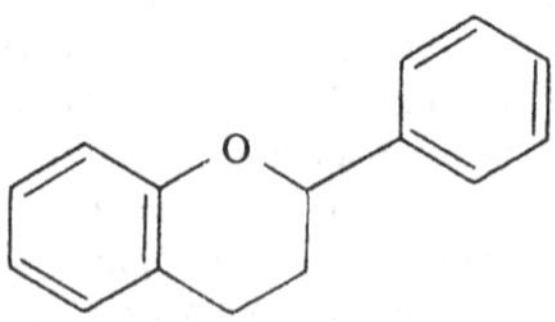                    (4.2) Flavan

The linking $C_3$ unit which completes the heterocyclic ring determines the class to which an individual flavonoid belongs. This unit can exist in several different oxidation states, each corresponding to a different flavonoid class, as illustrated in fig. 4.1. The most important of these classes are the flavones (**4.3**) and their 3-hydroxy derivatives the flavonols (**4.4**) and the antho-cyanidins (**4.5**) in which the heterocyclic ring in the compounds as isolated from acidic solution is in the form of a flavylium salt. 3-Deoxyanthocyanidins (**4.6**) also occur but are rare. Classes derived from flavan are the flavanones (**4.7**), the flavan-3-ols or catechins (**4.8**), the flavanonols or dihydroflavonols (**4.9**) and the flavan-3,4-diols or proanthocyanidins (**4.10**). Some related classes, isoflavones (**4.11**), chalcones (**4.12**) and aurones (**4.13**) are also illustrated in fig. 4.1. Although these last structures do not contain the 2-phenylchromone skeleton and therefore are not strictly flavonoids, they are nevertheless so closely related chemically and biosynthetically that they are always included in the flavonoid group. Note that the different classes of true flavonoid compounds are generally named as flavone or flavan deriva-tives, though the old name anthocyanidin persists for one group and trivial names are often used for certain other groups, *e.g.* catechins. Other old terms, anthoxanthin, anthochlor and chymochrome, are now, fortunately, rarely used.

### 4.2.2   *Ring substitution patterns ( fig. 4.2)*

Within each class individual compounds are characterised by the number and positions of substituents on the aromatic rings; commonly these substituents are hydroxy-groups, which may be methylated or glycosylated. The location of some of the hydroxy-groups is a consequence of the general biosynthetic pathway (see §4.6). Thus in most flavonoid compounds ring A has hydroxy-groups either at C-7 or at both C-5 and C-7. These are rarely methylated. Ring B is virtually always hydroxylated at C-4' and commonly also at C-3' and C-5'; hydroxy-groups at these latter two positions are often methylated.

### 4.2.3   *Glycosylation*

*In vivo* the flavonoids exist largely, perhaps entirely, as glycosides. In those flavonoid classes that contain a C-3 hydroxy-group, especially antho-cyanidins and flavonols, this is the favoured position for glycosylation, but glycoside residues at C-7, C-4' or C-5 are also encountered quite frequently. It is not uncommon for sugars to be present at more than one position, or for the sugars to be di- or trisaccharides. The natural pigments are usually β-glycosides, with D-glucose as the most common monosaccharide, but L-rhamnose and D-galactose are also found often, and other sugars more rarely. In some examples the sugar may be acylated with a phenolic acid, frequently a hydroxycinnamic acid.

Fig. 4.1. Basic skeletons of the main classes of flavonoids.

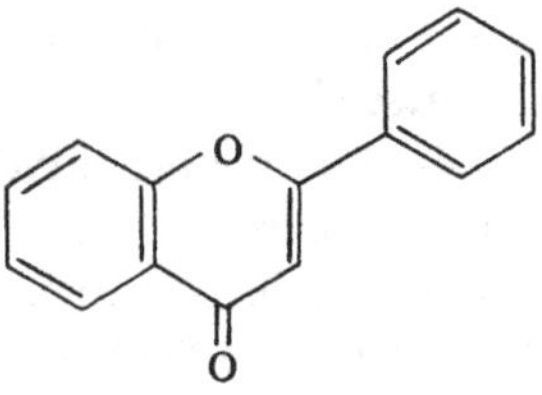

(4.3) Flavone

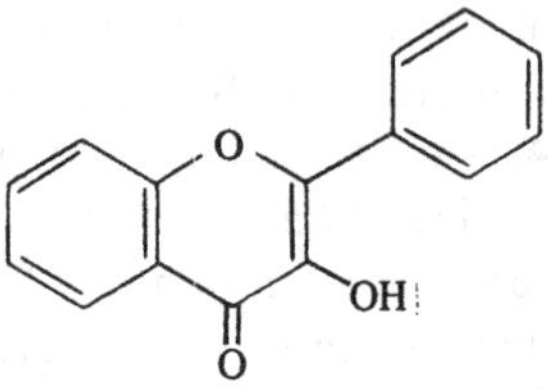

(4.4) Flavonol

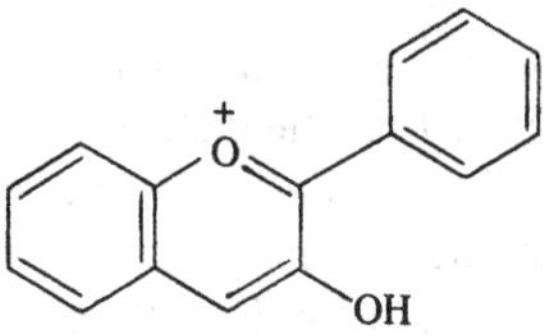

(4.5) Anthocyanidin

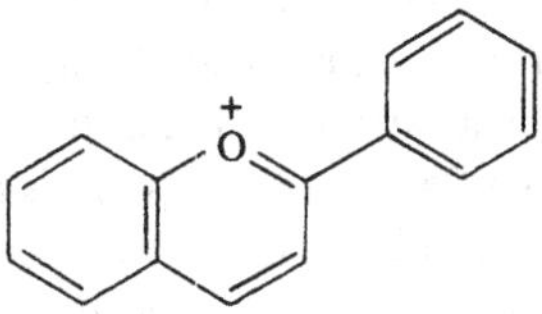

(4.6) Deoxyanthocyanidin

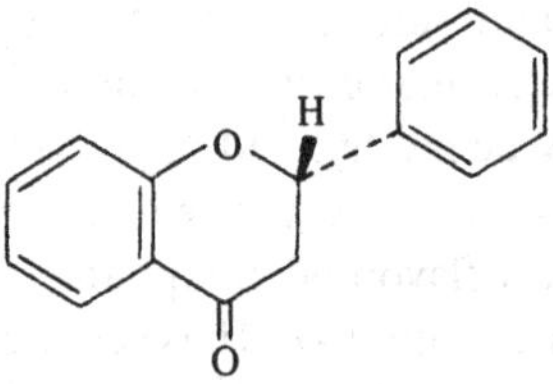

(4.7) Flavanone (2*S*)

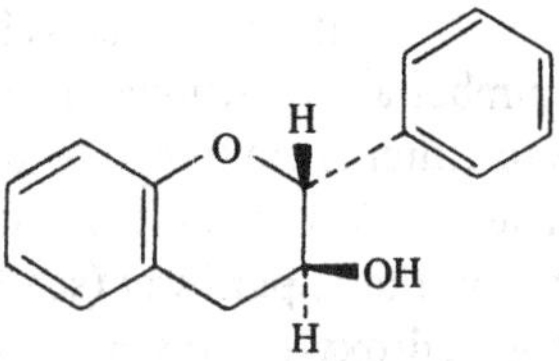

(4.8) Flavan-3-ol
(2*R*, 3*S*) = catechin
(2*R*, 3*R*) = epicatechin

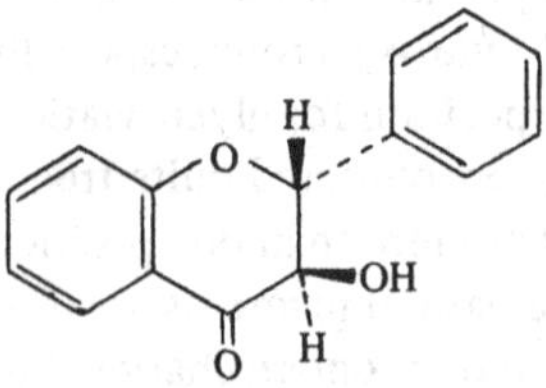

(4.9) Flavanonol (2*R*, 3*R*)

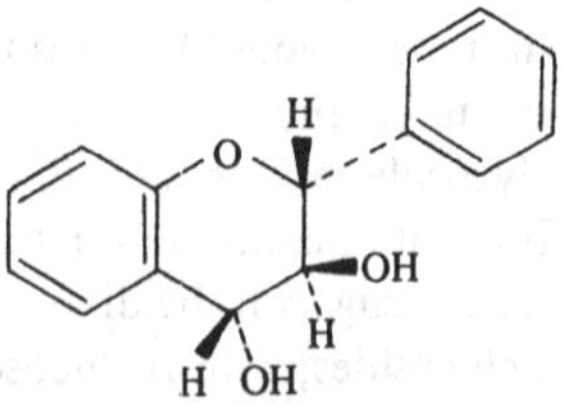

(4.10) Flavan-3, 4-diol
(2*R*, 3*R*, 4*R*) and epimers

Fig. 4.1. *contd.*

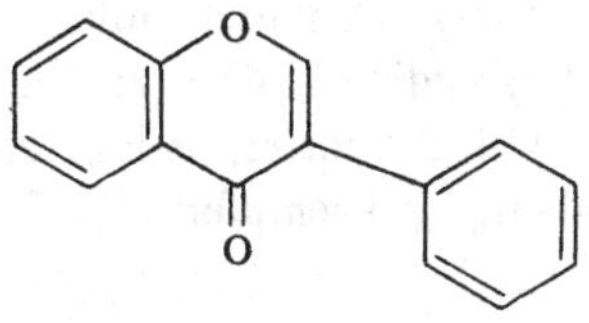

**(4.11)** Isoflavone
(3-phenylchromone)

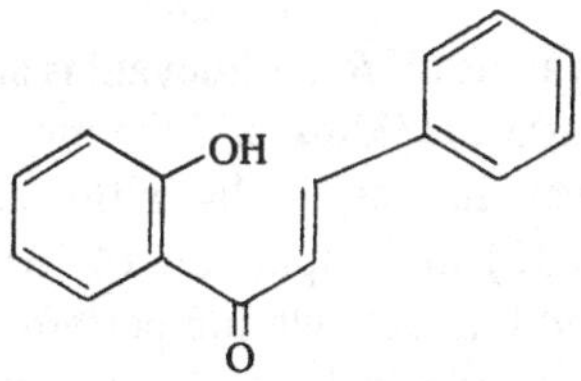

**(4.12)** Chalkone

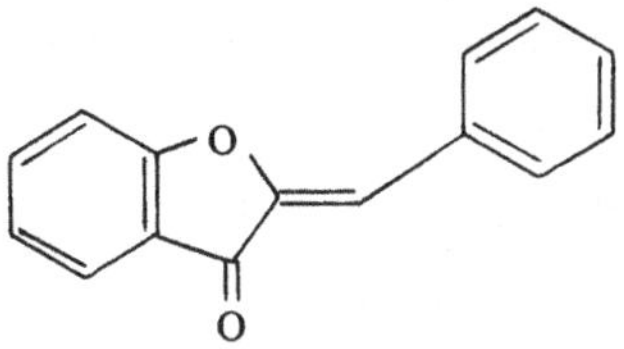

**(4.13)** Aurone

Fig. 4.2. Most common positions of substituent groups in flavonoids.

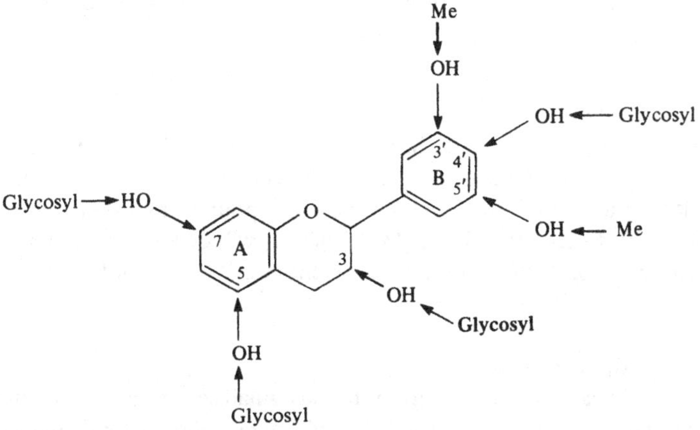

It is not unusual for a free flavonoid (aglycone) to be present in an extract. However, the free flavonoids that are isolated may not have been present *in vivo* but may have been formed rapidly during extraction by the action of glycosidase enzymes, some of which function even in the presence of high concentrations of organic solvents.

### 4.2.4     Some important examples

The most important flavonoid plant pigments *in vivo* are the antho-cyanins (*N.B.* **anthocyanins** are **glycosides**, the corresponding **aglycones** are **anthocyanidins**) and the most common and widely distributed antho-cyanidins, especially in flower petals, are pelargonidin (**4.14**), cyanidin (**4.15**) and delphinidin (**4.16**). The flavonols with the corresponding ring A and B hydroxylation patterns are also abundant, *e.g.* kaempferol (**4.17**), quercetin (**4.18**) and myricetin (**4.19**).

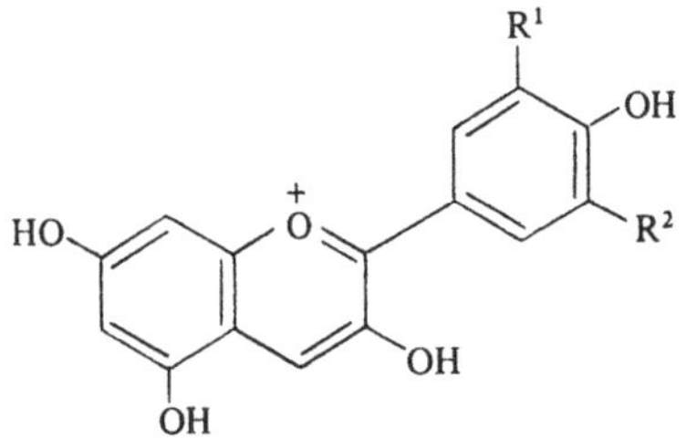

Common anthocyanidins

(**4.14**)  R$^1$ = R$^2$ = H: pelargonidin
(**4.15**)  R$^1$ = OH, R$^2$ = H: cyanidin
(**4.16**)  R$^1$ = R$^2$ = OH: delphinidin

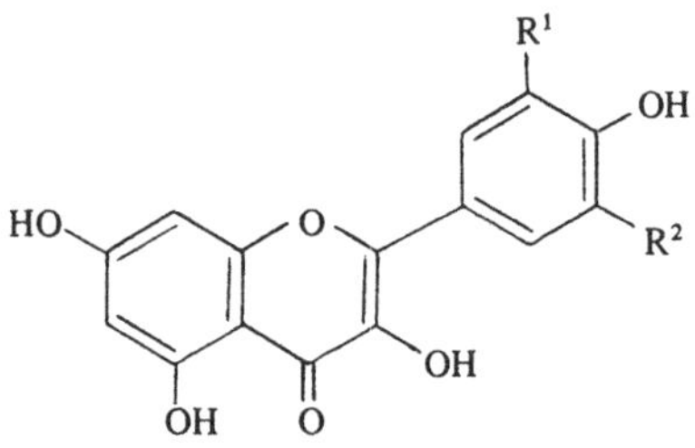

Common flavonols

(**4.17**)  R$^1$ = R$^2$ = H: kaempferol
(**4.18**)  R$^1$ = OH, R$^2$ = H: quercetin
(**4.19**)  R$^1$ = R$^2$ = OH: myricetin

The number of individual flavonoid aglycones is high, *e.g. ca* 300 flavones and flavonols, 50 chalkones, 20+ anthocyanidins, so the number of possible glycosides and acylated glycosides of these is obviously enormous. In this book the number of individual flavonoid compounds discussed and hence of trivial names encountered (see §*4.2.5* below) will be kept to a minimum. Whenever possible the information presented will be relevant to all members of a flavonoid class.

### 4.2.5     Nomenclature

As can be seen even from the very small selection of compounds encountered in the previous section, trivial names abound, and are usually derived from the name of the plant source from which the compound was first isolated or with which it is usually associated, *e.g.* delphinidin from the delphinium. This can lead to difficulty and confusion since very closely related compounds, *e.g.* different glycosides of the same aglycone, may have entirely unrelated names. In the most extreme cases two entirely different names have been used for the same compound.

## 4.3 Properties

### 4.3.1 Chemical properties

The chemical properties of flavonoids were studied intensively in the days of classical organic chemistry, and several general synthetic routes were devised. Flavonoids generally undergo the reactions characteristic of their substituent groups, *e.g.* hydroxyls. The $C_3$ link of the heterocyclic ring can undergo reduction and oxidation, thus permitting limited interconversion of some of the flavonoid classes. Degradation with alkali, often requiring forcing conditions, cleaves the flavonoid molecule into the two benzene ring fragments. This approach may be useful for identifying ring substitution patterns, although such information is now usually obtained by spectroscopic methods (u.v., m.s., n.m.r.).

### 4.3.2 Physical properties

Natural flavonoid glycosides are appreciably water-soluble and may be extracted from plant tissues with water or more commonly aqueous alcohols. Anthocyanins bear a positive charge and their extraction requires mildly acidic conditions. Generally flavonoids are more stable towards light and moderate heat and pH changes than are most other main groups of pigments.

*Purification.* Separation and purification of flavonoids are readily achieved by paper or thin-layer chromatography, and two-dimensional chromatography is commonly used to give a very rapid indication of the flavonoids present in a plant tissue. Even normally colourless compounds are easily detected by examination of chromatograms under u.v. light, especially in the presence of ammonia vapour, when characteristic fluorescence colours are observed.

### 4.3.3 Light absorption

As outlined in chapter **1**, the wavelength of light absorbed by a molecule is a function of the ease with which electronic transitions to higher energy levels can be brought about. In a given molecular type, the longer the conjugated chain or chromophore and the greater the number of contributing functional groups or auxochromes, then the greater will be the stabilisation of the excited state and the easier this excited state will be to produce, *i.e.* excitation will require less energy and absorption will occur at longer wavelengths. The flavonoid field illustrates this very well.

*Flavan derivatives* (fig. 4.3). In the hydroxylated flavan derivatives, *e.g.* flavan-3-ols (**4.8**) and -3,4-diols (**4.10**) the C-3,4 single bond effectively isolates the two benzene rings. These compounds therefore exhibit only transitions of the isolated benzene ring chromophores, and absorb in the u.v. region at 275–280 nm, just as do the corresponding simple phenols. Flavanones (**4.7**) and isoflavones (**4.11**) have ring A conjugated with the C-4 carbonyl group, and therefore have an absorption maximum similar to that of the corresponding hydroxyacetophenone.

*Flavone derivatives* (fig 4.4). In the flavones (**4.3**) and flavonols (**4.4**) conjugation between the C-4 carbonyl group and ring B occurs, and the spectra have

Fig. 4.3. U.v.-light absorption spectra of flavan derivatives in ethanol; a flavan-3,4-diol (———) and a flavanone (– – – –).

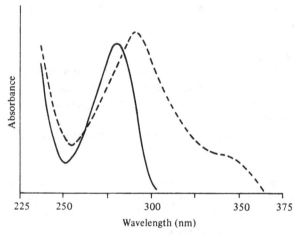

Fig. 4.4. Light absorption spectra of a flavone (– – – –) and a flavonol (———) in ethanol.

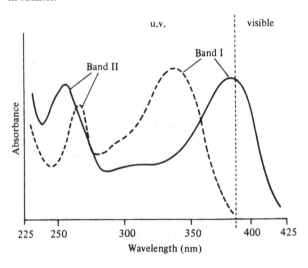

two strong, well-separated absorption bands. Band I, at longer wavelength, is associated with the conjugated ring B–carbonyl group chromophore, and the shorter wavelength band II with ring A. With the flavonols, the vinylic hydroxy-group at C-3 contributes to electron availability. Band I moves to longer wavelength than that of the corresponding flavone, and there may be sufficient absorption in the blue region of the spectrum for the flavonol to appear yellow.

*Chalkones and aurones* (fig. 4.5). The greater coplanarity possible in the chalkones (**4.12**) and even more so in the aurones (**4.13**) increases the ease of electronic excitation and therefore the $\lambda_{max}$. Chalkones are thus yellow and aurones orange.

Fig. 4.5. Light absorption spectra of a chalkone (– – – –) and an aurone (———) in ethanol.

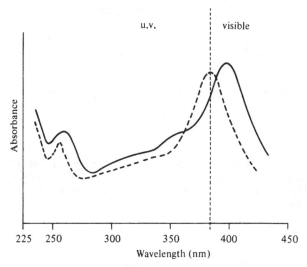

*Anthocyanidins* (fig. 4.6). The anthocyanidins and anthocyanins are usually isolated in acidic solution as flavylium salts (**4.20**) and the electrons of the heterocyclic oxygen atom contribute to the $\pi$-bonding in the heteroaromatic ring, so that the chromophore effectively extends over the entire molecule. The anthocyanidins hence absorb at the longest wavelengths of all the flavonoids and are orange, red, purple or blue. As the pH of an anthocyanidin solution is raised towards neutrality, however, more and more of the colourless pseudo-base form (2-hydroxychromene, **4.21**) is produced. At pH values above 7, the quinonoid base structure (**4.22**) is formed in small amounts (absorption at longer wavelength, blue in colour) but is unstable in solution

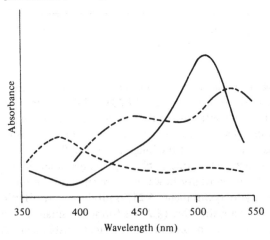

(4.20)  Flavylium salt (cyanidin)

(4.21)  Pseudo-base (2-hydroxychromene)                    'Chalkone form'

Fig. 4.6. Light absorption spectrum of an anthocyanidin; (*a*) at pH 1.0, flavylium salt (**4.20**) predominant (——); (*b*) at pH 4.0, pseudo-base form (**4.21**) predominant (– – – –); (*c*) at pH 7.5, anhydrobase form (**4.22**) predominant (– – – –).

(4.22) Quinonoidal base

in water. This quinonoid base structure is, however, the anthocyanin form which seems to be present and stable *in vivo*, especially in flower petals which frequently are more blue and more deeply coloured than would be expected from the absorption spectra of the isolated anthocyanins (in the flavylium salt form).

*Effect of substituent groups.* With all the flavonoid classes, hydroxy substituents contribute non-bonding electrons and increase the extent of charge delocalisation and hence stabilisation of the excited state, thus facilitating electronic excitation. The effect of increasing hydroxylation on light-absorption properties is readily seen from the spectra of the three common anthocyanidins (fig. 4.7). Pelargonidin (**4.14**), cyanidin (**4.15**) and delphinidin (**4.16**), with one, two and three ring B hydroxy-groups, respectively, are

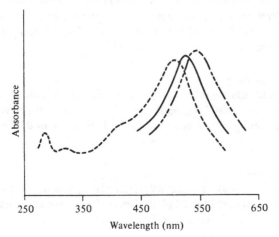

Fig. 4.7. Effect of hydroxy-substituents on the light absorption spectra of anthocyanins. Spectra (in ethanol–HCl) of glycosides of pelargonidin (**4.14**) (– – – –), cyanidin (**4.15**) (———) and delphinidin (**4.16**) (– – –—).

orange, red and purple. Similar effects are seen with other flavonoid classes, e.g. the flavonols kaempferol (**4.17**), quercetin (**4.18**) and myricetin (**4.19**) have $\lambda_{max}$ (in ethanol) at 368, 374 and 378 nm, respectively. Hydroxy-groups in other positions do not produce this effect. Methylation, glycosylation or acylation of the hydroxy-group usually diminishes or removes the bathochromic effect.

*Effects of pH and chelation.* The effect of variations in pH on the ionisation and absorption spectra of anthocyanidins has already been mentioned. All the flavonoid classes, however, are phenols and their spectra are therefore altered markedly by high pH (above pH 9) which brings about ionisation of the phenolic hydroxy-groups, and also by chelation with metal ions. (In the case of anthocyanidins and anthocyanins, high pH levels bring about irreversible breakdown.) The absorption maxima shift to longer wavelengths and many of the flavonoids become strongly coloured. Such ionisation, brought about by exposure to ammonia vapour, is frequently used to reveal flavonols, *etc.*, on chromatograms.

## 4.4    Distribution
### 4.4.1    In animals
Reports of the presence of flavonoids in animals are few. Flavones and flavonols are widely, though by no means commonly, found in larvae and adults of the Lepidoptera, especially white or cream butterflies. They are certainly obtained from the plants that form the diet. This is also the case with the larva of the pugmoth *Eupithecia oblongata* which may be blue or red depending on the anthocyanin content of the flowers on which it feeds. This adaptation to the background colour is an obvious advantage to the animal in terms of camouflage. Elsewhere flavonoids or flavonoid-like compounds have been identified in a land snail, *Helix pomatia* (a flavone?), a marine hydroid, *Sertularella*, and the fresh-water diving beetle *Dytiscus* (an aurone). The presence of flavonoid in the last-named species, a carnivore, is unexpected and unexplained.

### 4.4.2    In microorganisms
Only recently has microbial synthesis of a flavonoid been demonstrated conclusively. The mould *Aspergillus candidus* produces the flavonoid flavonin (**4.23**) apparently by the same biosynthetic route as used by higher plants.

### 4.4.3    In higher plants
Flavonoids are almost exclusively higher plant pigments. They may be synthesised and accumulated in all plant tissues – leaves, wood, roots, fruit, seeds and all parts of flowers, especially the petals. The native glyco-

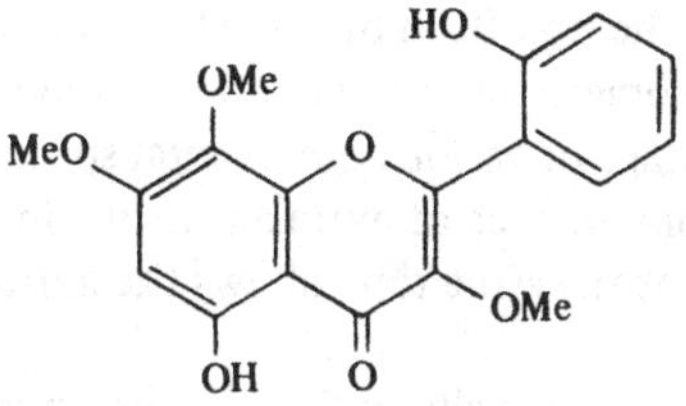

(4.23) Flavonin

sides are water-soluble and are found generally in the cell sap or vacuoles, although recent reports suggest that small amounts of flavonoids may be normal constituents of higher plant chloroplasts.

Very extensive systematic surveys have been made of the flavonoid compositions of higher plant species and cultivars. Two-dimensional paper chromatography has been used extensively as a 'finger-printing' method to give a rapid indication of the flavonoid composition of plant species. Flavonoid patterns have proved extremely valuable taxonomic markers both for establishing close correlations within plant families and also as an aid to distinguishing between closely related species. Both the classes of flavonoid present and the substitution patterns of the individual compounds may be highly characteristic.

## 4.5    Contribution to plant colours

### 4.5.1    Anthocyanins

Of all the flavonoids, it is the anthocyanins that make the greatest and most obvious contribution to plant colours. These compounds are strongly coloured orange, red, purple or blue, and are responsible for almost all the red-blue colours of flowers. A famous example is the red rose, coloured by derivatives of cyanidin. A clear correlation has been established between the colour of flowers and the structures of the anthocyanins that they contain. In an extensive survey it was found that most orange flowers contained pelargonidin, most red-magenta flowers cyanidin and most purple-blue ones delphinidin.

Anthocyanin colours are also extremely familiar in many red fruits, such as strawberry, raspberry, cherry, apple, where anthocyanin content is a sign of ripeness. Most 'black' fruits, *e.g.* blackberry, black grape, are in fact very deep red or purple due to the presence of very high concentrations of anthocyanin. This is nicely illustrated by the fact that the black grape gives rise to red wine, in which the anthocyanin content is much lower. Foods also provide examples of anthocyanins colouring other plant tissues such as leaves (red cabbage) and stems (rhubarb).

Anthocyanins are often produced in large amounts in young shoots and leaves, which therefore become red in colour, in contrast to the green of mature leaves. The deep-red colour of the first flush of rose stems and leaves in the spring is a familiar example. In some cases the red anthocyanin persists into maturity, giving rise to the red foliage of various ornamental species. The reds of autumn leaves too may be a consequence of extra anthocyanin synthesis. The destruction of chlorophyll at the same time allows the anthocyanin colour to be seen more easily.

The colour produced by anthocyanin, especially in flowers and fruit where the quinonoid form (4.22) predominates, may be highly dependent on factors such as pH, metal chelation and copigmentation. The effect of pH changes on the wavelength of light absorption by anthocyanins has already been mentioned (see §4.3.3). Anthocyanins, especially those with *ortho*-dihydroxy substitution in ring B, are able to chelate metal ions, the usual result being a shift in $\lambda_{max}$ to longer wavelength, *i.e.* the colour becomes more blue. The case of *Hydrangea macrophylla* is well known. The anthocyanin present in the sepals of both pink and blue flowers is the same, delphinidin-3-glucoside, but in the blue flowers it is chelated with aluminium and molybdenum ions. Enrichment of the soil with suitable metal ions can cause a previously pink shrub to produce blue flowers. Similarly, mineral deficiency can cause reversion from blue to pink.

Many horticultural breeding programmes are directed towards producing varieties of flowers with 'better', *i.e.* deeper or unusual, colours or patterns. In biochemical terms, the target is to achieve qualitative or quantitative changes in anthocyanin compositions. The search for a blue rose is an example. The aim here is to produce a blue colour either by suitable copigmentation or through a change in the main anthocyanidin from cyanidin to delphinidin, clearly an extremely difficult biochemical change to achieve since the rose does not normally possess the necessary enzymes to bring about the additional hydroxylation.

### 4.5.2    Other flavonoids

The contribution made to plant colours by other classes of flavonoids is usually less obvious, though chalkones and especially aurones are sometimes responsible for yellow flower colours, *e.g.* in the snapdragon *Antirrhinum majus*.

Flavones and flavonols have virtually no light absorption in the visible region of the spectrum, but are nonetheless essential to give 'body' to white and cream flowers which without them would be almost transparent. It must also be remembered that flavones and flavonols absorb light in the near u.v. region which is 'visible' to bees and other insects. Thus although these compounds and the flowers containing them are white or colourless to us they will appear 'coloured' to bees and thus help to attract the insects. Flavones and flavonols are almost always present in leaves, though because of the chlorophyll present they make no contribution to the colour.

Monomeric flavans, flavanones and their hydroxy-derivatives are not involved directly in plant coloration. Dimers, oligomers and polymers, however, especially of catechins and proanthocyanins, give rise to the brown colours of autumn and dried leaves and the dark colours of many heartwoods. Such compounds ('condensed tannins') also provide the warm brown colour of tea.

### 4.5.3 Copigmentation

Although colourless or nearly so, hydroxyflavans, flavones and flavonols do make an important contribution to many flower colours by copigmentation. These compounds frequently occur together with anthocyanins and form complexes which absorb more intensely and at longer wavelengths than the anthocyanins alone. Many blue flowers probably owe their colour to the presence of such copigmentation complexes, some of which have recently been isolated. For example, the pigment of the blue iris flower contains the quinonoidal base of an acylated delphinidin 3,5-diglucoside, stabilised by copigmentation with $C$-glycosyl flavones. In other examples, the copigmentation complex is stabilised by metal ions such as $Mg^{2+}$. Several blue flowers, however, *e.g. Lobelia*, do not employ copigmentation or metal complexing. In these the blue quinonoidal base itself is greatly stabilised by interactions with caffeoyl groups from an acylated sugar substituent.

Other aspects of the coloration of plant tissues by flavonoids and other pigments will be discussed in chapter 8.

### 4.6 Biosynthesis
### 4.6.1 Introduction

Flavonoid biosynthesis occurs on a very large scale. It has been estimated to account for the consumption of almost 2% of the total carbon fixed by photosynthesis in higher plants. The broad outline of the flavonoid biosynthetic pathway is well understood, but many of the details, especially at the enzyme level, remain to be elucidated.

It is convenient to consider the biosynthetic pathway in stages: (i) the formation of the basic $C_6$-$C_3$-$C_6$ skeleton, involving as it does the two major biosynthetic routes to phenols, the polyketide and shikimate pathways; (ii) the ways in which the several flavonoid classes may be produced from the basic $C_6$-$C_3$-$C_6$ precursor and the possible interrelationships between the different flavonoid classes; and (iii) the final modifications such as hydroxylation, methylation and glycosylation, which give rise to the many individual flavonoids within each class.

### 4.6.2 Formation of the basic $C_6$-$C_3$-$C_6$ skeleton

It has long been realised that ring A of the flavonoid molecule is derived from acetate units, whereas ring B and the linking three carbon atoms constitute a phenylpropanoid unit derived from shikimate.

Fig. 4.8. Deamination of phenylalanine or tyrosine to *trans*-cinnamic acid or *p*-coumaric acid.

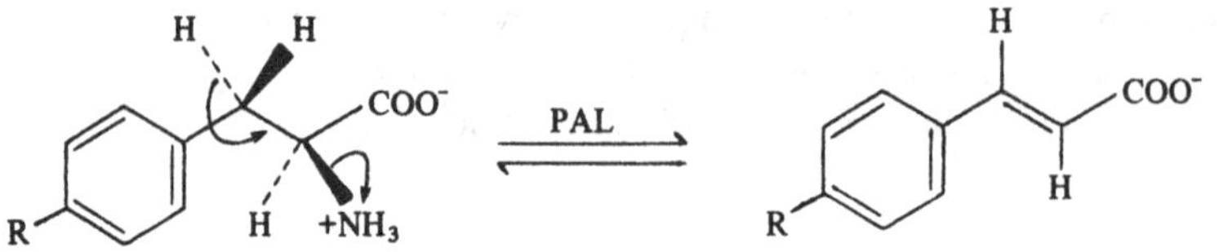

(4.24) R = H: phenylalanine
(4.26) R = OH: tyrosine

(4.25) R = H: *trans*-cinnamic acid
(4.27) R = OH: *p*-coumaric acid

Details of the shikimate pathway leading to phenylalanine have been given in chapter 3. Of fundamental importance in flavonoid biosynthesis is the reaction, catalysed by the enzyme phenylalanine-ammonia lyase (PAL), in which phenylalanine (4.24) is deaminated by an antiperiplanar elimination reaction (fig. 4.8) to give *trans*-cinnamic acid (4.25). In some cases similar deamination of tyrosine (4.26) gives *p*-coumaric acid (4.27), but the latter is usually obtained by hydroxylation of cinnamic acid with cinnamic acid 4-hydroxylase, a mixed-function oxidase enzyme which utilises NADPH and molecular $O_2$. This reaction involves an interesting intramolecular hydrogen migration, the 'NIH shift' (fig. 4.9).

The cinnamic acids are then activated as their coenzyme A thioesters by a reaction (fig. 4.10) analogous to acetyl-CoA synthesis from acetate. The cinnamoyl-CoA (4.28) then acts as 'starter' in a polyketide mechanism (see §3.6.3), condensing with three molecules of malonate as the CoA or related thioester. Cyclisation of the polyketide system (4.29) gives the hydroxylated ring A and completes the $C_6$-$C_3$-$C_6$ skeleton of a chalkone (4.30) (fig. 4.11). The formation of a chalkone from *p*-coumaroyl-CoA and malonyl-CoA by an enzyme from cell suspension cultures of parsley (*Petroselinum hortense*) has been demonstrated.

### 4.6.3    *Formation of different classes of flavonoids*

It is now well established that chalkones play the central role in the biosynthesis of all other classes of flavonoids. This is illustrated in fig. 4.12 which summarises currently accepted views of the possible biogenetic relationships among the different classes of flavonoids. In general, however, much more work, especially at the enzyme level, is needed before many of these possible interconversions can be regarded as proved.

*Flavanones.* Spontaneous isomerisation of chalkones to the corresponding flavanones occurs in solution, but in Nature this reaction is enzymic, catalysed by chalkone-flavanone isomerase, and gives the (2*S*)-flavanone

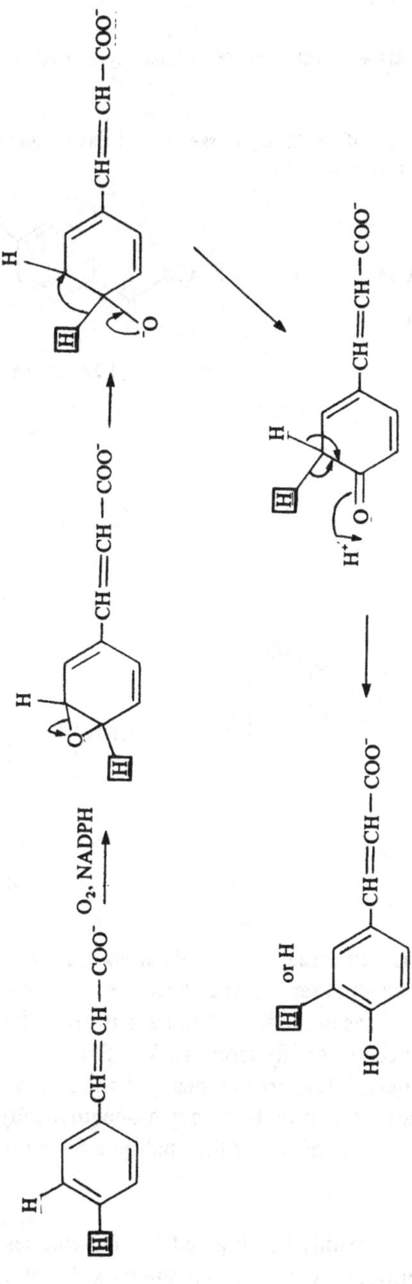

Fig. 4.9. Proposed mechanism for hydroxylation of cinnamic acid.

Fig. 4.10. Formation of cinnamoyl-CoA thioesters.

Cinnamate + ATP ⟶ Cinnamoyl-AMP + PP$_i$

Cinnamoyl-AMP + CoASH ⟶ Cinnamoyl-S . CoA + AMP

Fig. 4.11. Origin of the C$_6$-C$_3$-C$_6$ skeleton of flavonoids: mechanism of chalkone formation.

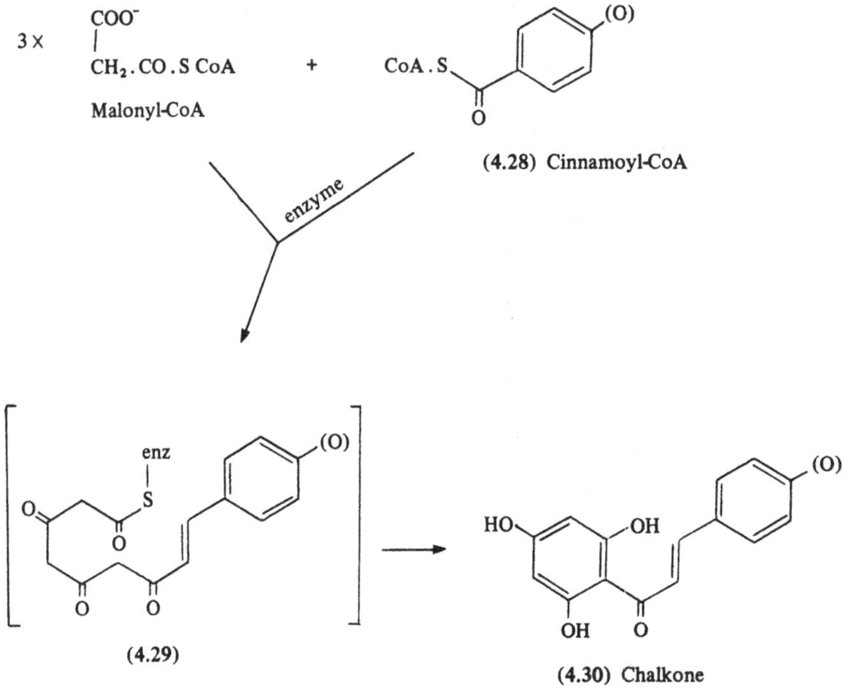

(4.28) Cinnamoyl-CoA

(4.29)

(4.30) Chalkone

(4.7). This process is quite freely reversible *in vivo*, so it has proved very difficult to determine whether it is the chalkone or the corresponding flavanone which serves as immediate precursor to the other flavonoid classes. The available evidence, especially from work with cell-suspension cultures of parsley and other species, favours the idea that the flavanones are direct precursors. This matter will only be resolved unequivocally in other cases by enzyme studies in systems where the chalkone-flavanone isomerase enzyme is inoperative.

*Aurones.* Aurones can readily be obtained from chalkones by chemical oxidation. Their formation *in vivo* by enzymic oxidation of chalkones has also been demonstrated but the mechanism of this reaction is not clear.

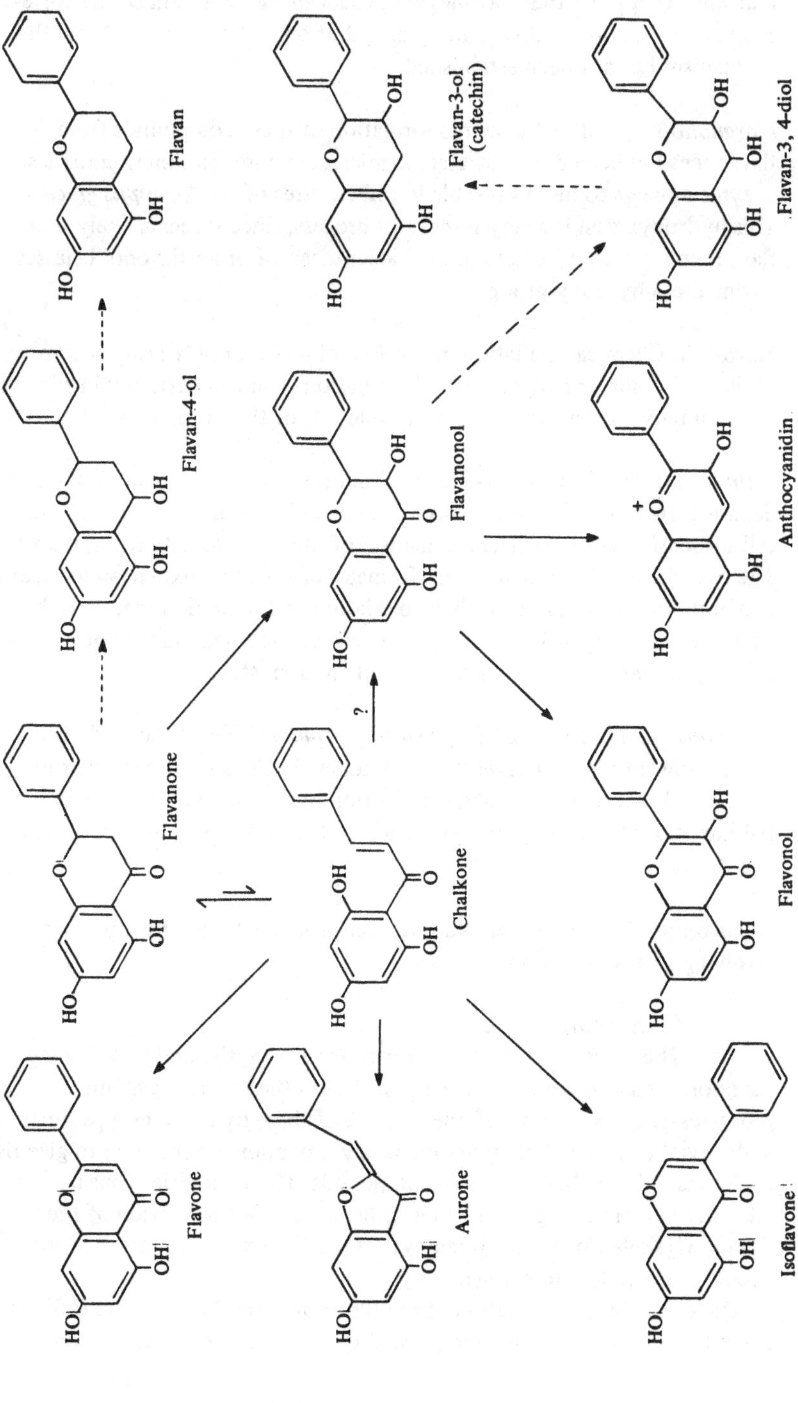

Fig. 4.12. Overall scheme of flavonoid biosynthesis illustrating demonstrated (⟶) and postulated (– – –⟶) biogenetic relationships among the various flavonoid classes.

*Flavones*. It appears that flavones are produced by oxidation of the corresponding flavanone, rather than by dehydration of flavanonols. Again the mechanism has not been established.

*Flavanonols*. The direct enzymic formation of these compounds from flavanones has been demonstrated. A microsomal mixed-function oxidase enzyme appears to be responsible in cell cultures of *Haplopappus gracilis*. This hydroxylation is a very important process, since it seems likely that flavanonols are intermediate in the biosynthesis of other flavonoid classes having the 3-hydroxy-group.

*Flavonols*. Chemical oxidation of flavanonol to flavonol is easily accomplished, and similar enzymic dehydrogenation is almost certainly involved in flavonol biosynthesis, though direct evidence for this is still lacking.

*Anthocyanidins*. The conversion of flavanonols into anthocyanidins has been demonstrated in several plant species, and confirmed in experiments with cell suspension cultures. The mechanism of the transformation is not known. The two classes of compounds are formally of the same oxidation level, and a hypothetical route, not involving oxidation and reduction reactions, has been proposed (fig. 4.13). The biosynthesis of the small number of 3-deoxy-anthocyanidins that occur in Nature remains a mystery.

*Catechins and flavan-3,4-diols (proanthocyanidins)*. These flavan derivatives are at a lower oxidation level than chalkones. Little is known of their biosynthesis, but a scheme involving reduction of the chalkone-flavanone to a flavene (**4.31**) has been proposed to account for their possible interrelation (fig. 4.14).

*Isoflavones*. These compounds are formed from chalkones by a process involving ring B migration.

### 4.6.4    Final modifications
The presence of hydroxy-functions at C-5, C-7 and C-4′ in most flavonoids is a consequence of the basic biosynthetic pathway. Other processes, such as removal of one or more of these hydroxy-groups, further hydroxylation, *O*- or *C*-methylation or glycosylation are required to give rise to the many individual flavonoid compounds. These modifications usually take place at a late stage in the biosynthesis, *i.e.* after formation of the basic $C_6$-$C_3$-$C_6$ skeleton, although removal of the C-5 oxygen function probably occurs at the polyketide stage.

The enzymic hydroxylation of the flavonoid ring B at C-3′ and C-5′ has been demonstrated, though the possibility that in some cases *p*-coumaric

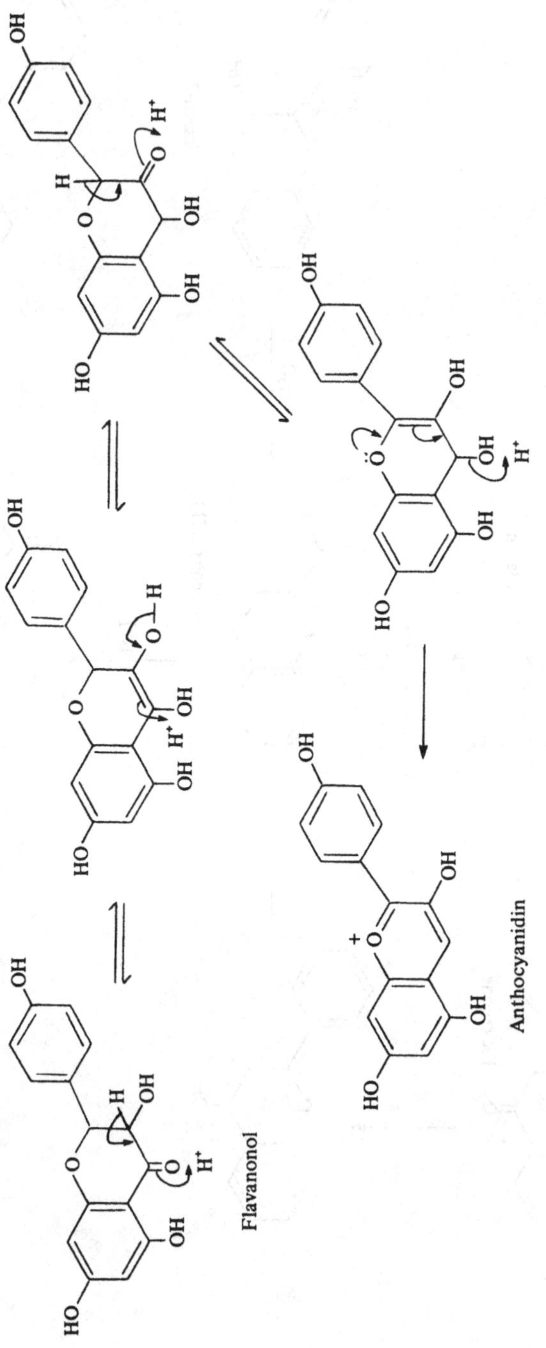

Fig. 4.13. Proposed mechanism of anthocyanidin biosynthesis.

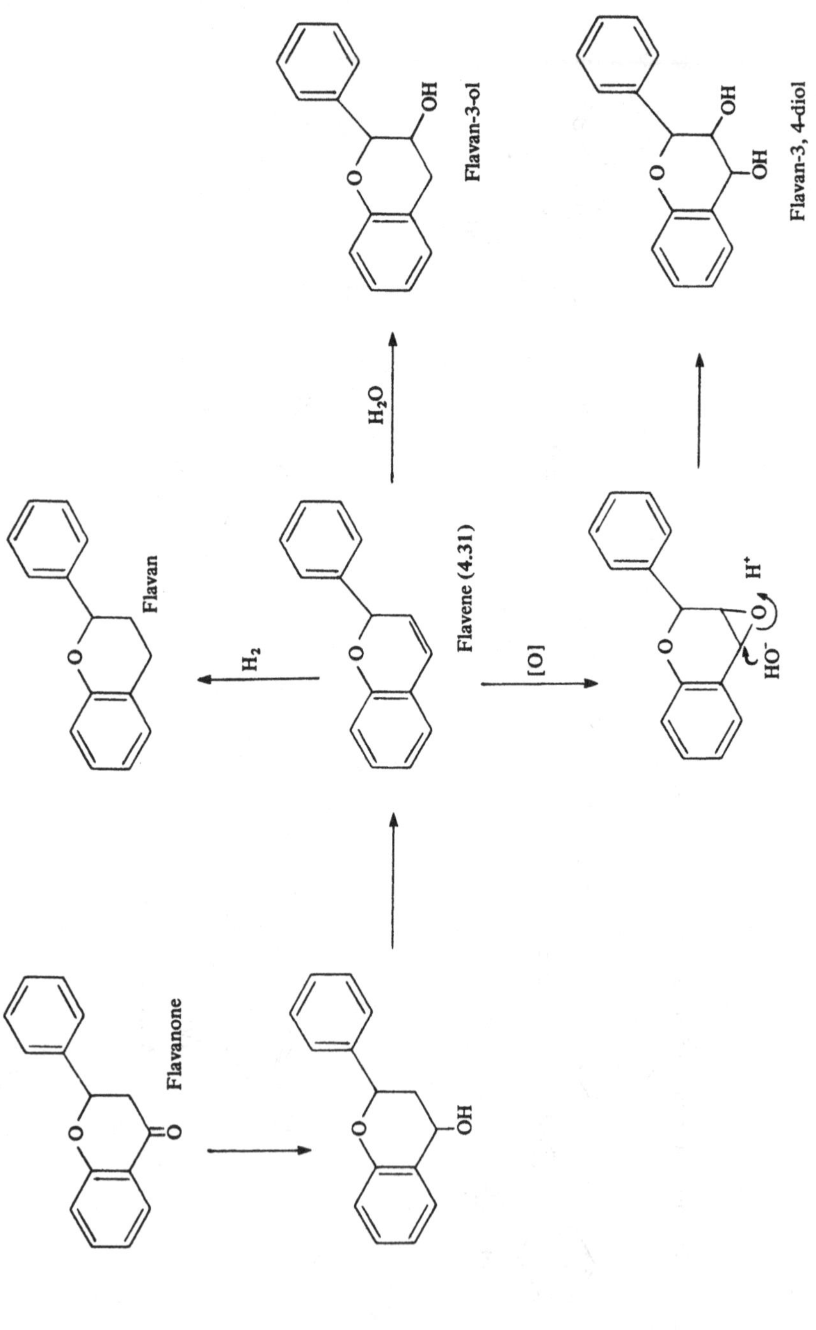

Fig. 4.14. Postulated mechanism for biosynthesis of flavan derivatives (ring hydroxy-substituents are not shown).

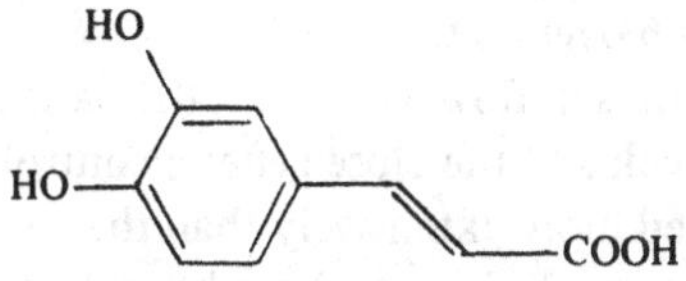

(4.32) Caffeic acid

acid is hydroxylated to caffeic acid (**4.32**) before being incorporated into the flavonoid molecule has not been ruled out entirely. Additional hydroxylation can apparently occur at virtually all levels of oxidation of the flavonoid skeleton.

*O*-Methylation utilises *S*-adenosylmethionine, and the enzymes responsible are very specific both with respect to the position of methylation and also the class of compound methylated. Normally the flavonoid methyl transferases catalyse methylation only at the flavone and flavonol levels, not of anthocyanidins.

*O*-Glycosylation appears to be a straightforward process, monosaccharide units being added from the activated UDP-derivatives.

Little is known of the mechanisms of *C*-methylation and *C*-glycosylation, but the same donors, *S*-adenosylmethionine and UDP-sugars, appear to be used as with *O*-substitution.

## 4.6.5   Oxidative polymerisation of flavonoids

The normally colourless hydroxylated flavans (catechins, flavan-3,4-diols) can give rise to coloured products by oxidative dimerisation and polymerisation processes. The familiar warm brown colour of tea is due to products of this kind, *e.g.* theaflavin (**4.33**).

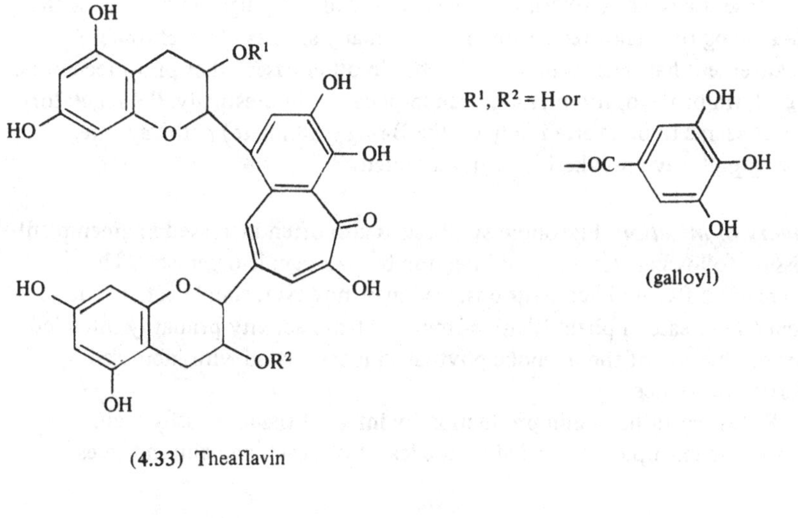

$R^1, R^2 = H$ or

(galloyl)

(4.33) Theaflavin

### 4.6.6    Genetic control of flavonoid biosynthesis

The inheritance of anthocyanins in flowers formed the basis of the classical genetics experiments of Mendel, and the close genetic control of flavonoid biosynthesis has been studied more extensively than that of any other plant constituents. Many single gene differences have been described which result in large qualitative and quantitative variations in flavonoid patterns. Many biochemical effects, *e.g.* total production of flavonoids (*i.e.* coloured *versus* albino phenotypes), concentrations of different flavonoid classes or individual compounds, structural modifications such as hydroxylation, methylation and glycosylation, and flavonoid distribution in different plant parts, have been correlated with specific genetic factors. This genetic work has helped considerably in the elucidation of flavonoid biosynthetic pathways, and also forms the basis of many of the extensive plant breeding programmes being undertaken in the search for new horticultural varieties.

### 4.6.7    Other factors affecting flavonoid biosynthesis

Many internal and environmental factors influence or regulate flavonoid biosynthesis. Among the most important are light and stress conditions such as wounding or infection. Most of the studies of these effects, however, have been physiological rather than biochemical.

*Light.* The effect of light has been studied most extensively. Generally light stimulates synthesis of flavonoids, especially anthocyanins, largely by influencing the level and activity of the biosynthetic enzymes. The enzymes behave as two groups in their responses to light induction. PAL and the enzymes which convert cinnamate into *p*-coumaroyl-CoA are induced most rapidly, whereas the later enzymes form a second group which show a slower response. Enzyme synthesis *de novo* occurs following light induction, with PAL being the major regulation point. In many species phytochrome involvement has been demonstrated, but in other cases other photoreceptors, *e.g.* flavin or flavoprotein, have been implicated. Interestingly, the regulatory effect seems to be exerted only on the B-ring (shikimate) pathway; the A-ring pathway (polyketide) is not affected.

*Injury or infection.* Flavonoid synthesis is also often increased in green plant tissues following wounding or infection by pathogenic organisms. The increase in flavonoid content observed may, however, simply reflect the general increase in phenylalanine-ammonia lyase activity primarily intended for production of the phenolic phytoalexins associated with increased disease resistance.

Excessive anthocyanin production by infected tissue is easily seen. A striking example is provided by the leaves of peach and almond trees infected with the peach leaf-curl fungus. These leaves take on the appearance of brilliant orange-red seed-pods or fruits. Another illustration is given by fruit such as apples. Immature fruit infested with insect larvae commonly synthesise increased amounts of anthocyanin and seem to be prematurely ripe. They can thus readily be distinguished from the healthy fruit on the tree.

*Age*. The production of flavonoids in green tissues also seems to depend on factors such as age and the general stage of growth. Young tissues characteristically contain greater concentrations of flavonoids than do mature tissues. In particular, anthocyanins are often produced in large amounts in new shoots and leaves, which may thus be red in contrast to the green of mature leaves.

## 4.7 Metabolism of flavonoids by animals

All herbivorous animals take in large amounts of flavonoids in their diets. Only in a few cases (see § *4.4.1*) is flavonoid material utilised by the animal for pigmentation or other purposes. In most cases the flavonoids, like other foreign phenolic compounds, are converted into sulphates or glucuronates and excreted, or they may be broken down into smaller phenolic acids (fig. 4.15*a*). There are no reports that flavonoids in general may have any harmful effects towards animals.

## 4.8 Microbial degradation of flavonoids

Little work has been done on the microbial degradation of flavonoids, although some fungi are known to degrade the $C_3$ unit and produce simple phenols from rings A and B (fig. 4.15*b*). It also seems likely that those soil microorganisms which are capable of oxidising and cleaving the aromatic rings of simple phenols should similarly degrade flavonoids.

## 4.9 Functions of flavonoids in plants

The main function that flavonoids perform in plants is as pigments, giving colour to the tissues that synthesise and accumulate them (see chapter 8). There have been many suggestions, however, that at least some flavonoids may have other important roles to play. Their strong light absorption in the u.v. region has led to the suggestion that flavones, flavonols, and anthocyanins could act as a protective screen against harmful u.v. radiations.

Other forms of protection may also be afforded. It is claimed that flavonoids in leaves may serve to deter insects from feeding, and may thus in the long term protect the plant from excessive insect damage. Flavonoids have also been implicated, along with other plant phenols, in the resistance of plants to disease and infection.

Fig. 4.15. Metabolism of quercetin glycosides, *e.g.* rutin, (*a*) by animals, (*b*) by microbes.

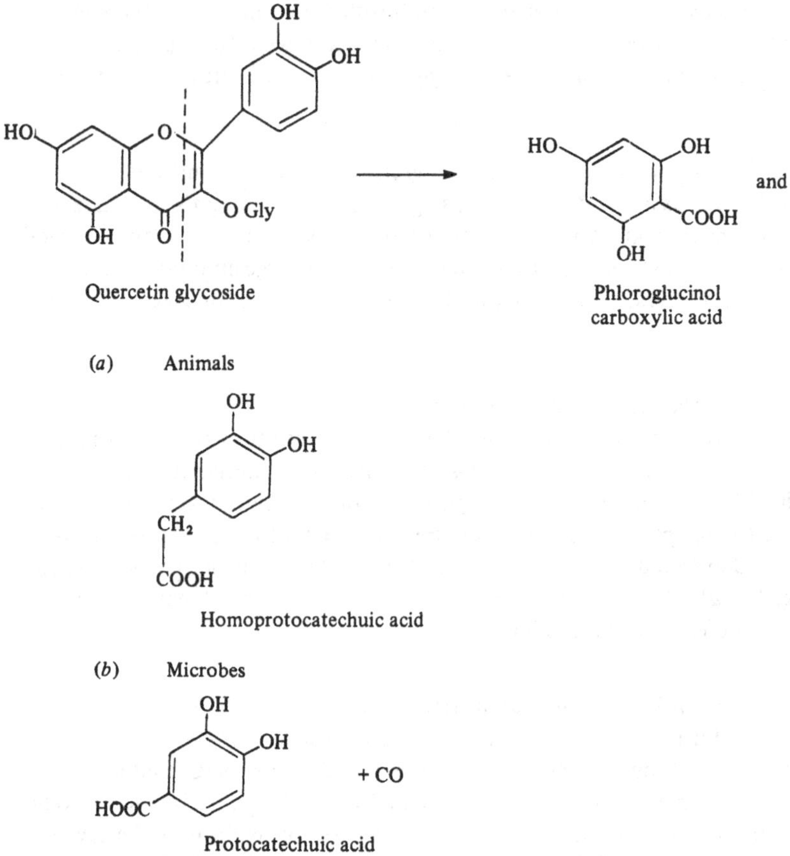
OH
OH
HO
O
O Gly
OH
O
Quercetin glycoside
HO
OH
COOH
OH
Phloroglucinol carboxylic acid
and
(*a*) Animals
OH
OH
$CH_2$
COOH
Homoprotocatechuic acid
(*b*) Microbes
OH
OH
HOOC
+ CO
Protocatechuic acid

Finally the recent detection of flavonoids in the chloroplasts of all species so far examined suggests that the flavonoids may have an important, though as yet unknown, function in that organelle.

## 4.10 Use of anthocyanins as food colorants

The use of anthocyanins as food colour additives has largely been confined to the employment of extracts of highly pigmented fruit sources, such as black grapes and bilberries. These preparations are very crude mixtures containing a wide variety of water-soluble substances. Pure anthocyanin preparations cannot yet be used as food colorants; the instability of their colours at pH values above 4 is a great disadvantage. Methods are being developed, however, by which the flavylium salt or the quinonoid base form can be stabilised at the pH range usually found in drinks and food. Thus self-association between molecules in the cation form produces a disproportionately more intense and stable colour at high concentrations. Copigmentation can be used, especially with the inclusion of acetaldehyde, which gives highly coloured and stable pigments, even at pH 6.

**4.11     Conclusions and comments**
     After the heterogeneity of the quinones, we return with flavonoids to a group of compounds which all have the same carbon skeleton. Different oxidation states give the different flavonoid classes, each with its own peculiar light absorption properties. One class, the anthocyanins, probably in a quinonoid form, provides the most striking colours in flowers and fruits. Anthocyanins are therefore of especial interest to the horticulturalist, and changes in anthocyanin composition are at the heart of many plant breeding programmes to produce new colour varieties of flowers.
     Flavonoid patterns, including the variety of glycosidic derivatives, have been and will continue to be useful markers for taxonomic correlation. Such systematic surveys are likely to provide novel flavonoid structures for the organic chemist. Much work remains to be done on flavonoid biosynthesis. Direct proof of the proposed enzymic interconversions of the various flavonoid classes is in many cases still lacking, the mechanisms of many of the individual reactions are not fully understood, and there has been little detailed work on the enzymes responsible, except with cell suspension cultures from a few species, *e.g.* parsley. Although the physiological and environmental factors (*e.g.* light) which regulate flavonoid biosynthesis as a whole are recognised, the mechanisms that regulate flavonoid patterns and the differential biosynthesis of flavonoids, especially anthocyanins, in differently coloured areas of flowers and other plant tissues are largely unknown. Such knowledge would be of particular interest to the horticulturist, as would a greater understanding of the mechanisms of copigmentation and other phenomena which modify the basic colour produced by anthocyanins *in vivo*. The flavonoid field may be one in which plant tissue cultures will prove particularly useful for studies of biosynthesis and its regulation and also of the mechanism of colour enrichment.
     In addition to these major areas, many other problems remain unsolved. If small amounts of flavonoids are present in plant chloroplasts, what is their function there? Are they really absent from algae, even if present and essential in plant chloroplasts? Many animals, including humans, ingest large amounts of flavonoids in their food. How are they metabolised? Have they any harmful effects? Do the animals use them for any useful purpose?
     As regulations on food additives becomes stricter, interest in flavonoids, especially anthocyanins, as industrial food colours from natural sources will surely increase, and these questions about their metabolic fate in animals, particularly humans, will become even more important. Methods will be

sought for the industrial synthesis of certain flavonoids, or for large-scale biological production, possibly including the use of tissue cultures, and ways of stabilising the anthocyanin colour in food preparations will be essential.


## 4.12    Suggested further reading

The definitive book on flavonoids is the large volume edited by Harborne, Mabry and Mabry (1975) which gives comprehensive information about all aspects of the chemistry and biochemistry of these compounds. The older book by Geissman (1962) on flavonoid chemistry remains useful. A more biochemical or systematic approach is adopted by Harborne (1967) in *Comparative biochemistry of the flavonoids*. Another Harborne volume (1964) on the biochemistry of phenolic compounds in general also contains useful material. Two recent review articles by Swain (1976*a*) and Wong (1976) describe, respectively, the nature and properties of flavonoids and their biosynthesis. Flavonoid functions in plants are discussed by Harborne (1976) in the same volume. The article by Haslam (1979) on the shikimate biosynthetic pathway includes flavonoid biosynthesis, and reaction mechanisms are considered in as much detail as experimental evidence permits. An article by Grisebach (1980) provides up-to-date information on selected aspects of flavonoid biosynthesis. A very recent review by Timberlake (1980) provides insight into the problems associated with the use of anthocyanins as food colorants.

Methods in general use in flavonoid biochemistry are outlined by Swain (1976*b*) and more details are included in yet another book by Harborne (1973).


## 4.13    Selected bibliography

Geissman, T. A. (ed.) (1962) *The chemistry of flavonoid compounds*. New York: Macmillan.

Grisebach, H. (1980) Recent developments in flavonoid biosynthesis, in *Pigments in plants*, 2nd edition, ed. F.-C. Czygan, p. 187. Stuttgart and New York: Gustav-Fischer.

Harborne, J. B. (ed.) (1964) *Biochemistry of phenolic compounds*. London and New York: Academic Press.

Harborne, J. B. (1967) *Comparative biochemistry of the flavonoids*. London and New York: Academic Press.

Harborne, J. B. (1973) *Phytochemical methods*. London: Chapman and Hall.

Harborne, J. B. (1976) Functions of flavonoids in plants, in *Chemistry and biochemistry of plant pigments*, 2nd edition, vol. 1, ed. T. W. Goodwin, p. 736. London, New York and San Francisco: Academic Press.

Harborne, J. B., Mabry, T. J. and Mabry, H. (eds) (1975) *The Flavonoids*. London: Chapman and Hall.

Haslam, E. (1979) Shikimic acid metabolites, in *Comprehensive organic chemistry*, vol. 5, ed. E. Haslam, p. 1167. Oxford and London: Pergamon.

Swain, T. (1976*a*) Nature and properties of flavonoids, in *Chemistry and biochemistry of plant pigments*, 2nd edition, vol. 1, ed. T. W. Goodwin, p. 425. London, New York and San Francisco: Academic Press.

Swain, T. (1976*b*) Flavonoids, in *Chemistry and biochemistry of plant pigments*, 2nd edition, vol. 2, ed. T. W. Goodwin, p. 166. London, New York and San Francisco: Academic Press.

Timberlake, C. F. (1980) Anthocyanins – occurrence, extraction and chemistry, *Food Chem.*, 5, 69.

Wong, E. (1976) Biosynthesis of flavonoids, in *Chemistry and biochemistry of plant pigments*, 2nd edition, vol. 1, ed. T. W. Goodwin, p. 464. London, New York and San Francisco: Academic Press.

# 5    Tetrapyrroles

## 5.1    Introduction

The $N$-heterocyclic compound pyrrole (**5.1**) is a very stable heteroaromatic system, but simple monopyrroles are seldom encountered in Nature
(see, however, the *Wallemia* pigments). Di- and tripyrroles are also rare,
although the red bacterial pigment prodigiosin is now known to be a linear
tripyrrole (**5.2**). In contrast, some of the most familiar of all natural pigments
have cyclic tetrapyrrole structures, including such important substances as
chlorophyll, the green light-harvesting pigment of plants, and haem, which
forms the basis of the oxygen-transporting red blood pigments. Related to
these are the bilins, which are linear tetrapyrroles. This group includes the
animal bile pigments, the phycobilin accessory photosynthetic pigments of
some algae, and the chromophore of the plant photoregulatory pigment,
phytochrome.

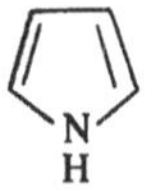

(5.1) Pyrrole 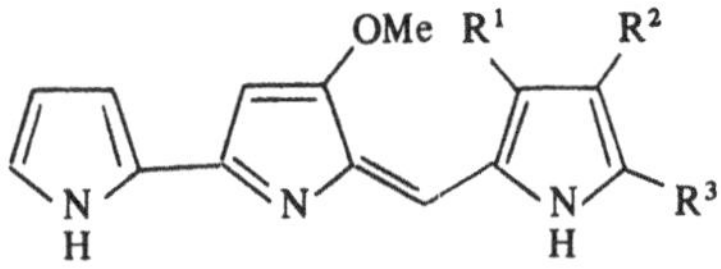            (5.2) Prodigiosin: $R^1$, $R^2$, $R^3$ = H or alkyl

## 5.2    General structural features

The most important natural tetrapyrrole pigments are **porphyrins**,
containing a '**super-ring**' or **macrocycle** in which the four pyrrole residues are
linked by single bridging carbon atoms. The basic structure is that of **porphin**,
which is illustrated in fig. 5.1, with the commonly used Fischer numbering
system. The four pyrrole rings are designated A–D or I–IV, and the linking
**methine** (or **methene**) **bridge** or **meso** carbon atoms $\alpha$–$\delta$. The peripheral
pyrrole carbon atoms of the 'super-ring' are numbered 1–8. In the various
natural pigments these carbons bear additional carbon sidechains (see below).

In 1960 an IUPAC commission recommended a new nomenclature and numbering system (fig. 5.2). Although the familiar Fischer system has retained much of its popularity, the IUPAC scheme will be used in this book.

The basic porphin structure (5.3) is present in haem, whereas chlorophylls and bacteriochlorophylls, respectively, contain the dihydroporphin ($\equiv$ chlorin, 5.4) and tetrahydroporphin ($\equiv$ bacteriochlorin, 5.5) systems. In the natural cyclic tetrapyrrole pigments, the pyrrole *N*-atoms chelate a metal ion, usually $Mg^{2+}$ (in chlorophylls) or $Fe^{2+}$ (haem).

The biosynthetic process by which the porphin ring is formed (see §5.9.4) gives a first cyclic tetrapyrrole in which each pyrrole ring carries both

Fig. 5.1. Basic structure and Fischer numbering scheme for the porphin ring system.

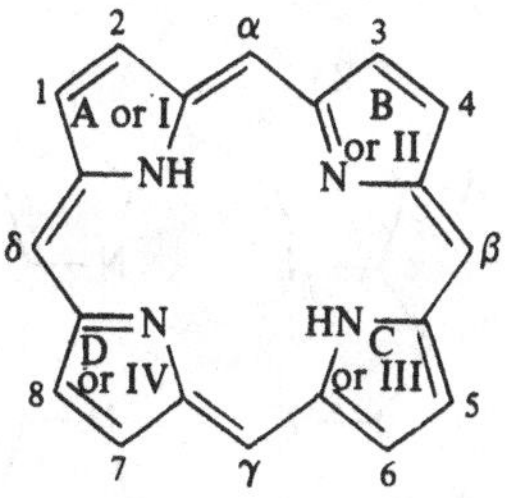

Fig. 5.2. IUPAC numbering system for porphin.

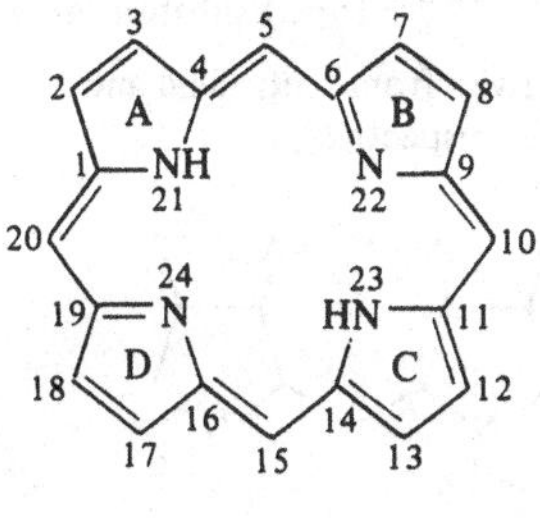

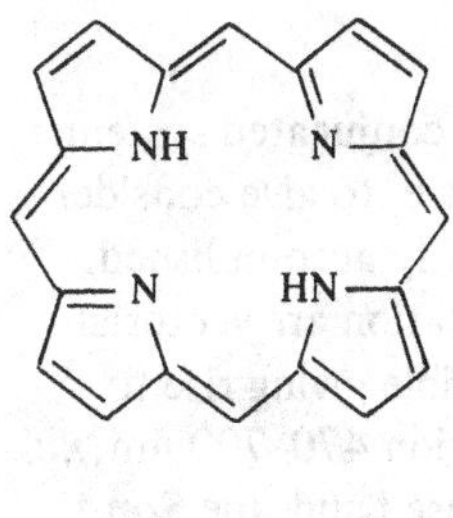

(5.3) Porphin 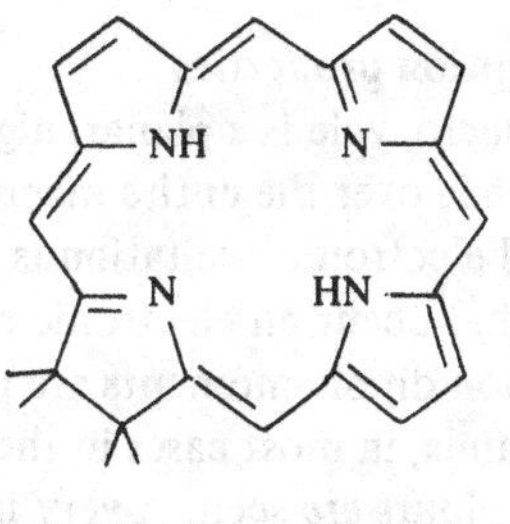    (5.4) Dihydroporphin (chlorin) 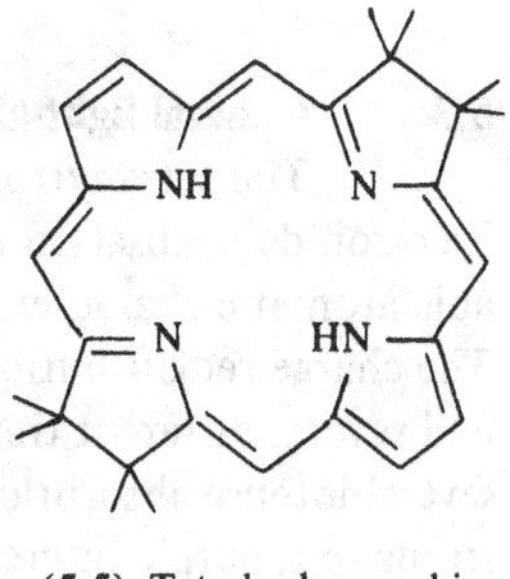    (5.5) Tetrahydroporphin (bacteriochlorin)

a $C_2$ (acetic acid) and a $C_3$ (propionic acid) sidechain. Many dispositions of
the sidechains are theoretically possible, but almost all naturally occurring
tetrapyrroles are derived from a precursor with the sidechains located as
illustrated in **5.6**. This is known as the **Type III** substitution pattern. Occasionally compounds are encountered with an alternative (**Type I**) substitution
pattern (**5.7**). Compounds possessing other arrangements of the sidechains
do not appear to occur naturally.

The simpler linear tetrapyrroles found in Nature have the basic structure
(**5.8**). These compounds, known as **bilins**, or formerly as bile pigments, are
formed by plant and animal systems from porphyrin structures normally
by cleavage of the α-meso carbon bridge (IUPAC C-5). Structure **5.8** illustrates
the sidechain substitution pattern for a bilin formed by α-cleavage of a Type
III porphyrin.

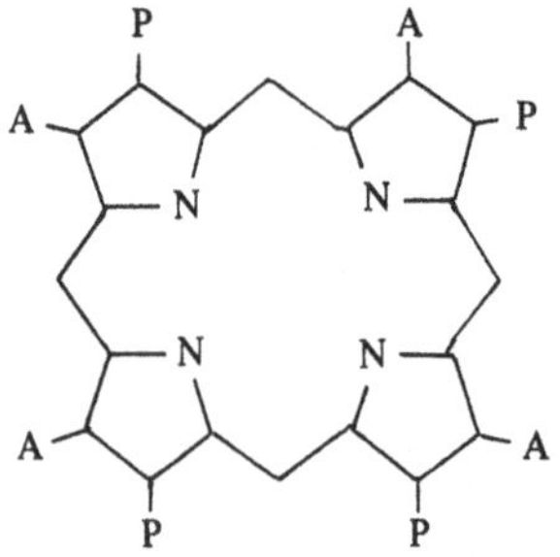

(5.6) Type III substitution pattern 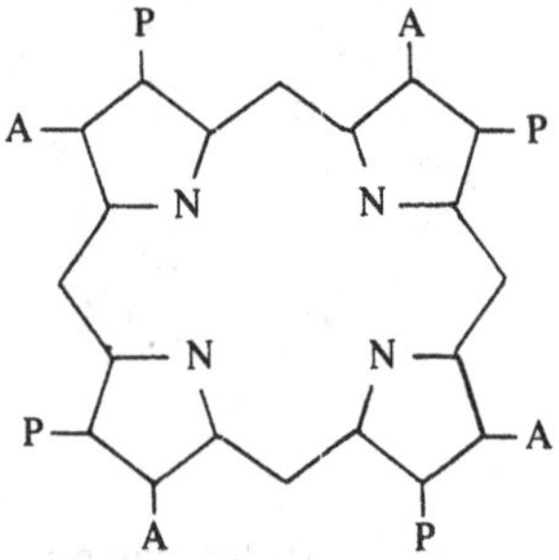          (5.7) Type I substitution pattern

(A and P represent substituents derived from acetic acid and
propionic acid sidechains, respectively)

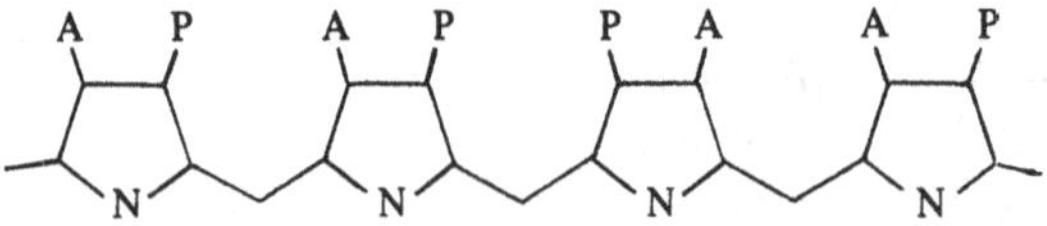

(5.8) Basic bilin skeleton

## 5.3    General light-absorption properties

The tetrapyrrole macrocycle is a planar, highly conjugated system.
Electron delocalisation extends over the entire macrocycle, to give considerable aromatic character, and electronic excitation is readily accomplished.
The charge redistributions that occur on electronic excitation are vectorial
and several different transition dipole moments are possible giving rise to
several intense absorption bands, in most cases in the region 470–700 nm, *i.e.*
strong red, purple or green colours are seen. A very intense band, the **Soret
band**, which occurs at *ca* 400 nm, is a consequence of the symmetrical

disposition of the four pyrrole *N*-atoms, and is highly characteristic of the tetrapyrrole macrocycle.

Changes in hydrogenation level obviously affect the extent and polarisation of the conjugated chromophore and consequently the position of the absorption maxima. The spectrum is also influenced by the substitution pattern of the pyrrole rings. These effects are of great diagnostic value in the identification of structural types. Ionisation brought about by acidic or alkaline conditions also causes changes in the spectra.

The linear tetrapyrroles show a simpler pattern of strong absorption in the visible region. Increased polarisation results in the shifting of the longest wavelength band to even longer wavelength, but the Soret band characteristic of the macrocyclic structures is no longer present. Saturation, especially of the *meso* carbon atoms, occurs readily, decreasing the length of the conjugated chromophore and hence the wavelength of maximum absorption.

In the following sections of this chapter, some of the most important natural tetrapyrroles will be described, and in many cases details of the absorption spectra will be given.

## 5.4    Chlorophylls
### 5.4.1    *Structures and distribution*

The contribution that chlorophylls make to the external colour of the plants that contain them is clearly seen in the pervading green of natural vegetation. All green tissues of higher plants contain in their photosynthetic organelles (chloroplasts) two chlorophylls, *a* and *b*. The nucleus of these compounds is phorbin (**5.9**), which is basically a chlorin structure but

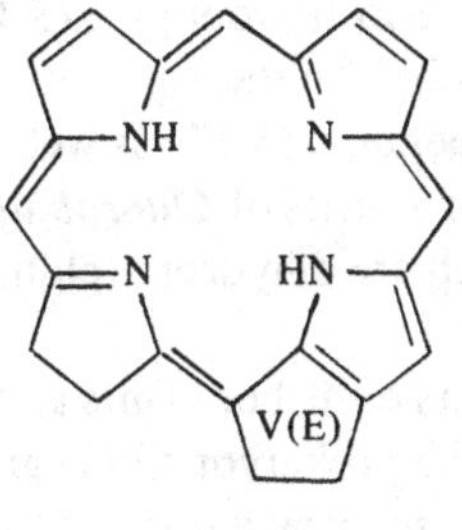

(**5.9**) Phorbin

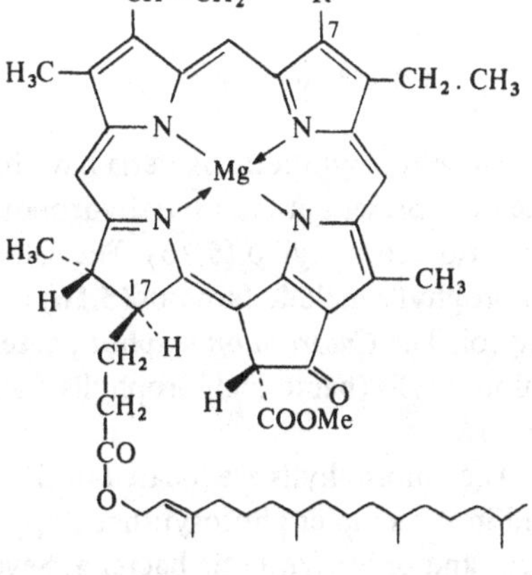

(**5.10**) Chlorophyll *a*: R = CH$_3$

(**5.11**) Chlorophyll *b*: R = CHO

contains an additional ring, the **isocyclic ring, ring E** or **ring V**, derived biosynthetically (see §*5.9.8*) by oxidation and cyclisation of the C-13 propionic acid substituent on to the γ-methine bridge carbon (C-15). The full structures of chlorophylls *a* and *b* are as shown (**5.10, 5.11**). These differ only in the substituent at C-7 which is —CH$_3$ in chlorophyll *a* but —CHO in chlorophyll *b*. Other notable features are a chelated Mg$^{2+}$ ion and esterification of the C-17 propionic acid substituent by the isoprenoid alcohol phytol. The existence of chlorophylls with geranylgeraniol as the esterifying alcohol has recently been demonstrated.

Chlorophyll *a* is also present in all algae, alone in the Cyanophyta and accompanied by chlorophyll *b* in the Chlorophyta and Euglenaphyta. Other algal classes also contain chlorophyll *a*, but have additional chlorophylls with slightly different substitution patterns. For example, the Chrysophyta, Pyrrophyta and Phaeophyta have chlorophyll *c* (**5.12**) and the Rhodophyta chlorophyll *d* (**5.13**).

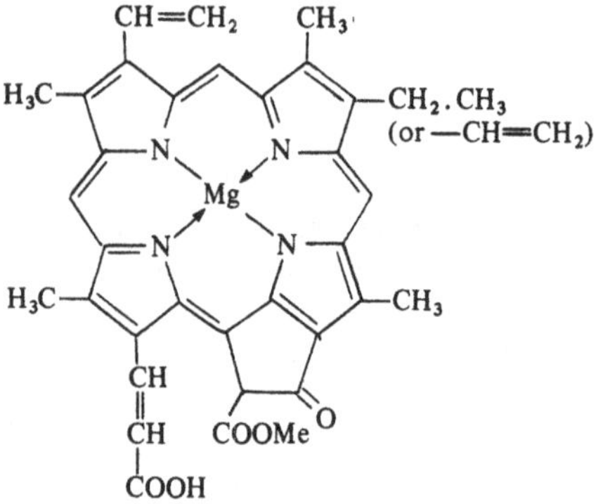

(5.12) Chlorophyll *c*

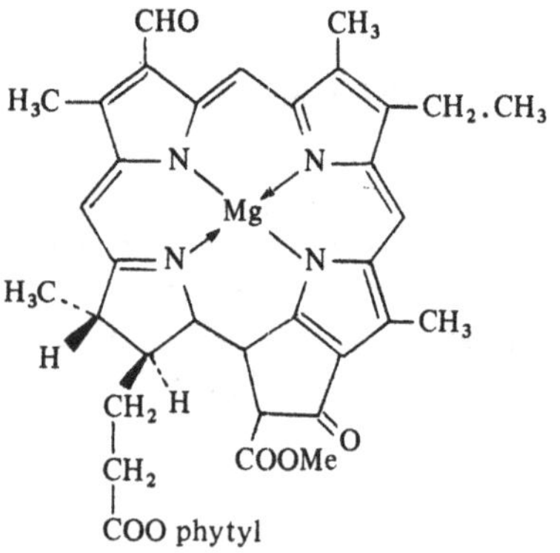

(5.13) Chlorophyll *d*

The photosynthetic bacteria have bacteriochlorophylls. In most species, the ones present are the tetrahydroporphins, bacteriochlorophyll *a* (**5.14**) and bacteriochlorophyll *b* (**5.15**). The esterifying alcohols in these bacteriochlorophylls include farnesol (**5.16**) and geranylgeraniol (**5.17**) as well as phytol. The *Chlorobium* sulphur bacteria contain a series of *Chlorobium* chlorophylls (bacteriochlorophylls *c* and *d*) which are dihydroporphins, such as **5.18**.

The chlorophylls are located in the chloroplasts of higher plants and algae and in the simpler photosynthetic apparatus of the prokaryotic blue-green algae and photosynthetic bacteria. Several spectroscopically distinct forms of chlorophyll have been recognised *in vivo*, indicating different functions in photosynthesis (chapter **10**). The spectroscopic differences are caused by

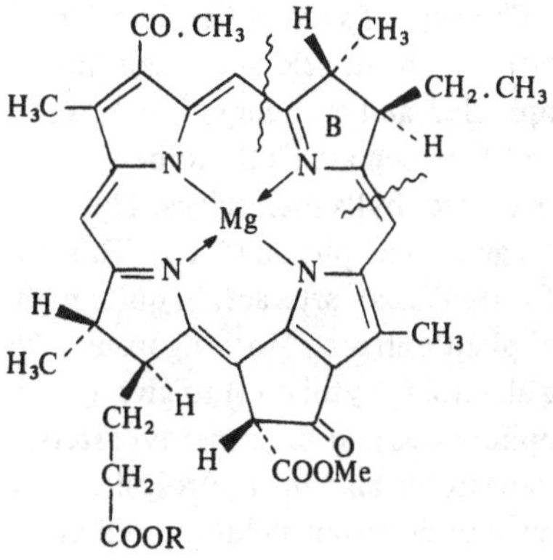

(5.14) Bacteriochlorophyll *a*
(R = farnesyl or geranylgeranyl)

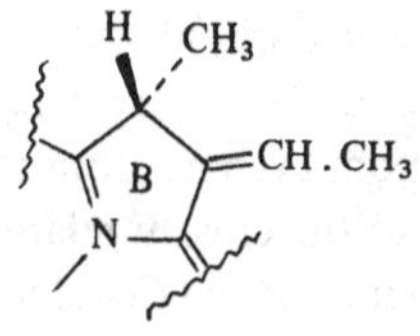

(5.15) Bacteriochlorophyll *b*

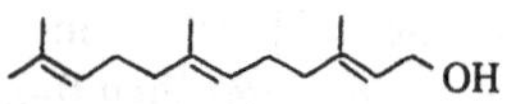

(5.16) Farnesol

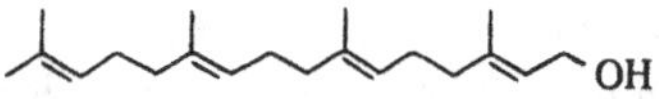

(5.17) Geranylgeraniol

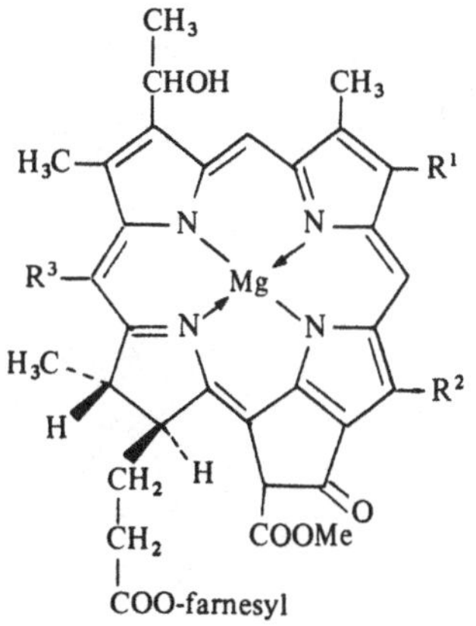

$R^1 = CH_2.CH_3$, $CH_2.CH_2.CH_3$
or $CH_2.CH(CH_3)_2$

$R^2 = CH_3$ or $CH_2.CH_3$

$R^3 = H$, $CH_3$ or $CH_2.CH_3$

(5.18) *Chlorobium* chlorophylls (bacteriochlorophylls *c* and *d*)

differences in the microenvironment of the chlorophyll molecules, *e.g.*
association with protein, molecular stacking.

## 5.4.2   General properties

The chlorophylls are esters and are soluble in most organic solvents.
They may be extracted from the tissues in which they occur by polar organic

solvents, particularly acetone and alcohols. Chlorophylls are rather unstable compounds and are readily destroyed by light, $O_2$, heat, acid and alkali. In solution, even at room temperature, chlorophylls *a* and *b* undergo isomerisation to the closely related chlorophylls *a'* and *b'* which are likely to be epimeric (at the COOMe of ring E) with the chlorophylls themselves. The magnesium is removed very easily by acid to give phaeophytin (**5.19**). This occurs so readily that phaeophytin is usually seen, as an artefact, in quite high concentration on chromatograms of natural plant extracts. More vigorous acid treatment also removes the esterifying alcohol (phytol *etc.*) to give a water-soluble phaeophorbide (**5.20**). Phaeophorbides and their methyl esters, and chlorophyllide (**5.21**) are produced by anaerobic alkaline hydrolysis. Saponification therefore provides a useful way of destroying chlorophyll in order to facilitate work with other plant lipids. The water-soluble chlorophyll degradation products are efficiently washed out of the extract.

Some of these derivatives have been used widely in radioisotope ($^{14}$C, $^{3}$H) and stable isotope ($^{13}$C) labelling studies of chlorophyll biosynthesis. The degradation products are relatively stable, easy to handle and purify, and give n.m.r. and other spectra which are much simpler to interpret than those of the chlorophylls themselves.

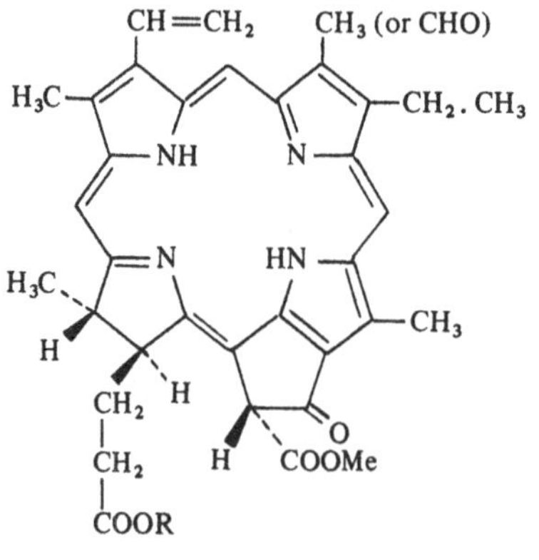

(**5.19**) Phaeophytin (R = phytyl)
(**5.20**) Phaeophorbide (R = H)

*5.4.3    Spectroscopic properties*
    The absorption spectra of chlorophylls *a* and *b* in diethyl ether solution are illustrated in fig. 5.3. The Soret bands are seen at 430 and 455 nm, respectively, and the longest wavelength or α-bands at 662 and 641 nm, respectively. The general features of the spectra of chlorophylls *c* and *d* (fig. 5.4) and of the *Chlorobium* chlorophylls (fig. 5.5) are similar to these, but the absorption maxima occur at different wavelengths.

Fig. 5.3. Light absorption spectra of chlorophyll *a* (———) and chlorophyll *b* (– – – –) in diethyl ether.

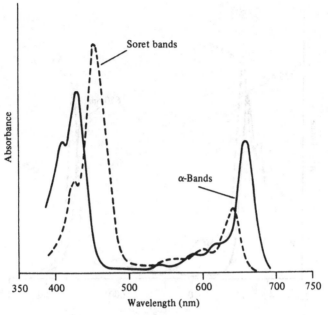

Fig. 5.4. Light absorption spectra of chlorophyll *c* (– – – –) and chlorophyll *d* (———) in diethyl ether.

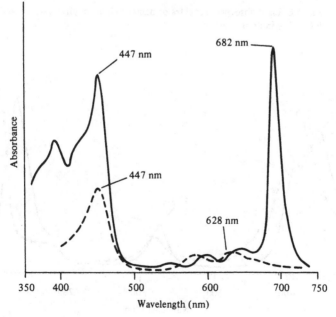

Fig. 5.5. Light absorption spectra of *Chlorobium* 'chlorophyll-650' (——)
and 'chlorophyll-660' (– – – –) in diethyl ether.

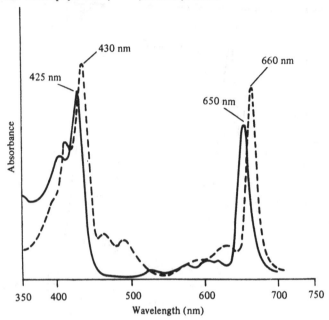

Fig. 5.6. Light absorption spectra of bacteriochlorophylls *a* (——) and
*b* (– – – –) in diethyl ether.

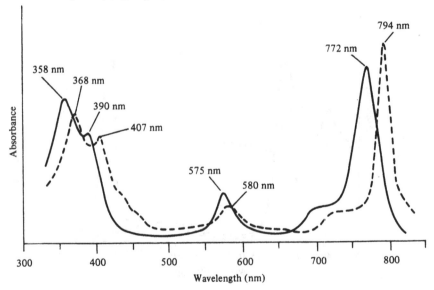

Bacteriochlorophylls *a* and *b* (fig. 5.6) also give a Soret band at *ca* 400 nm (358 and 368 nm, respectively) but the greater polarisation of the tetrahydroporphin chromophore causes the α-bands to appear at much longer wavelength (772 and 794 nm, respectively), *i.e.* in the infrared. Bacteriochlorophylls are thus not intensely coloured.

### 5.4.4    Chlorophyll and derivatives in animals

Free chlorophyll is not normally found in animals, except in the gut of herbivores. Some animal species however, particularly aquatic inverte- brates, contain commensal or symbiotic unicellular algae. An interesting example is a tropical didemnid ascidian, which was found to contain a hitherto unknown kind of prokaryotic green alga, *Prochloron*, now con- sidered by many to be a 'missing link' in the evolution of higher plants and their chloroplasts. Other animals are able to salvage and accumulate chloroplasts, still in functional condition, from dietary plant material. One example is a sacoglossan mollusc, *Elysia viridis*, which may contain as many as $10^8$ chloroplasts per animal. The chloroplasts are derived from the siphonalean alga *Codium fragile*, and those in the animal show rates of photo- synthesis comparable with those of the intact alga. All these animals must obviously contain chlorophylls and are therefore usually green in colour.

Chlorophyll breakdown products, including phaeophorbides, chlorins and phylloerythrin (**5.22**) are rarely found naturally in plants (though they are common as artefacts), but are used by some animals, especially certain poly- chaete and echiurid worms, as integumental pigments. These compounds may be produced by the animal from dietary chlorophyll, or they could be obtained direct from the decaying plant detritus on which the animals feed. In herbivorous or omnivorous mammals, including man, the main breakdown product of chlorophyll is phylloerythrin, which is present in the alimentary canal and in the bile.

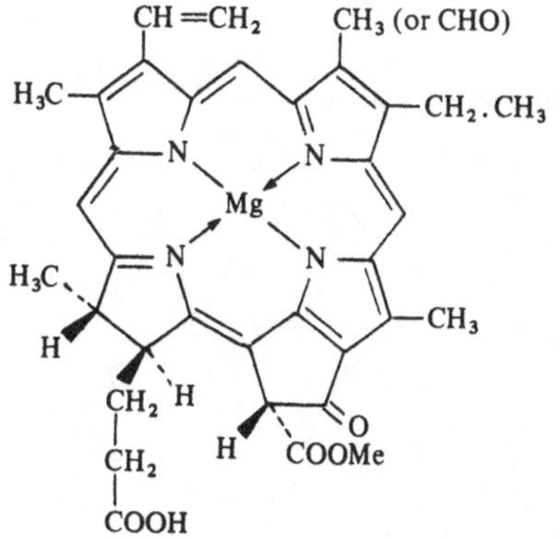

(5.21) Chlorophyllide

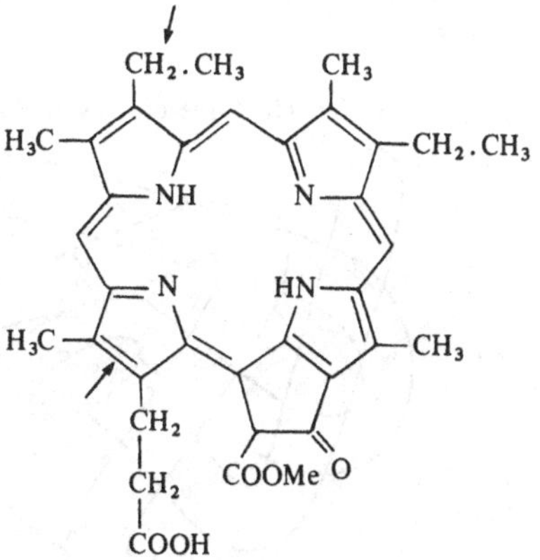

(5.22) Phylloerythrin

## 5.5    Haem and haemoproteins

Just as the most familiar plant pigment, chlorophyll, is a porphyrin, so too is the prosthetic group of the best-known animal pigment, the red blood protein haemoglobin, and the closely related myoglobin. These substances are haemoproteins, *i.e.* proteins which have as prosthetic group an iron-chelated porphyrin or **haem**. Other haemoproteins of great biological importance include the cytochromes and some enzymes such as peroxidase and catalase. In haemoglobin and myoglobin, the iron remains in the ferrous form, $Fe^{2+}$; the ferric, $Fe^{3+}$, form is inactive. In contrast to this, the functioning of the cytochromes depends on the freely reversible interconversion of the oxidised and reduced forms.

### 5.5.1    *Haemoglobin and myoglobin: structures and properties*

The vital, oxygen-transporting blood pigment of most animals, including mammals, is haemoglobin, a haemoprotein which has as its prosthetic group protohaem (**5.23**), the Fe-chelate of protoporphyrin IX. Muscle contains a structurally and functionally similar pigment, myoglobin. These two proteins were the first to have their three-dimensional structures elucidated by X-ray crystallography. Myoglobin has a single polypeptide chain of 153 amino acid residues and molecular weight 17 800. Its three-dimensional structure is illustrated in fig. 5.7. The folding of the peptide chain is such that the molecule is very compact. About three-quarters of the chain is in an α-helical form, with eight different helical segments. The outside of the molecule has both polar and non-polar residues, but the inside consists almost exclusively of non-polar residues which provide a hydrophobic environment for the single haem group. The iron atom of the haem is always in the ferrous state, $Fe^{2+}$, and is capable of complexing with six ligands, four of which are the four pyrrole *N*-atoms of the porphyrin. The fifth site is used for binding the haem to the protein by coordination with an imidazole nitrogen of a

Fig. 5.7. The three-dimensional structure of myoglobin.

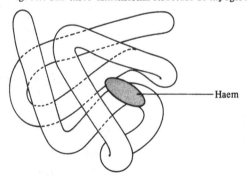

Haem

(5.23) Protohaem IX

histidine residue of the globin (known as the proximal histidine). In
**deoxymyoglobin**, the sixth site remains vacant and available for oxygen
uptake. A second histidine residue (the distal histidine) is very near to the
oxygen-binding site, but not bonded to the haem. Oxygen readily binds to
the sixth coordination site to give **oxymyoglobin**. Oxidation of the iron to
the $Fe^{3+}$ form, *i.e.* ferrihaem, can be achieved but the resulting ferrimyo-
globin is able to bind only water, not oxygen, at the sixth site and therefore
cannot function in oxygen transport.

The protein structure of haemoglobin is more complex. Mammalian
(including human) haemoglobin has a molecular weight of 64 500 and
consists of four polypeptide chains, each with its own haem group. The
predominant form of haemoglobin in human adults, haemoglobin A, has two
pairs of polypeptide chains, the α-chains, each consisting of 141 amino acid
residues, and the β-chains, with 146 residues each. In a minor adult haemoglobin, $A_2$, and in foetal haemoglobin F, other variants replace the β-chains.
Although the amino acid sequences of these haemoglobin peptide chains
differ greatly from that of myoglobin, the three-dimensional structures are
strikingly similar, and the haem groups again occupy hydrophobic clefts
within the folded protein. The proximal and distal histidine residues are
among the only nine amino acids to be conserved in all myoglobins and
haemoglobins from a range of animal species.

In haemoglobin, the four polypeptide chains are associated in an approximately tetrahedral array to give a nearly spherical molecule. Each α-chain is in
contact with the two β-chains, but there are few interactions between the two
α-chains or the two β-chains. Each of the four haem groups in haemoglobin
can take up one molecule of oxygen. The oxygen-containing form is called
**oxyhaemoglobin** (5.24b), the oxygen-free form **deoxyhaemoglobin** (5.24a).

142    *Tetrapyrroles*

Several changes occur in the three-dimensional structure when deoxyhaemoglobin takes up oxygen, notably a movement of the $Fe^{2+}$ into the plane of the
haem ring system (see below and fig. 5.8). As before, oxidation to the ferric
state gives a form, methaemoglobin, which cannot take up molecular oxygen.

*5.5.2    Haemoglobin and myoglobin: functioning in oxygen transport*
The ability to complex reversibly with oxygen renders haemoglobin
of vital importance as an oxygen transporter in animals. In mammals, haemoglobin is present in the red blood cells, and is responsible for carrying oxygen
from the lungs through the arteries, arterioles and capillaries to the various
body tissues. It also assists in carrying carbon dioxide on the return journey
from the tissues to the lungs. The oxygen-carrying capacity of arterial blood
is some 70 times greater when haemoglobin is present than in its absence. The
oxygen affinity of myoglobin is considerably greater than that of haemoglobin, so myoglobin can therefore accept oxygen from haemoglobin for use
or storage in muscle cells.

Fig. 5.8. Oxygen binding by haem in haemoglobin.

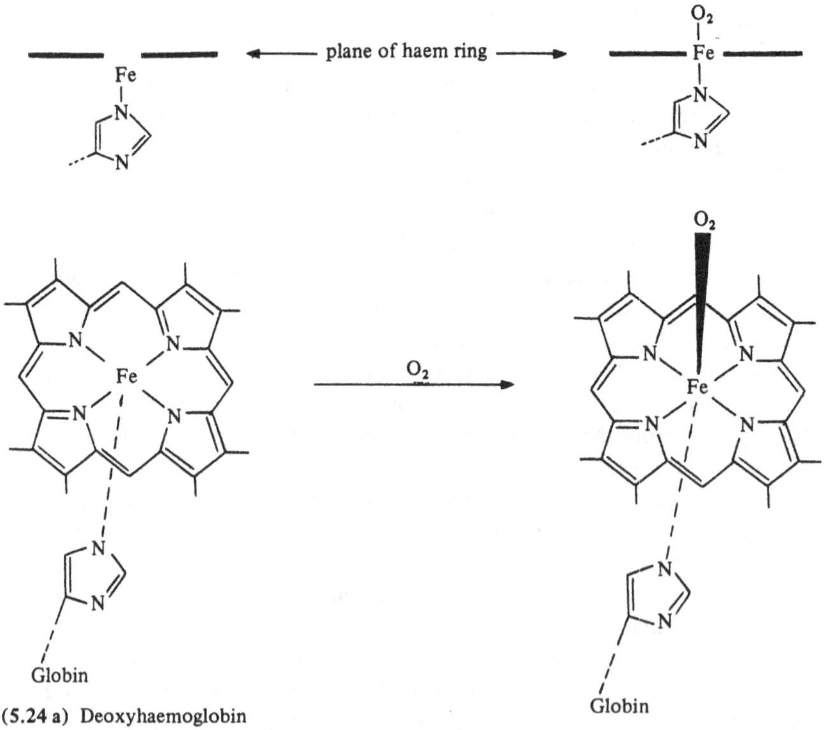

(5.24 a) Deoxyhaemoglobin

(5.24 b) Oxyhaemoglobin

Details of the mechanism of oxygen binding by haemoglobin are complex but well understood. In addition to being extremely important physiologically, this process provides an excellent illustration of allosteric interactions and regulation. The peculiarities of oxygen binding by haemoglobin may be summarised as follows:

(i) The shape of the oxygen-dissociation curve of haemoglobin is sigmoidal (that of myoglobin is hyperbolic). This indicates that the binding of oxygen to haem is cooperative, *i.e.* the binding of oxygen to one haem facilitates binding to the other haems.

(ii) The affinity of haemoglobin for oxygen is dependent on pH and affected by $CO_2$.

(iii) Organic phosphates, especially 'DPG' (2,3-diphosphoglycerate or 2,3-bisphosphoglycerate) also affect the oxygen affinity of haemoglobin.

*Cooperative oxygen binding.* If haemoglobin is dissociated into its constituent peptide chains, the isolated chains behave very much like myoglobin; they exhibit hyperbolic oxygen-dissociation curves and their oxygen-binding properties are not affected by $CO_2$, DPG or pH. The allosteric properties of haemoglobin clearly arise from interactions between the four subunits. The primary factor involved is the movement of the iron atom of haem when oxygen is bound. As mentioned previously, the haem iron is displaced (by 0.75 Å) out of the plane of the porphyrin in deoxyhaemoglobin. Binding of oxygen pulls the iron atom into the plane in oxyhaemoglobin, and the proximal histidine is pulled towards the haem ring (fig. 5.8). This movement leads to further small changes in the tertiary structure of that subunit; in particular a tyrosine residue and the adjacent *C*-terminal amino acid undergo quite large displacements. As a result, some of the interactions between subunits are removed and the quaternary structure is consequently destabilised. The transmitted effects on the conformation of the other subunits make oxygen binding by the other subunits easier. Conversely, when oxygen is 'unloaded' from one haem site, the resulting changes in conformation and subunit interactions facilitate unloading from the other subunit sites.

*Effect of $CO_2$ and pH (Bohr effect) (fig. 5.9).* Increasing acidity (lower pH) enhances the release of $O_2$ by haemoglobin. Increased concentrations of $CO_2$ also lower the oxygen affinity. This is extremely important physiologically, since rapidly metabolising tissues such as muscle produce much $CO_2$ and acid. These high $CO_2$ and $H^+$ levels promote the release of $O_2$ from haemoglobin, and thus the need for more oxygen in the metabolically active tissues can be met. When the oxygen is released the deoxyhaemoglobin takes up $H^+$ and $CO_2$. The higher oxygen concentration in the alveolar capillaries of the lung then drives off the $H^+$ and $CO_2$ as the deoxyhaemoglobin takes up oxygen

again. The structural transformations that take place during these changes
have been identified. Conformational changes on going from oxyhaemoglobin
to deoxyhaemoglobin place acidic amino acid residues in closer proximity to
certain histidine and terminal $NH_2$-groups. The change in the local charge
environment raises the pK of the residue, thus increasing the affinity for $H^+$.
$CO_2$ also binds much more readily to deoxyhaemoglobin than to oxyhaemoglobin. Binding occurs to the terminal $\alpha$-$NH_2$-group of each chain to give
carbamino-derivatives.

Fig. 5.9. Equation summarising the Bohr effect (Hb = haemoglobin).

$O_2$—Hb                in actively metabolising                $O_2$
                        tissue, especially muscle

                        _____→

$CO_2$                  ←_____        Hb—$CO_2$
                        in alveolar capillaries of lung         $H^+$

$H^+$

*Effect of DPG.* In man, DPG reduces the oxygen affinity of haemoglobin by
a factor of 26. This is very significant physiologically since without this
effect it would be difficult for haemoglobin to unload much oxygen in the
tissue capillaries. The DPG effect arises because the molecule is able to bind
to deoxyhaemoglobin but not to oxyhaemoglobin. One molecule of DPG is
bound per haemoglobin tetramer, in the central cavity in close proximity to
all four subunits. The binding of DPG and $O_2$ is mutually exclusive. On
oxygenation, conformational changes ensure that the central cavity becomes
much smaller and the DPG molecule is extruded. However, the need to break
the DPG–protein interactions makes it more difficult for oxygen binding to
be achieved. Uptake of DPG requires breaking of the haemoglobin–$O_2$
binding, so DPG thus facilitates oxygen unloading. The functioning of DPG is
represented simply in the equation in fig. 5.10.

Fig. 5.10. Equation representing the effect of bisphosphoglycerate (DPG) on
haemoglobin (Hb).

$$Hb—DPG + 4 O_2 \rightleftharpoons Hb (O_2)_4 + DPG$$

*Effects of other substances.* The affinity of haemoglobin for carbon monoxide
is very much greater than for oxygen; CO can therefore displace oxygen from
oxyhaemoglobin. The carboxyhaemoglobin produced cannot function as an
oxygen carrier, hence the efficacy of carbon monoxide as a poison. As
described below, carboxyhaemoglobin is cherry-red in colour. This characteristic colour is clearly seen in the complexion of people who are suffering from
carbon monoxide poisoning, which is thus easy to diagnose.

Drugs can also seriously impair the functioning of haemoglobin. Metabolic products of acetanilide, phenacetin and several other drugs are known to induce the oxidation of haemoglobin to the $Fe^{3+}$ form, methaemoglobin, resulting in a serious reduction of the oxygen-carrying capacity of the blood.

*Foetal haemoglobin.* In common with many other animals, man has different types of haemoglobin in the blood at different stages in life. The light absorption and electrophoretic properties of foetal blood haemoglobin differ from those of the adult pigment. A third haemoglobin type is present in the blood of the early embryo. Foetal haemoglobin F has a higher oxygen affinity than adult haemoglobin A. Thus optimal transfer of oxygen is possible from haemoglobin A of the mother to haemoglobin F of the foetus. Haemoglobin F has a higher oxygen affinity because it binds DPG less strongly than does haemoglobin A.

*Genetic defects in haemoglobin structure and functioning.* Many genetic variations in human haemoglobin have been recognised. The best known of these is found in the condition 'sickle-cell anaemia', a single gene mutation which, when homozygous, deforms the red blood cell into a sickle shape. Sickle-cell haemoglobin S differs from normal haemoglobin in one amino acid residue only, in the β-chains. The substitution, however, is of a polar amino acid, glutamic acid, by a non-polar one, valine, and drastically reduces the solubility of deoxyhaemoglobin S, though the solubility of oxyhaemoglobin S is normal. The deoxyhaemoglobin S forms a fibrous precipitate that causes deformation and destruction of the red cell, and consequently a chronic haemolytic anaemia.

More than 100 mutant haemoglobins have now been recognised. Some of the substitutions are harmless 'surface' substitutions, but others which affect the oxygen-binding sites, the tertiary structure or the quaternary subunit interactions and the consequent allosteric effects, may drastically affect the binding of oxygen.

### 5.5.3  Spectroscopic properties of haem and haemoglobin

Like the chlorophylls, haem and haemoproteins exhibit an intense Soret band in the 400 nm region, and further intense absorption peaks between 500 and 600 nm. The absorption maxima for deoxyhaemoglobin (*ca* 425 and 560 nm) and oxyhaemoglobin (*ca* 414, 543, 578 nm) are distinctive and characteristic (fig. 5.11). Haemoglobin is thus purple and oxyhaemoglobin is orange-red. In the CO-adduct carboxyhaemoglobin (cherry-red) the maxima are shifted slightly to the blue compared with those of oxyhaemoglobin. The spectra of oxy- and deoxymyoglobin are very similar to those of the respective haemoglobins. The exact absorption maxima of haemoglobins from different animal species are highly characteristic and depend on the

Fig. 5.11. Light absorption spectra of haemoglobin (———) oxyhaemoglobin
(– – – –) and carboxyhaemoglobin (+++++).

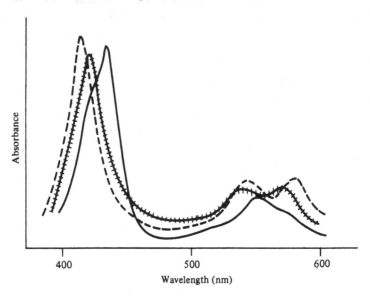

properties of the protein entity. This has proved very useful for taxonomic
correlations. A haem may be removed from its protein and complexed
through its remaining two coordination sites to other nitrogenous molecules
such as pyridine. These products, known as haemochromogens, have very
characteristic absorption spectra and are very useful for the identification of
haem prosthetic groups.

### 5.5.4    Distribution of haemoglobin

Haemoglobin is the blood pigment of virtually all vertebrate and also
many invertebrate animals. It is, however, thought not to occur in the
Porifera, Coelenterata, Rotifera, Sipuncula, Polyzoa, Brachiopoda, Onycho-
phora, Tardigrada, Chaetognatha, and Protochordata, and is rather rare in the
Arthropoda. In vertebrates the haemoglobin is located in the red blood cells.
In the invertebrates it may be present in corpuscles but more commonly exists
in solution in the blood or haemolymph. Haemoglobin may be found in
almost all body tissues.

There is very wide variation in the size of the haemoglobin complex. Thus,
as already mentioned, mammalian haemoglobin is a tetrameric form, relative molecular mass 64 500. Elsewhere in the animal kingdom monomeric and
dimeric forms (relative molecular mass 17 000 and 34 000, respectively)
have been found, whereas in many invertebrates the composite molecules are
much larger and may have relative molecular masses up to 3 000 000.

There is also considerable variation in the amino acid compositions and sequences in the haemoglobins of different kinds of animal, leading to differences in light absorption maxima, solubility, isoelectric point, oxygen affinity and resistance of the isolated pigment towards acid, alkali and heat. Even amongst the mammals there are considerable differences in the stability and crystalline form of the isolated haemoglobins. These structural differences are restricted to the protein part of the molecule; all forms of haemoglobin contain the same protohaem prosthetic group.

Besides differences in haemoglobin between families, genera, species or even sub-species, there may also be different haemoglobins in one individual at successive periods in life or even at the same time. As outlined above, man, for example, has different types of haemoglobin in his blood at different stages in life. The same situation obtains with other species; the haemoglobins of chick and hen, or of tadpole and frog are different. In each case the pigment of the immature form has the higher oxygen affinity.

Many animals have more than one blood haemoglobin component to serve, in some instances, different functions. The two haemoglobin components of the Pacific salmon, *Onchorhynchus keta*, differ widely in oxygen affinity; this may be related to the requirement for this fish to spend part of its life in salt water and part in fresh.

Although haemoglobin is usually thought of as purely an animal product, one form, leghaemoglobin, has been identified in leguminous plants. Its presence is confined to root-nodule cells containing symbiotic nitrogen-fixing bacteria (*Rhizobium* spp.). Haemoproteins with properties of haemoglobins have also been detected in some fungi and protozoa.

### 5.5.5 Contribution of haemoglobin to animal colours

Haemoglobin in red blood corpuscles in the capillary blood vessels of the dermis gives the pink skin tint to the 'white' races of man. In most other vertebrates this coloration is concealed by hair, feathers, scales or other skin pigments. Colour due to haemoglobin can, however, be seen in the pink tint of the tongue, inside of the ears and tip of the nose of many mammals. Certain specific tissues or body areas may be coloured red by haemoglobin, to serve as warning or sexual signals (see chapter 8) *e.g.* the bare neck of some vultures, the wattles of the turkey and other birds, and the buttocks of the baboon.

Among the invertebrates, coloration by haemoglobin is seen in many polychaete and oligochaete annelids (*e.g.* bloodworms) and, under certain circumstances, cladoceran and phyllopod Crustacea. Generally, however, few animals, vertebrate or invertebrate, are coloured by haemoglobin.

### 5.5.6 Chlorocruorin

Closely related to haemoglobin is another haemoprotein, chlorocruorin or chlorohaemoglobin, which serves as the oxygen-carrying pigment

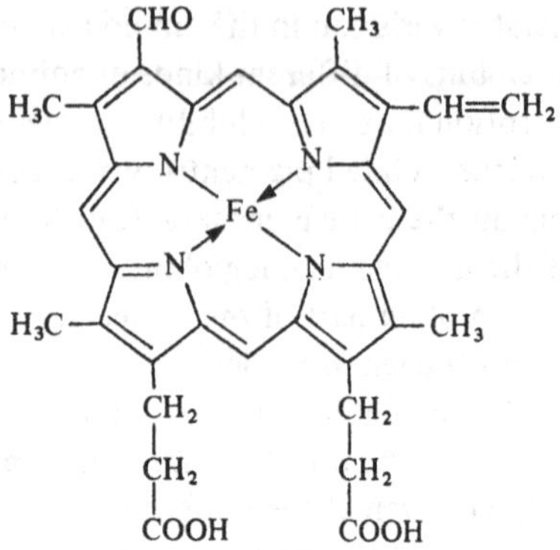

(5.25) Chlorocruorohaem

in the green blood of a very restricted group of polychaete worms. Chlorocruorin has as its prosthetic group chlorocruorohaem (5.25) which differs
from protohaem only by having the C-3 vinyl sidechain replaced by a formyl
or aldehyde group. Otherwise the complexing of this haem with the globin
protein, and the oxygen affinity of the complex, are very similar to those
described above for haemoglobin. The absorption spectrum of oxychloro-
cruorin exhibits a Soret band at 430 nm and $\alpha$- and $\beta$-absorption bands at 604
and 558 nm, *i.e.* at longer wavelengths than those of oxyhaemoglobin. The
colours of the oxy- and deoxy- forms of chlorocruorin differ very little. This
pigment is, however, highly dichroic, and its colour changes from red to green
on dilution. Only in very few species does chlorocruorin impart a greenish
colour to the body.

### 5.5.7    Haemocyanin, haemerythrin and haemovanadin

These compounds occur as blood or respiratory pigments in small
numbers of invertebrates. They are metalloproteins and are included here to
stress that despite their names they are not haem or porphyrin derivatives.
For example, the haemocyanins of gastropod molluscs such as the snail
*Helix pomatia* are giant proteins ($9 \times 10^6$ daltons). The functional unit, which
binds one molecule of $O_2$, is a pair of copper atoms, immediately surrounded
by a compact folded polypeptide, relative molecular mass 50 000. Some
seven to nine of these functional domains constitute a $4-5 \times 10^5$ dalton unit,
and the haemocyanin is a composite of 20 of these units.

### 5.5.8    Electron transport cytochromes

The cytochromes are a group of small haemoproteins which, unlike
haemoglobin and myoglobin, readily undergo reversible oxidation and reduction of the haem iron atom. This property renders them of extreme biological
importance in electron transport. Cytochromes are present in all animals,

plants and aerobic microorganisms, and a large number of individual cyto-chromes have been identified in the many animal, plant and microbial species examined. Structurally they fall into four main groups designated cyto-chromes *a*, *b*, *c* and *d*. Individual cytochromes are assigned to one of these groups according to the nature of the haem prosthetic group present, and the way that this is bound to the protein.

The prosthetic group of cytochrome *b* is protohaem, (5.23) as in haemo-globin. The haem group of cytochrome *a* is designated haem *a* (5.26) and differs from protohaem in that the C-18 methyl group is replaced by CHO and the vinyl group at C-3 has been modified by the addition of a $C_{15}$ (farnesyl) isoprenoid chain. The title cytochrome *d* is applied to those cyto-chromes with a dihydroporphyrin (chlorin)–iron prosthetic group (5.27). The sidechain substituents may vary. The cytochrome *c* group includes all cyto-chromes with covalent thioether links between the haem sidechains and the protein, *e.g.* (5.28).

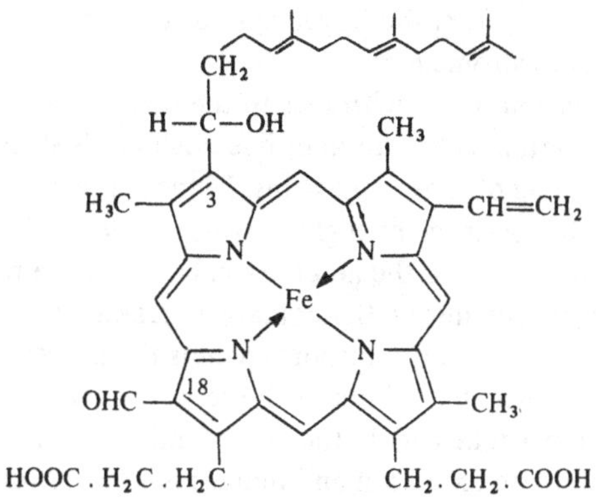

(5.26) Haem *a*

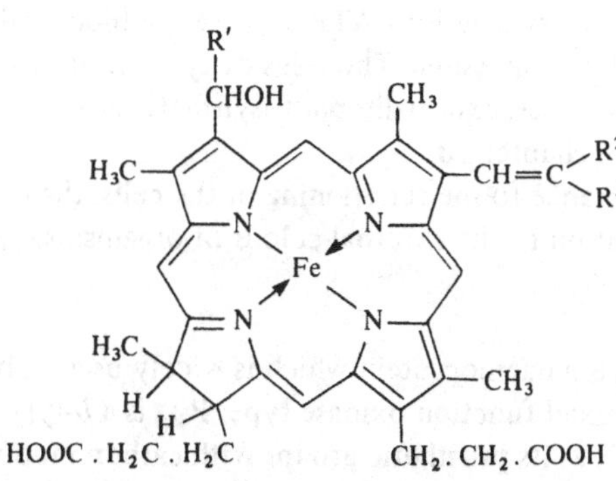

(5.27) Cytochrome *d* prosthetic group
($R^1$, $R^2$, $R^3$ not certain)

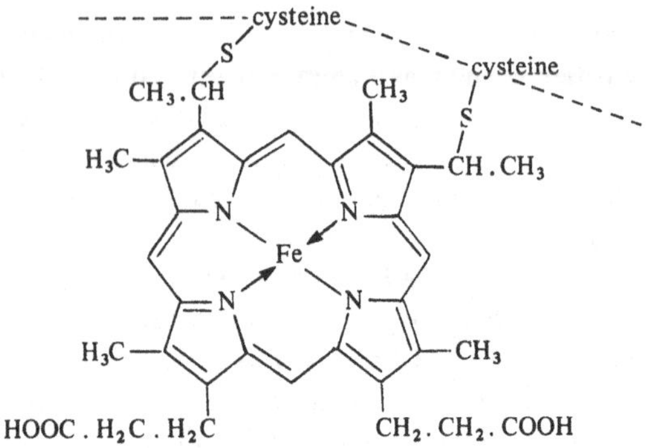

(5.28) Cytochrome *c*—protein linkage

Individual cytochromes in these groups are identified by subscript numbers, *e.g.* cytochrome $b_6$, or from the wavelength of the $\alpha$-band in the absorption spectrum, *e.g.* cytochrome *b*-550.

The quantitative contribution of cytochromes to the total tetrapyrrole content of organisms that contain chlorophyll or haemoglobin is slight, but they are vital to the functioning of these organisms. In the mitochondria of eukaryotic cells they form the basis of the highly organised electron transport chain (fig. 5.12) which is used for the aerobic oxidation of the reduced coenzymes (*e.g.* NADH) produced during the oxidative breakdown of nutrients. Associated with this electron transport chain is the process of oxidative phosphorylation by which the bulk of the respiring cell's ATP is generated. Several cytochromes take part in the chain, and the haem iron atom of each undergoes alternate oxidation and reduction as electrons are passed along the series ultimately to molecular oxygen. Cytochromes also function in a similar way in photosynthetic electron transport. Without these electron transport systems the synthesis of ATP to drive the biochemical reactions of the cell would be impossible. The roles of cytochromes and other pigments in these processes, expecially photosynthetic electron transport, are considered in chapter **10**.

Although of vital importance to the functioning of the cells, the cytochromes make no contribution to the external colour of organisms.

### 5.5.9   Cytochrome $P_{450}$

Cytochrome $P_{450}$ is a haemoprotein which is widely used in biological oxidations of the mixed-function oxidase type. $P_{450}$ is a *b*-type cytochrome with protohaem IX as its prosthetic group. With carbon monoxide, the $Fe^{2+}$ form gives a stable product, $\lambda_{max}$ 450 nm, from which '$P_{450}$' derives its name.

Mixed-function oxidase reactions catalyse the insertion of one atom of an oxygen molecule into an organic molecule RH to give a hydroxy product ROH. The other oxygen atom is reduced to water. A second substrate (coenzyme), usually NAD(P)H, is used as an electron donor. The whole system takes the form of a small electron transport chain, including a flavo-protein and $P_{450}$, which accepts electrons from the reduced flavin in two one-electron steps and passes these electrons on to molecular oxygen. The substrate RH seems to be bound to $P_{450}$ during the reaction, a possible mechanism for which is outlined in fig. 5.13. Such hydroxylations character-istically occur with retention of configuration. Examples of mixed-function

Fig. 5.12. The role of cytochromes in the mitochondrial electron transport system.

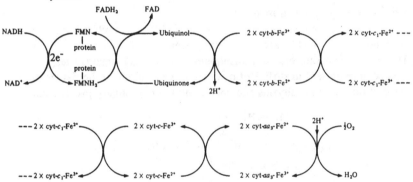

Fig. 5.13. A mechanism for the functioning of cytochrome $P_{450}$ in mixed-function oxidase oxidation of a substrate RH to ROH.

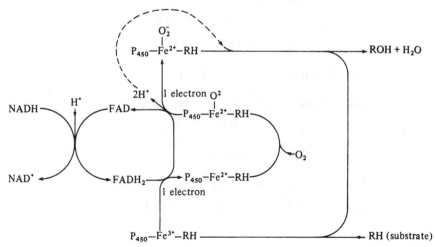

oxidase reactions are found in the hydroxylation of steroids in liver micro-somes, and in the hydroxylation of drugs (detoxication). $P_{450}$ is inducible by the presence of many foreign organic compounds.

### 5.5.10   Other haemoprotein enzymes

Besides the cytochromes, several other enzymes rely on a haem prosthetic group for their catalytic action. Such haemoproteins include peroxidases and catalases from various plant and animal sources. The porphyrin is normally protohaem. For example, horseradish peroxidase, relative molecular mass 44 000, contains one haem which catalyses the oxidation of phenolic compounds by $H_2O_2$. Catalase (from beef liver) has relative molecular mass 248 000 and four haem groups. It catalyses the breakdown of $H_2O_2$ to water with an extremely high turnover rate.

## 5.6   Free porphyrins in animals

Free porphyrins – protoporphyrin (**5.29**), uroporphyrin (**5.30**) and coproporphyrin (**5.31**) – are commonly present in animal urine and faeces. They are also encountered not infrequently in various animal tissues, but are very seldom present in sufficient quantity to give colour to the tissues. In the invertebrates free porphyrins and haems and their secondary protein conju-

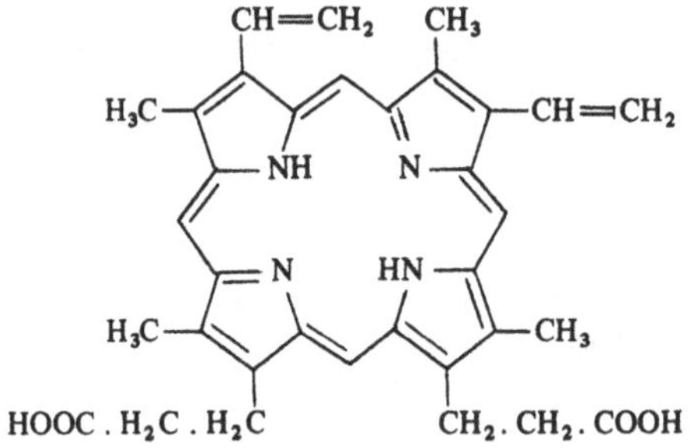

(5.29) Protoporphyrin

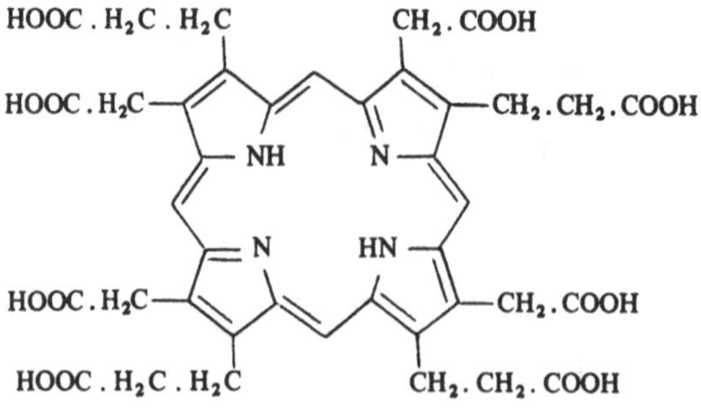

(5.30)  Uroporphyrin III

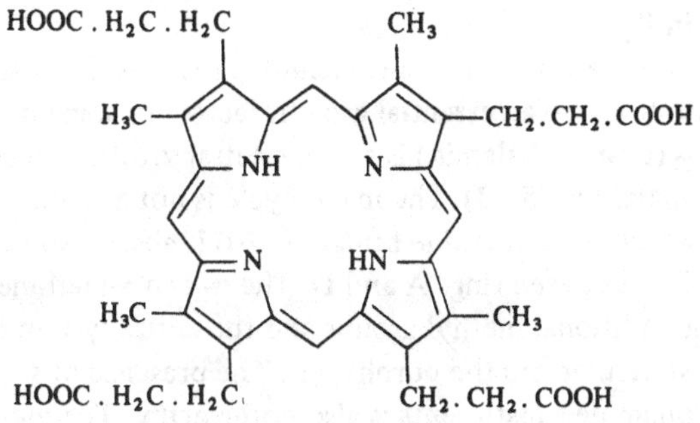

(5.31) Coproporphyrin III

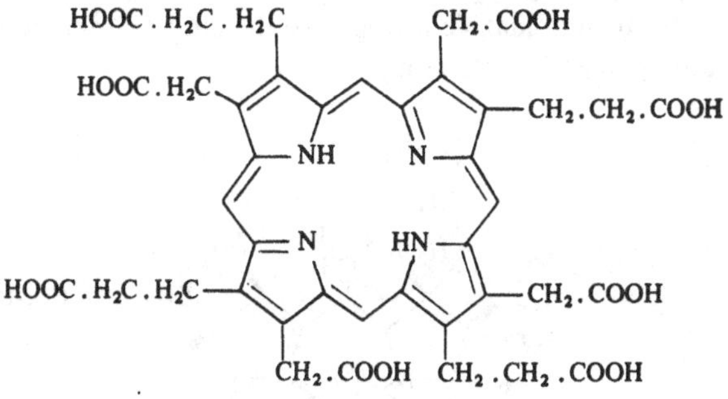

(5.32) Uroporphyrin I

gates are distributed sporadically, but are most common in worms and molluscs. A familiar example of a porphyrin giving integumental colour is in the earthworm; the purplish colour of the anterior dorsal integument is due to protoporphyrin. The shells of some molluscs contain quite large amounts of urophorphyrin, often the uroporphyrin I isomer (5.32). Free porphyrins are very rare or unknown in protozoa, coelenterates, arthropods and protochordates.

In the vertebrates, especially mammals, free porphyrins are rare but are occasionally present in some internal tissues. There are, however, many examples of protoporphyrins occurring in and contributing, as with molluscs, to the brown colours and intricate patterns of the egg shells of many birds. Porphyrins readily bind to minerals, and the shell colours produced are usually remarkably stable. Also in bird feathers it is not uncommon for coproporphyrin III to be present, sometimes in large amounts. The striking bright-red colour of wing feathers of the touraco bird is due to the presence of a copper chelate of uroporphyrin III. This pigment is highly water-soluble, though fortunately only under alkaline conditions, so the bird is not at risk of losing its colour in its natural habitat, tropical rain forests.

In general the contribution of free porphyrins to animal colours is slight.

### 5.7    Vitamin B$_{12}$

B$_{12}$ is important to mammals, including man, as the anti-pernicious anaemia vitamin. It is also an essential growth factor for some microorganisms. Structurally, B$_{12}$ (cyanocobalamin) is a cyclic tetrapyrrole with cobalt, Co$^{+}$, as the chelated metal ion (**5.33**). The macrocycle is not a standard porphyrin but a **corrin** in which the δ-methine bridge (C-20) is absent, so that there is a direct covalent link between rings A and D. The α- and γ-methine bridges (C-5, C-15) bear additional methyl groups and the entire system is in a more highly reduced state than are the porphyrins. The presence of several acetamide and propionamide substituents is also noteworthy. The most unusual feature is the inclusion of dimethylbenzimidazole ribosyl phosphate, linked both through the phosphate group to a ring D sidechain and also *via* a coordination link from an imidazole nitrogen to the cobalt. Adenine may replace the dimethylbenzimidazole group.

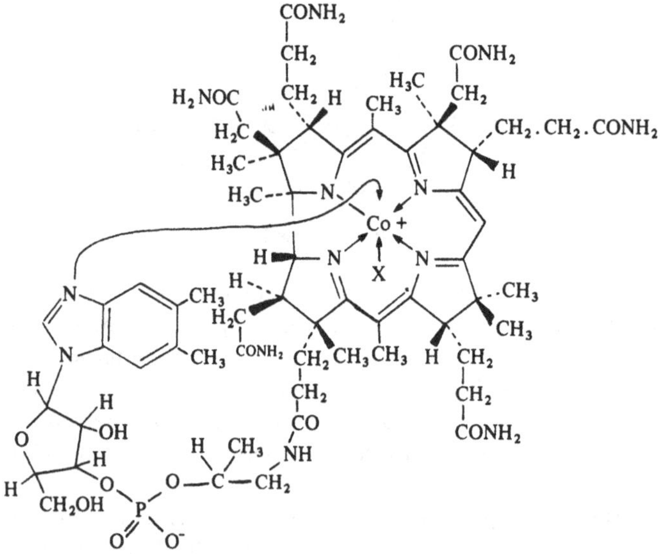

(5.33) Vitamin B$_{12}$

Cyanocobalamin: X = CN

B$_{12}$ coenzyme:
X =

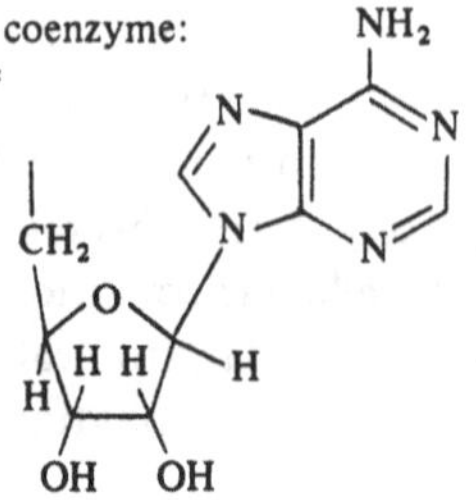

Although $B_{12}$ is usually isolated with CN as the sixth ligand of the $Co^+$, this is not the active form *in vivo*. The metabolically active form, $B_{12}$-coenzyme, has the CN replaced by adenosine, linked directly through the C-5 carbon of the ribose moeity. The coenzyme functions in rearrangement reactions such as the rearrangement of methylmalonyl-CoA to succinyl-CoA.

Vitamin $B_{12}$ is solely a microbial product and cannot be synthesised by animals or plants. Animals obtain their supplies mainly from gut microorganisms. $B_{12}$ is produced commercially from microbial cultures.

$B_{12}$ is present in such small amounts *in vivo* that its dark-red colour makes no contribution to external appearance.

## 5.8     Linear tetrapyrroles – bilins
### 5.8.1     Introduction
Bilins are distributed widely though irregularly in both the plant and animal kingdoms. They occur in animals as catabolic products of haem. In plants and algae they are also derived from porphyrins; in some algae they are produced in large amounts as protein conjugates which play an important role in photosynthesis. Structurally, bilins are linear or open tetrapyrroles with structures based on the skeleton **5.8**. The carbon bridges (a,b,c) linking the pyrrole rings can be either saturated ($-CH_2-$) or unsaturated ($-CH=$). If all three carbon bridges are saturated, the structure is a **bilane**, with one unsaturated $-CH=$ bridge, a **bilene**. **Bilidienes** and **bilitrienes** have two and three unsaturated $-CH=$ bridges, respectively. Thus, of the structures quoted below, urobilinogen (**5.37**) is an example of a bilane, urobilin (**5.39**) is a b-bilene, bilirubin (**5.35**) an a,c-bilidiene, and biliverdin (**5.34**) an a,b,c-bilitriene. All these compounds exist primarily in the diketo form, as illustrated.

### 5.8.2     Animal bilins ('bile pigments')
*Structures and formation.* In mammals, degradation of the haem of haemoglobin and other haemoproteins gives rise to open tetrapyrroles, long known as bile pigments. This catabolic process occurs mainly in the reticuloendothelial cells of spleen, liver, bone-marrow and to a lesser extent kidney. Haemoglobin of senescent erythrocytes provides the bulk of the haem that is catabolised in this way, though haem prosthetic groups from other proteins *e.g.* haem enzymes, are also utilised.

In the formation of bilins, enzyme action cleaves the haem porphyrin system at the α-methine bridge (C-5) to yield the first open tetrapyrrole biliverdin IXα (**5.34**), a bilitriene. This is the green bile pigment which colours the faeces of most birds and amphibians. In most mammals, including man, the central $-CH=$ group is reduced to $-CH_2-$ to give the orange bilidiene bilirubin (**5.35**). This is converted into the diglucuronide (**5.36**) which passes through the intestine. Intestinal bacteria then bring about further reductions to give colourless products such as the bilane urobilinogen (**5.37**). As well as

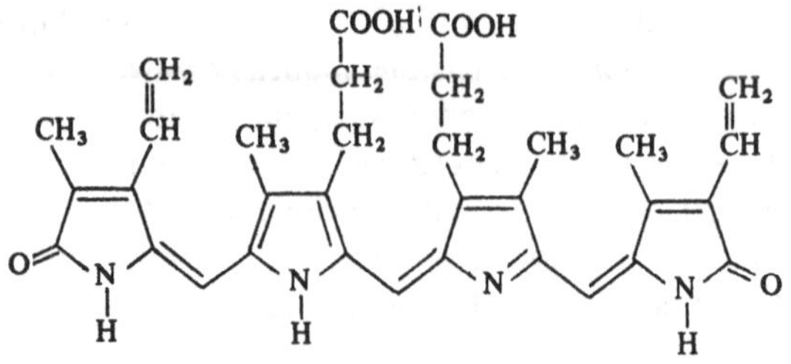

(5.34)  Biliverdin IXα

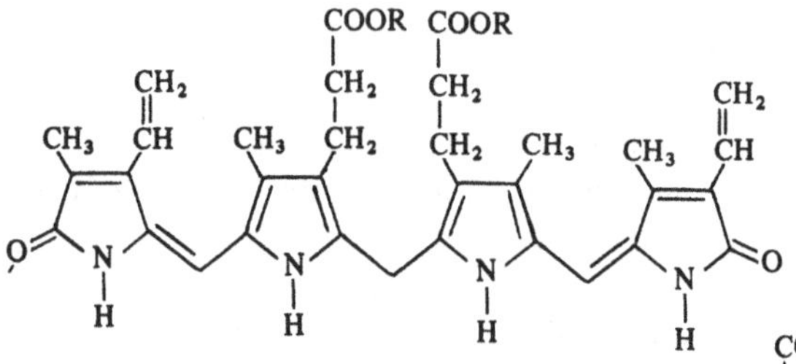

(5.35)  Bilirubin (R = H)
(5.36)  Bilirubin diglucuronide

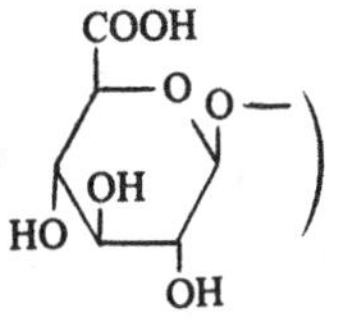

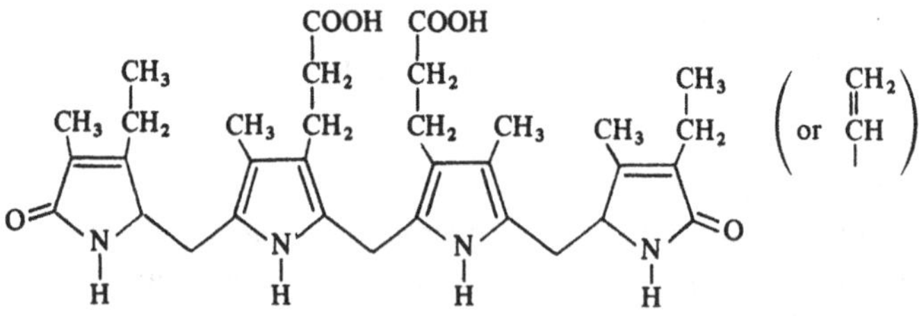

(5.37)  Urobilinogen

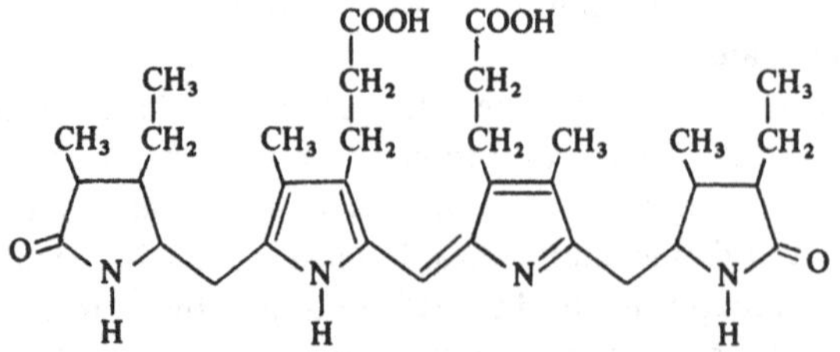

(5.38)  Stercobilin

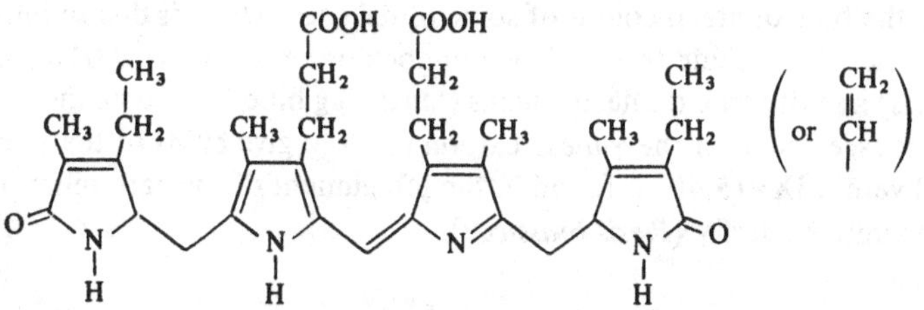

(5.39) Urobilin

reduction of the bridge —CH= groups, reduction of the vinyl sidechains to ethyl also occurs, to give '*meso*' structures. These products are oxidised in air to the yellow-brown bilenes stercobilin (5.38) and urobilin (5.39), which are largely responsible for the colour of faeces and urine. Further details of these processes are given when tetrapyrrole biosynthesis is discussed below (see §*5.9.14*).

*Properties.* Biliverdin and bilirubin are acids and are soluble in aqueous alkali. Salts with most other metal ions are insoluble; the calcium salt of bilirubin is the main component of gallstones. Opening of the porphyrin macrocycle renders both the pyrrole rings and the methine carbons more accessible to chemical attack and hence more reactive.

The bilins exhibit characteristic intense light absorption in the visible region, but no Soret band is given. Saturation of the methine-bridge carbon atoms shortens the chromophore and absorption maxima at shorter wavelengths are observed. Thus biliverdin has $\lambda_{max}$ at about 680 nm (in acid solution), bilirubin at around 450 nm, and urobilin about 490 nm. The bilanes, such as urobilinogen, have no absorption in the visible region.

*Occurrence and distribution.* Although the formation of bilins from haem has been studied most intensively in mammals, this process occurs widely throughout the animal kingdom. Bilins have been identified in most major groups of animals. They serve as integumental pigments in many invertebrates, notably worms (*e.g. Nereis diversicolor*, the usually bright-green worm of sea-shore rock pools) and insects. The familiar green colour of the blood and integument of many grasshoppers, caterpillars *etc.* is due to bilins of the biliverdin type and not to chlorophyll as was first thought. Glaucobilin (mesobiliverdin) in which the vinyl sidechains have been reduced to ethyl groups, seems to occur quite commonly in these invertebrates.

In vertebrates, bilins are familiar as the pigments responsible for the colour of the bile and faeces, but are rarely responsible for any external colours. In man, however, jaundice is characterised by the skin's taking on a yellow hue. The pigment responsible for this is bilirubin which is present in the blood is substantial quantities. The only other familiar example in verte-

brates is the blue or green colour of some birds' eggs, which is due to biliverdin. Examples include those of the dunnock or hedge sparrow (*Prunella
modularis*) and of some domestic ducks ('duck-egg blue'). In some insects,
haem cleavage occurs at the γ-*meso* carbon atom to give bilins of the γ-series.
Thus biliverdin IXγ (**5.40**) is found in the integument of the caterpillar of the
cabbage white butterfly (*Pieris brassicae*).

(5.40) Biliverdin IXγ

*Functions*. Despite intensive study, especially with vertebrates, no specific
physiological role has been established for bilins in any bodily function.

### 5.8.3     Phycobilins and phycobiliproteins

*Occurrence, properties and structures*. In the plant kingdom, bilins
occur as protein conjugates (phycobiliproteins) in three divisions of algae, the
Rhodophyta (red algae), Cyanophyta (blue-green algae) and Cryptophyta
(cryptomonads). These algal biliproteins are acidic, water-soluble, globular
proteins, originally thought to be of high relative molecular mass, but now
known to be made up of much smaller subunits (*ca* 20 000 daltons). The
phycobilin prosthetic groups are strongly, *i.e.* covalently, bound to the protein
and can only be removed by very drastic chemical methods. This has made the
structural characterisation of the phycobilins very difficult. The phycobiliproteins are further aggregated into particles called phycobilisomes (see
§*10.8.2* and fig. 10.17) which play an important accessory light-harvesting
role in photosynthesis.

Quantitatively, the two main groups of algal biliproteins are the red phycoerythrins and the blue phycocyanins. Most of the relevant algal species
contain both a phycoerythrin and one or more phycocyanins, with one
individual biliprotein usually predominating. It is usually phycoerythrin
which predominates in red algae and phycocyanin in blue-green, though
there are red phycoerythrin-predominant members of the Cyanophyta. The
relative amounts of phycocyanin and phycoerythrin present depend on the
spectral quality of the light available to the cells, *e.g.* in green light conditions,
synthesis of the green-absorbing red pigment phycoerythrin is favoured,

whereas in red light phycocyanin predominates. In addition to these two, a small amount of allophycocyanin occurs at the core of the phycobilisome structure.

Several types of phycoerythrin and phycocyanin can be distinguished spectroscopically. Typical absorption spectra of the pigments from one species are illustrated in fig. 5.14. Thus although phycoerythrin from all algal sources has a characteristic major absorption maximum at 560–570 nm (the absorption maxima of the cryptomonad pigments may be slightly outside the normal ranges), there are differences between the absorption spectra of the phycoerythrins from Cyanophyta, Rhodophyta and Cryptophyta, and also between samples from different species. The same is true of phycocyanin ($\lambda_{max}$ 610–620 nm). Allophycocyanin, which accepts energy passed on from phycoerythrin and phycocyanin in the phycobilisome, absorbs at even longer wavelength (650 nm).

In general, the prosthetic group chromophores of phycocyanin and phycoerythrin are phycocyanobilin and phycoerythrobilin, respectively. The situation is not always as simple as this, however; some of the biliproteins contain two different chromophores, *e.g.* Rhodophyta phycocyanin has both phycocyanobilin and phycoerythrobilin.

The characterisation of these phycobilin chromophores has proved extremely difficult. It is now clear that, like the animal bile pigments, they

Fig. 5.14. Light absorption spectra of a phycocyanin (———), a phycoerythrin (– – – –) and an allophycocyanin (–···–) from the blue-green alga *Chlorogloea*. Spectra in aqueous solution, pH 7.0.

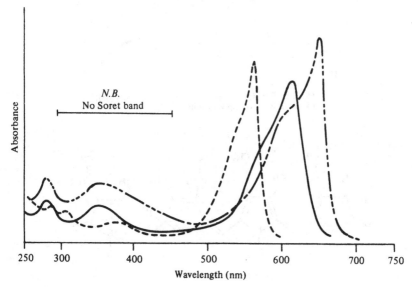

are all IXα isomers derived by cleavage of the α-methine bridge of a porphyrin. In early work, treatment of phycobiliproteins with boiling methanol released the modified chromophores 'phycoerythrobilin-690' (= phycobiliverdin, **5.41**) and 'phycoerythrobilin-590' (=phycoviolin, **5.42a**). The true chromophores, phycocyanobilin and phycoerythrobilin, are formed from these by thioether linkage between the ring A ethylidene group and a cysteine residue of the protein (fig. 5.15). A further example, phycourobilin from Rhodophyta phycoerythrin has been suggested to be formed from the chromophore (**5.43**) by thioether links with substituents on rings A and D. (*N.B.* For convenience these chromophores are all represented as linear structures, without regard to their true natural conformation.)

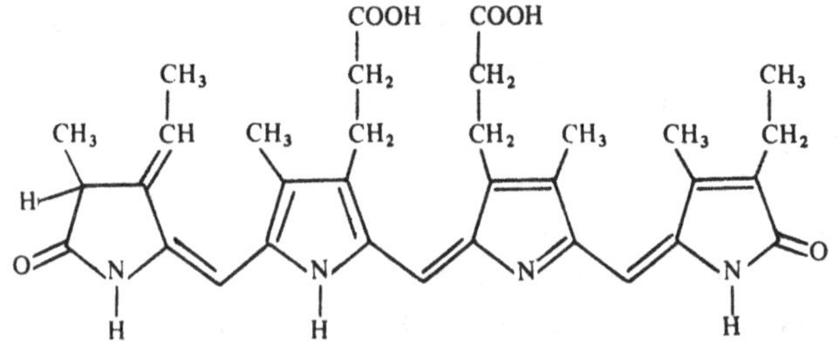

(5.41) Phycoerythrobilin-690 = phycobiliverdin

(5.42 a)  R = H; phycoerythrobilin-590 = phycoviolin
(5.42 b)  R = CH₃; aplysioviolin

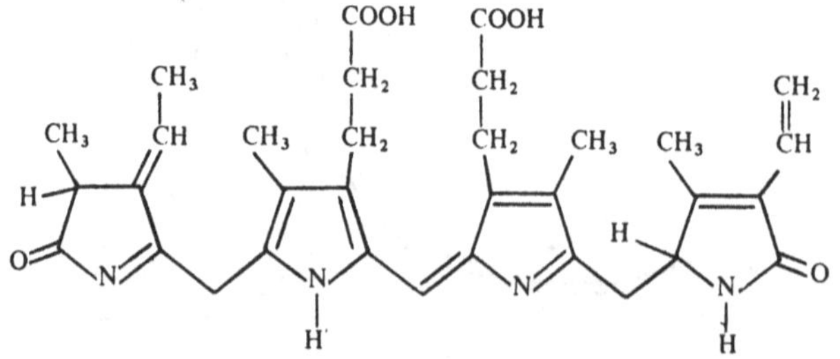

(5.43) Phycourobilin

Fig. 5.15. Proposed covalent linkage of phycoerythrobilin, phycocyanobilin and phycourobilin to protein.

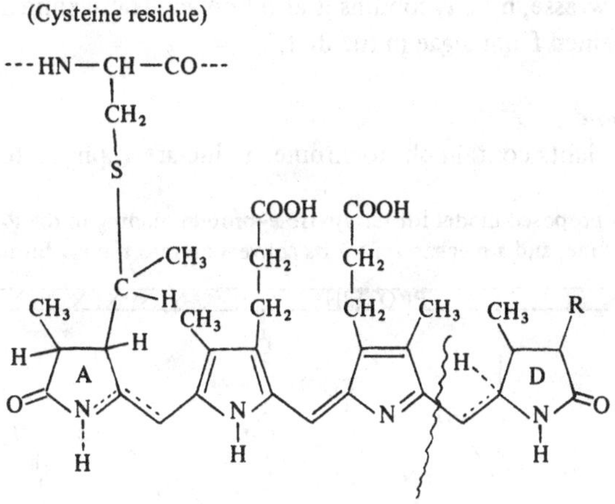

(Cysteine residue)

Proposed additional binding
of phycourobilin

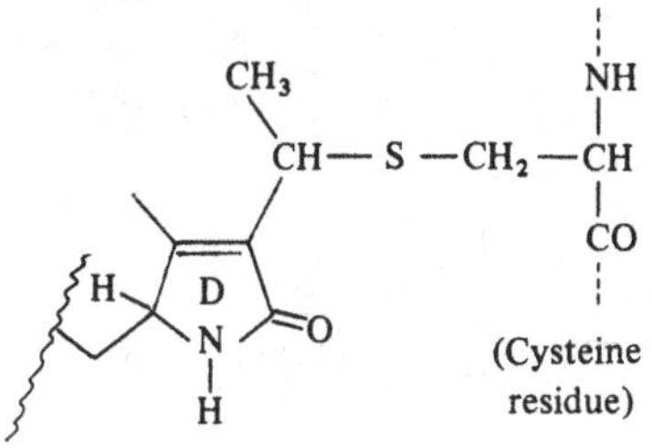

(Cysteine
residue)

*Distribution and function in plants.* In the plant kingdom the presence of phycobilins is restricted to the three algal classes. The concentration of biliprotein is usually high - values of up to 24% of algal dry mass have been recorded for phycocyanin. These high concentrations of biliprotein are responsible for the characteristic blue and red colours of most members of the Cyanophyta and Rhodophyta. In these algae, the biliprotein pigments are present in aggregates (bilisomes) in the photosynthetic structures where they play an important role in light harvesting. This functioning in photosynthesis will be discussed in chapter **10**.

*Occurrence in animals.* A number of instances have been reported of the presence of biliproteins like phycocyanin and phycoerythrin in invertebrate animals, particularly molluscs, some of which use these pigments as body and shell colours. A well-characterised example is the presence of substantial amounts of aplysioviolin (**5.42b**), the methyl ester of phycoerythrobilin, in the purple defensive secretion of the sea hare *Aplysia*. Coloration of some fishes, *e.g.* the blue wrasse, by phycobilins is also known. These animal phycobilins are obtained from algae in the diet.

### 5.8.4   *Phytochrome*

All higher plants contain phytochrome, a blue-green photochromic

Fig. 5.16. A proposed model for tetrapyrrole–protein binding in the $P_r$ form of phytochrome, and a mechanism for its conversion into the $P_{fr}$ form.

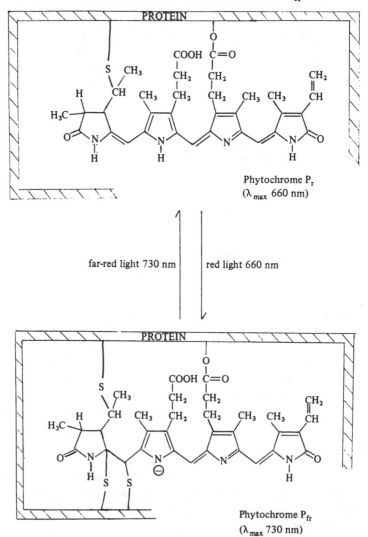

far-red light 730 nm     red light 660 nm

pigment which controls a very wide range of developmental and metabolic processes. Phytochrome is a 120 000 dalton protein with a linear tetrapyrrole or bilin as its prosthetic group chromophore. It exists and functions in two forms, $P_r$ which absorbs maximally in the red region of the spectrum (660 nm) and $P_{fr}$ with maximum absorption in the far-red (730 nm). The two forms are freely interconvertible; on absorption of red light, $P_r$ is converted into $P_{fr}$ which in turn gives $P_r$ on absorbing far-red light.

Details of the protein structure and ultrastructure of phytochrome have not been elucidated fully; different plant species appear to have different protein structures. The nature of the tetrapyrrole chromophore that can be isolated (**5.44**) and its very clear relationship with the algal phycobilins are now established. The exact nature of the binding of the chromophore to the protein is not yet known, but it is clearly different from the binding present in the phycobiliproteins. Fig. 5.16 illustrates one proposal for the tetrapyrrole-protein binding of the $P_r$ form of phytochrome and a mechanism for its conversion into the $P_{fr}$ form.

Further details of the functioning of phytochrome in plants are given in chapter **11**.

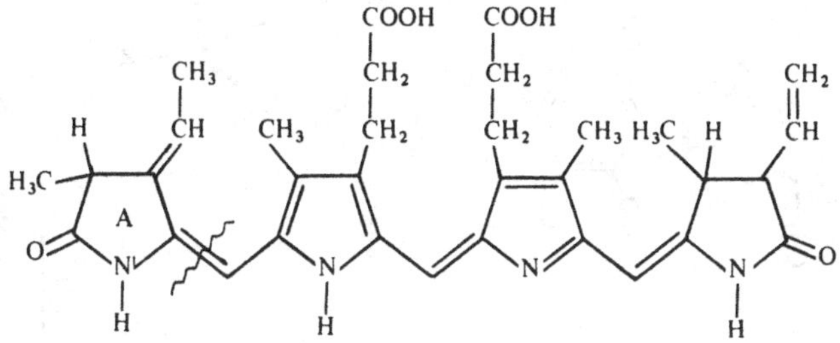

(**5.44**) Phytochrome chromophore ($P_r$)

## 5.9 Biosynthesis and metabolism of tetrapyrroles

### 5.9.1 Introduction

As far as is known, the same basic pathway is used by all forms of living organisms to make porphyrins, whether haem or chlorophyll is to be the final product. In both animals and plants the linear bilins are formed from porphyrin precursors.

It is convenient to consider porphyrin biosynthesis in stages:

(a) formation of δ-aminolaevulinic acid;
(b) formation of the monopyrrole, porphobilinogen;
(c) formation of uroporphyrinogen, the first tetrapyrrole macrocycle;
(d) sidechain modification to give protoporphyrinogen;

Fig. 5.17. ALA synthetase route for formation of δ-aminolaevulinic acid from glycine and succinyl-CoA.

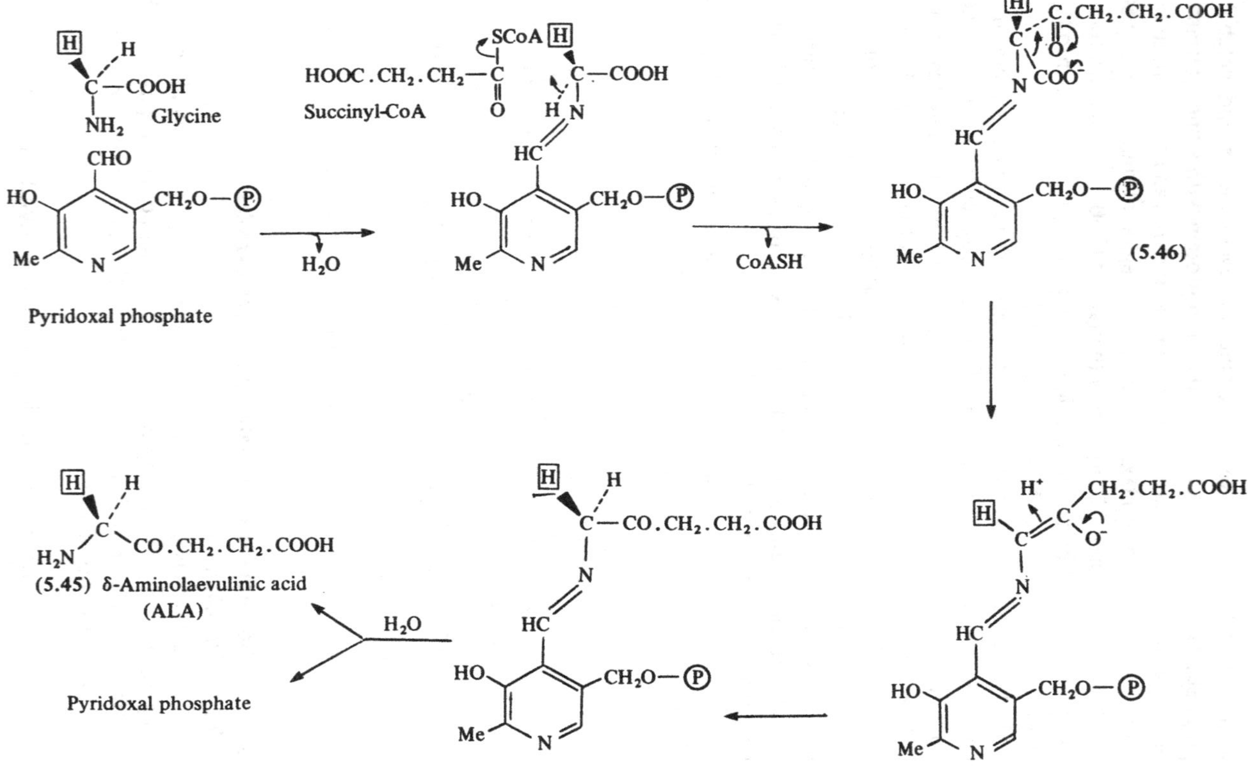

(e) dehydrogenation of the macrocycle to give protoporphyrin IX;
(f) metal chelation to give haem or the chlorophyll precursor magnesium protoporphyrin IX;
(g) further modifications leading to chlorophylls.

Stages (a) to (f) are common to haem and chlorophyll biosynthesis.

### 5.9.2    *Formation of δ-aminolaevulinic acid*

δ-Aminolaevulinic acid (ALA, **5.45**) may be produced by two alternative routes. The main route in animals and probably also in bacteria uses the enzyme ALA synthetase, which catalyses the formation of ALA from succinyl-CoA and glycine. The reaction requires pyridoxal phosphate and proceeds *via* a Schiff base intermediate (fig. 5.17). The intermediates such as α-amino-β-oxoadipic acid all remain bound as Schiff bases (**5.46**), and decarboxylation occurs before release of the ALA from the pyridoxal phosphate. The stereochemistry of the process has been determined. The 2-*pro-S* hydrogen atom of glycine is retained throughout, indicating that the reaction sequence occurs with one retention and one inversion of configuration.

In algae and higher green plants most, if not all, of the ALA is obtained by transamination (fig. 5.18). ALA transaminase catalyses the transfer of an amino group from an amino acid such as L-alanine to γ,δ-dioxovaleric acid (α-oxoglutaraldehyde, **5.49**), which is itself formed from L-glutamate (**5.47**) by transamination to α-oxoglutarate (**5.48**) and reduction of this.

Fig. 5.18. Formation of δ-aminolaevulinic acid by transamination.

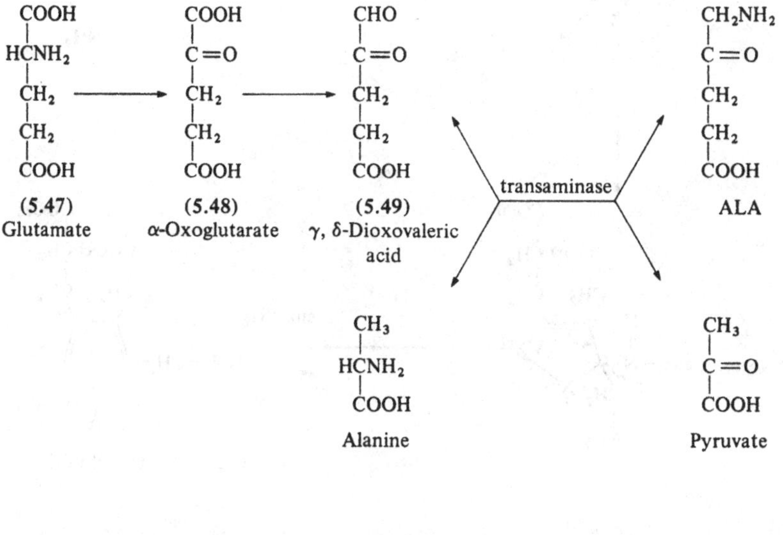

$$
\begin{array}{cccc}
\text{COOH} & \text{COOH} & \text{CHO} & \text{CH}_2\text{NH}_2 \\
| & | & | & | \\
\text{HCNH}_2 & \text{C=O} & \text{C=O} & \text{C=O} \\
| & | & | & | \\
\text{CH}_2 \longrightarrow & \text{CH}_2 \longrightarrow & \text{CH}_2 & \text{CH}_2 \\
| & | & | & | \\
\text{CH}_2 & \text{CH}_2 & \text{CH}_2 & \text{CH}_2 \\
| & | & | & | \\
\text{COOH} & \text{COOH} & \text{COOH} & \text{COOH} \\
(\mathbf{5.47}) & (\mathbf{5.48}) & (\mathbf{5.49}) & \text{ALA} \\
\text{Glutamate} & \text{α-Oxoglutarate} & \text{γ, δ-Dioxovaleric} & \\
& & \text{acid} &
\end{array}
$$

transaminase

$$
\begin{array}{cc}
\text{CH}_3 & \text{CH}_3 \\
| & | \\
\text{HCNH}_2 & \text{C=O} \\
| & | \\
\text{COOH} & \text{COOH} \\
\text{Alanine} & \text{Pyruvate}
\end{array}
$$

Fig. 5.19. The formation of porphobilinogen: a possible mechanism involving Schiff base formation between ALA and a lysine residue of the enzyme ALA dehydratase.

(a) Overall process

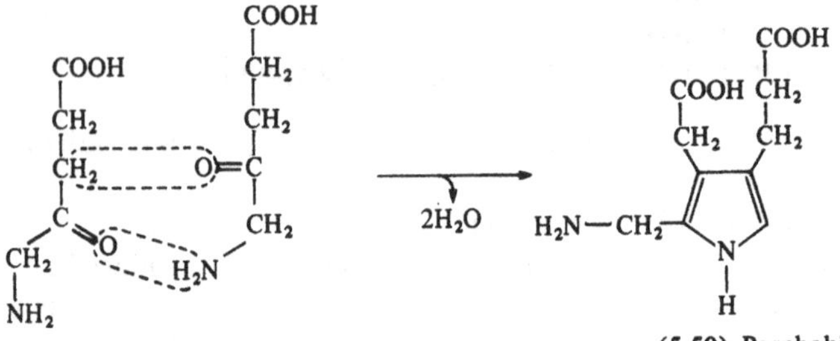

(5.50) Porphobilinogen (PBG)

(b) Possible mechanism

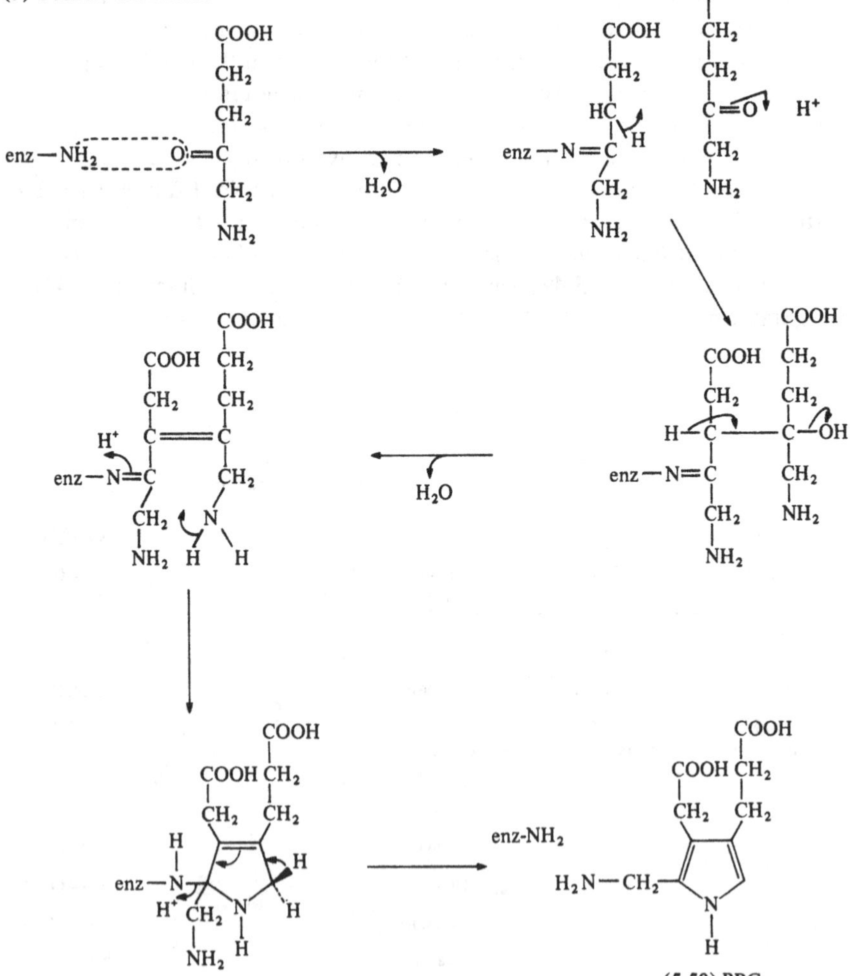

(5.50) PBG

*5.9.3    Formation of porphobilinogen*

The unsymmetrical Knorr condensation of two molecules of ALA occurs by a process (fig. 5.19) involving an aldol condensation, water elimination and Schiff base formation. A single enzyme, ALA dehydratase (or dehydrase) ($\equiv$ PBG synthetase) catalyses the reaction and the product is porphobilinogen (PBG, **5.50**). In the final step the loss of hydrogen from C-2 of the pyrrole is stereospecific, the hydrogen lost being that which originates as the *pro-R* hydrogen at C-5 of ALA.

*5.9.4    Formation of the first tetrapyrrole, uroporphyrinogen III*

This is the most interesting yet the most complex part of the whole pathway. The tetrapyrrole macrocycle is made from four identical mono-pyrrole units, in the form of PBG. The straightforward head-to-tail condensation of four molecules of PBG would yield the tetrapyrrole uroporphyrinogen I (**5.51**). The normal first cyclic tetrapyrrole that can be isolated as an intermediate in the biosynthetic pathway to chlorophylls and haem is a different isomer, uroporphyrinogen III (**5.52**) in which the positions of the acetic and propionic acid sidechain substituents on one of the pyrrole rings are formally reversed.

The formation of uroporphyrinogen III from four molecules of PBG is a complex process, still being studied intensively. It involves two enzymes, PBG deaminase ($\equiv$ uroporphyrinogen I synthetase) and uroporphyrinogen III cosynthetase, which appear to act in concert rather than strictly sequentially. If the cosynthetase is denatured, for example by heat (55–60°C), the first enzyme remains active and uroporphyrinogen I results. This cannot, however, be isomerised to uroporphyrinogen III. The likely sequence of events is illustrated in fig. 5.20. Four units of PBG are assembled stepwise, starting with ring A, into the linear bilane (**5.53**) by the PBG deaminase enzyme, which then catalyses formation of the key intermediate hydroxy derivative (**5.54**). This is then rapidly cyclised by the cosynthetase to give uroporphyrinogen III. An enzyme-stabilised methylenepyrrolenine intermediate (**5.55**) may be involved. During the cyclisation, an intramolecular rearrangement occurs involving ring D. A mechanism including a spiro-type intermediate (**5.56**) is considered likely. In the absence of the cosynthetase, the linear tetrapyrrole can readily be cyclised to uroporphyrinogen I.

*5.9.5    Conversion into protoporphyrinogen (fig. 5.21)*

The next step in the biosynthetic pathway is the sequential decarboxylation of the four acetic acid sidechains of uroporphyrinogen to four methyl groups. The propionic acid sidechains are not attacked at this stage. Uroporphyrinogen I, when it occurs, can be converted thus into copro-porphyrinogen I (**5.57**), but the normal biosynthetic route produces coproporphyrinogen III (**5.58**) from uroporphyrinogen III. A single enzyme, uroporphyrinogen III decarboxylase, catalyses the removal of all four

Fig. 5.20. A possible mechanism for the formation of uroporphyrinogen IH rather than the I-isomer (A = —CH₂.COOH; P = —CH₂.CH₂.COOH).

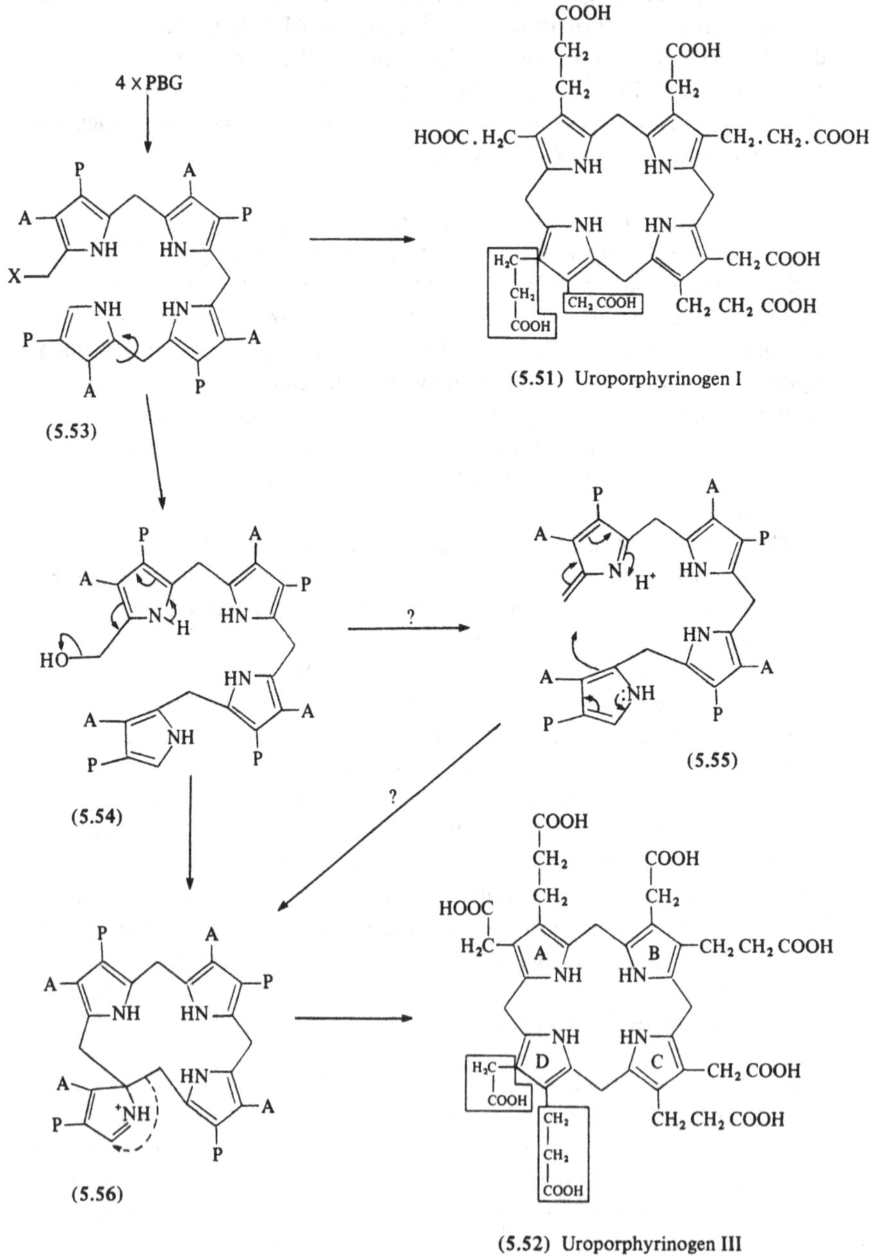

(5.53)

(5.51)  Uroporphyrinogen I

(5.54)

(5.55)

(5.56)

(5.52)  Uroporphyrinogen III

carboxy-groups. The intermediates do not normally seem to be released from the enzyme, but minute amounts of the possible intermediates have been obtained from rat faeces. Available evidence supports a decarboxylation sequence: ring D, A, B, C. The stereochemical course of decarboxylation has been determined; it proceeds with retention of configuration (fig. 5.21b).

Two of the propionic acid sidechains of coproporphyrinogen III (that on ring A, followed by that on ring B) are then modified into vinyl groups by

Fig. 5.21. Conversion of uroporphyrinogen into coproporphyrinogen and thence into protoporphyrinogen IX. (*a*) Pathway; (*b*) stereochemistry.

(*a*) Pathway

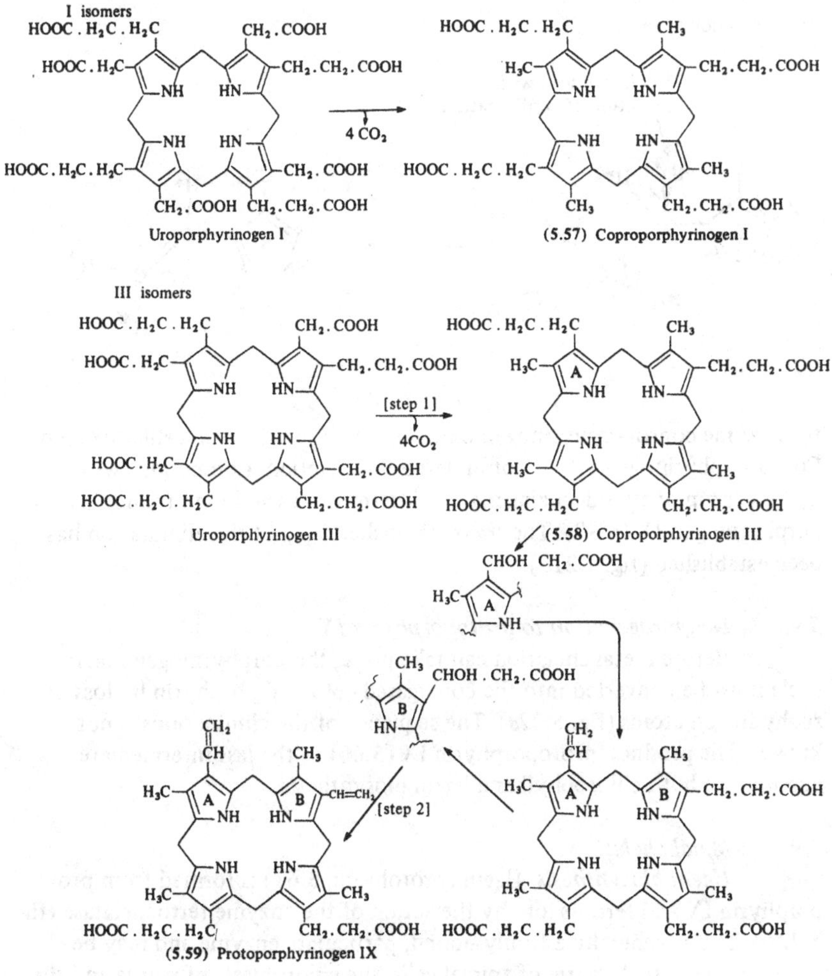

I isomers

Uroporphyrinogen I

(5.57) Coproporphyrinogen I

III isomers

Uroporphyrinogen III

(5.58) Coproporphyrinogen III

(5.59) Protoporphyrinogen IX

Fig. 5.21 – *continued*

(*b*) Sterochemistry
*i*) Decarboxylation (step 1)

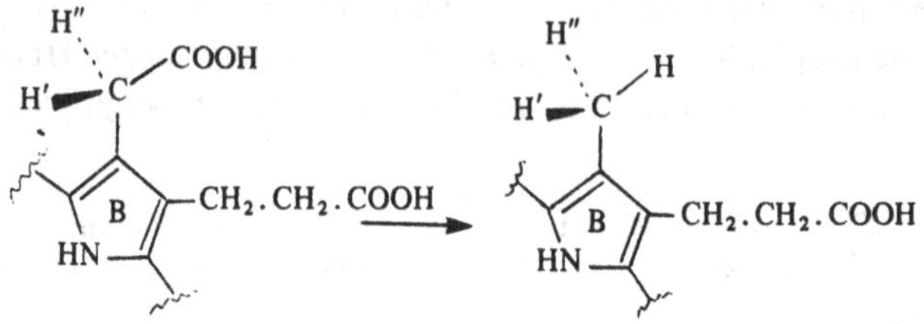

*ii*) Elimination (step 2)

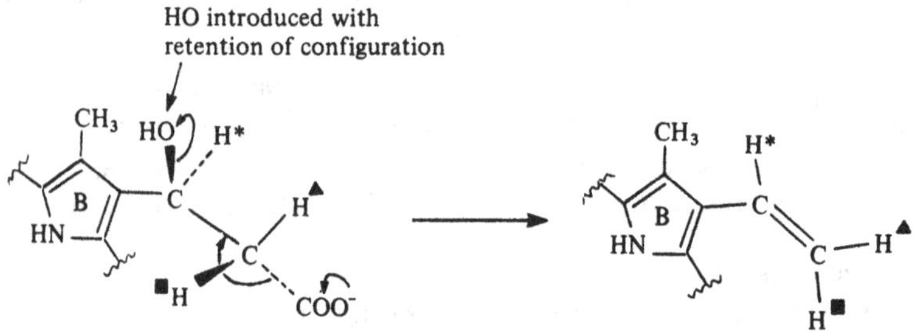

the enzyme coproporphyrinogen oxidase, which requires molecular oxygen. Coproporphyrinogen I does not undergo this reaction. Compounds with hydroxypropionate sidechains may be intermediates and the product is proto-porphyrinogen IX (**5.59**). The *trans* stereochemistry of the elimination has been established (fig. 5.21*b*).

### 5.9.6    *Dehydrogenation to protoporphyrin IX*
Before metal chelation can take place, the porphyrinogen macro-cycle must be converted into the conjugated coloured porphyrin by loss of six hydrogen atoms (fig. 5.22*a*). The sequence of the eliminations is not known. The product, protoporphyrin IX (**5.60**), is the last intermediate common to both chlorophyll and haem biosynthesis.

### 5.9.7    *Metal chelation*
*Haem biosynthesis.* Haem (protohaem, **5.61**) is formed from proto-porphyrin IX and ferrous ion, by the action of the enzyme ferrochelatase (fig. 5.22*b*). This is generally a firmly bound, particulate enzyme and may be found in the mitochondria of animal cells, the chloroplasts of plants and the

Fig. 5.22. (*a*) Aromatisation of protoporphyrinogen IX to give protoporphyrin IX. (*b*) Chelation with Fe$^{2+}$ to form haem. (*c*) Introduction of magnesium to give methyl magnesium protoporphyrin IX.

Protoporphyrinogen IX

(*a*)

6(H)

(5.60) Protoporphyrin IX

(*b*)

Fe$^{2+}$

(5.61) Haem (protohaem)

(*c*)

Mg$^{2+}$

(5.62) Magnesium protoporphyrin IX

[CH$_3$]

(5.63) Methyl magnesium protoporphyrin IX

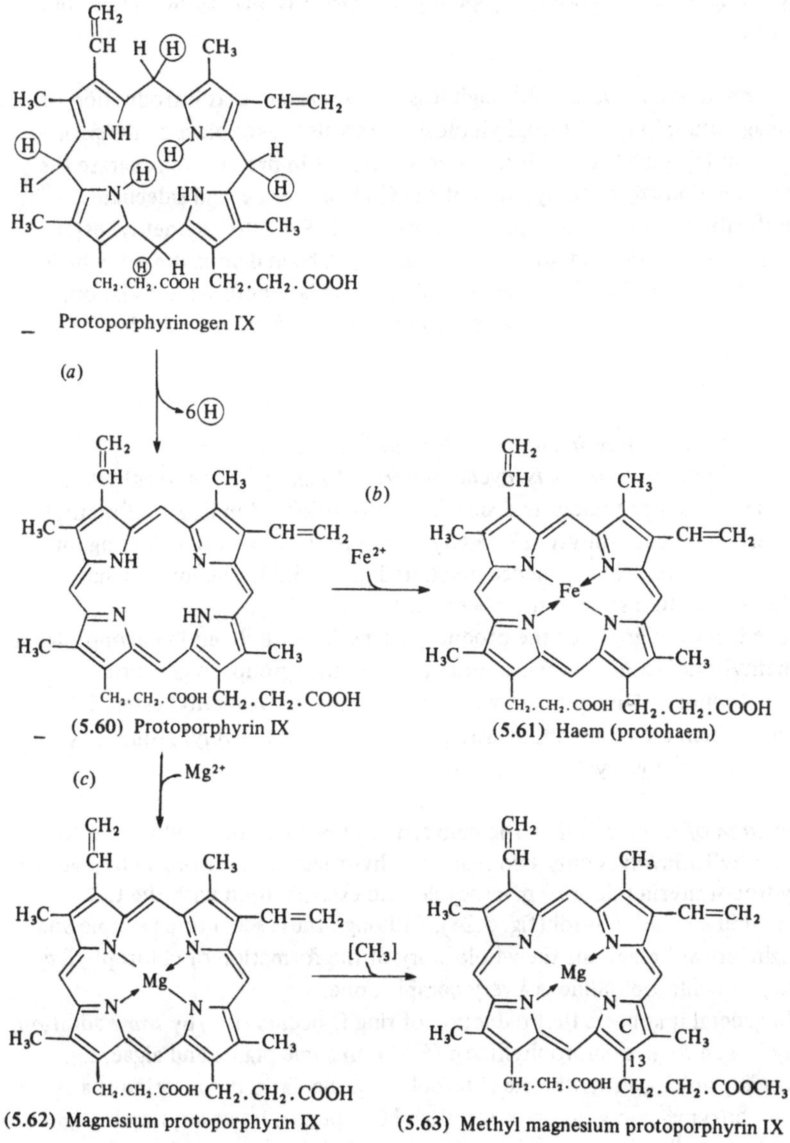

photosynthetic membranes of photosynthetic bacteria. In some cases, however, a soluble ferrochelatase is also present. Purified ferrochelatase is capable of inserting other metal ions, *e.g.* $Zn^{2+}$, $Co^{2+}$, and of utilising other porphyrins. Lipid, probably phospholipid, appears to play some role in the process.

*Chlorophyll biosynthesis.* Although it is almost certain that introduction of the magnesium ion in chlorophyll biosynthesis also takes place at the proto-porphyrin IX stage (fig. 5.22c), it is very difficult in practice to separate the metal chelation from methylation of the C-13 propionic acid sidechain. Esterification of magnesium protoporphyrin IX (**5.62**) to the methyl ester (**5.63**) with *S*-adenosylmethionine has, however, been demonstrated, which supports the idea that chelation normally takes place before methylation. Preliminary studies of the enzyme responsible for insertion of the $Mg^{2+}$ ion have been reported.

### 5.9.8     *Further steps in chlorophyll* a *biosynthesis*
*Formation of the isocyclic ring (ring E) and protochlorophyllide* a.
The methylated propionic acid sidechain at position 13 in ring C of methyl magnesium protoporphyrin IX (**5.63**) is used to form the isocyclic ring, or ring E, by the reaction sequence illustrated in fig. 5.23. The intermediates in this $\beta$-oxidation scheme have been identified.

The C-8 vinyl group of the product, magnesium 3,8-divinyl phaeoporphyrin $a_5$ methyl ester (**5.64**) is then saturated to an ethyl group, to give proto-chlorophyllide *a* ($\equiv$magnesium vinyl phaeoporphyrin $a_5$ methyl ester, **5.65**). There is much argument about whether reduction of the vinyl group may occur before the isocyclic ring is formed.

*Formation of chlorophyll* a. The conversion of protochlorophyllide *a* into chlorophyll *a* involves only two reactions, hydrogenation of ring D to give the dihydroporphyrin (chlorin) macrocycle, and esterification with the $C_{20}$ isoprenoid alcohol, phytol (fig. 5.24). Although these seem to be simple and straightforward reactions the whole story of the formation of chlorophyll *a* from protochlorophyllide is a very complex one.

In general it appears that reduction of ring D occurs first, by *trans* addition of hydrogen to give chlorophyllide *a* (**5.66**). In some plants and algae, *e.g. Chlorella*, which can synthesise chlorophyll in the dark, the reaction is a simple, enzyme-catalysed, dark reaction. Most plants, however, require light for chlorophyll synthesis, and the ring D saturation appears to be a photo-conversion, not of free protochlorophyllide *a* but of a protochlorophyllide-protein complex, known as the protochlorophyllide holochrome, which contains one molecule of protochlorophyllide per molecule of protein. Recent evidence indicates that the holochrome protein is a photo-enzyme, proto-chlorophyllide reductase. Formation of a protochlorophyllide–enzyme–NADPH complex occurs in the dark but light is required to bring about the actual reduction of ring D of the protochlorophyllide.

Fig. 5.23. Formation of the isocyclic ring and protochlorophyllide *a*.

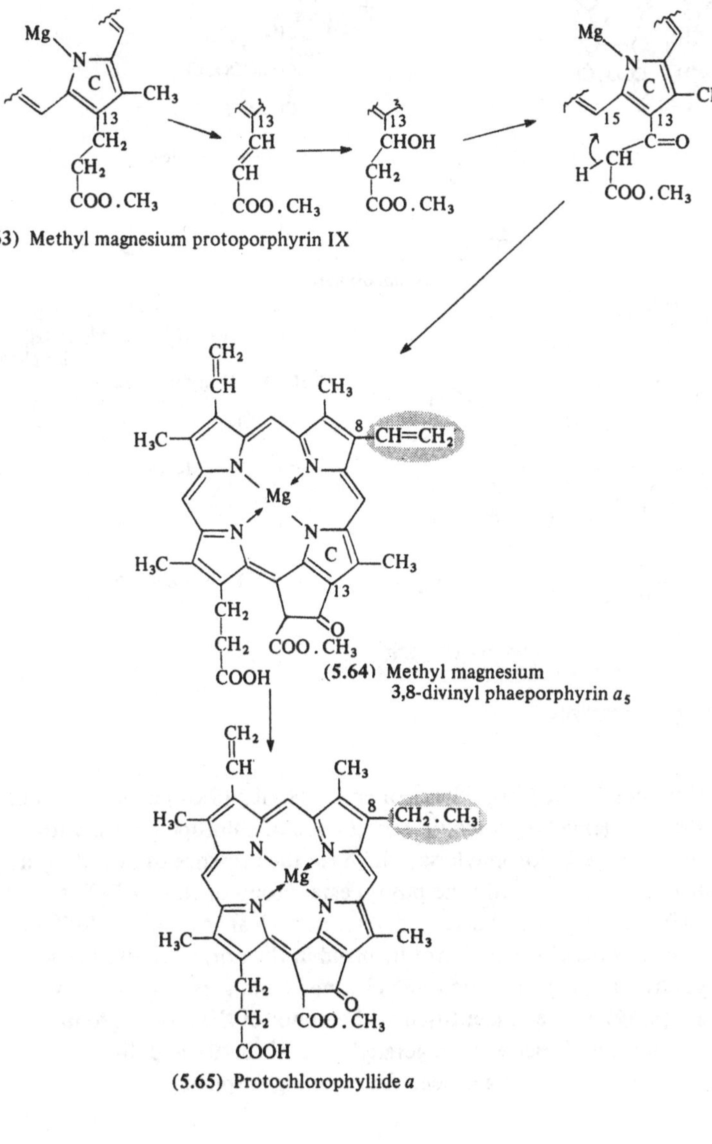

(5.63) Methyl magnesium protoporphyrin IX

(5.64) Methyl magnesium
3,8-divinyl phaeporphyrin $a_5$

(5.65) Protochlorophyllide *a*

Fig. 5.24. Conversion of protochlorophyllide *a* into chlorophylls *a* and *b*.

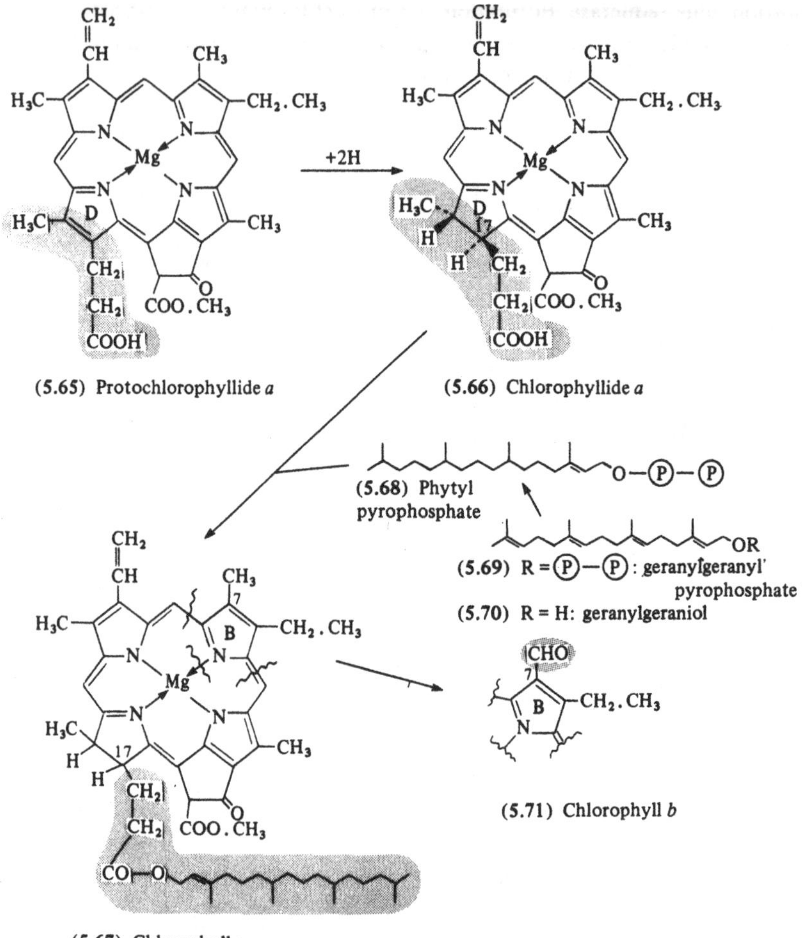

(5.65) Protochlorophyllide *a*

+2H →

(5.66) Chlorophyllide *a*

(5.68) Phytyl pyrophosphate

(5.69) R = $\circledP$—$\circledP$ : geranylgeranyl' pyrophosphate

(5.70) R = H: geranylgeraniol

(5.71) Chlorophyll *b*

(5.67) Chlorophyll *a*

The final step in the biosynthesis of chlorophyll *a* (5.67) is the esterification of the C-17 (ring D) propionic acid residue of chlorophyllide *a* with phytol. An enzyme, chlorophyllase, which (in the presence of 30% acetone!) will hydrolyse or transesterify the phytyl ester group of chlorophyll is well known. After many years of argument, it is now clear that chlorophyllase acting in the reverse direction is not involved in the normal biosynthetic pathway. Phytyl pyrophosphate (5.68) formed from geranylgeranyl pyrophosphate (5.69) has been identified as the donor of the phytol group. Recently chlorophyll species with geranylgeraniol (5.70), and di- and tetrahydrogeranylgeraniol have been found in higher plants, suggesting that geranylgeraniol could in some cases be the esterifying alcohol, and could perhaps be reduced to phytol after addition to the porphyrin.

### 5.9.9    *Formation of chlorophyll* b

It is most likely that chlorophyll *b* **(5.71)** is formed from chlorophyll *a* by oxidation of the C-7 methyl group to an aldehyde. Nothing is known about the mechanism of the presumed oxidation process involved. Alternative suggestions that the two chlorophylls are synthesised by separate or divergent pathways or that chlorophyll *a* is formed from *b* have been discounted.

Fig. 5.25. Possible routes for the formation of chlorophylls *c* and *d*.

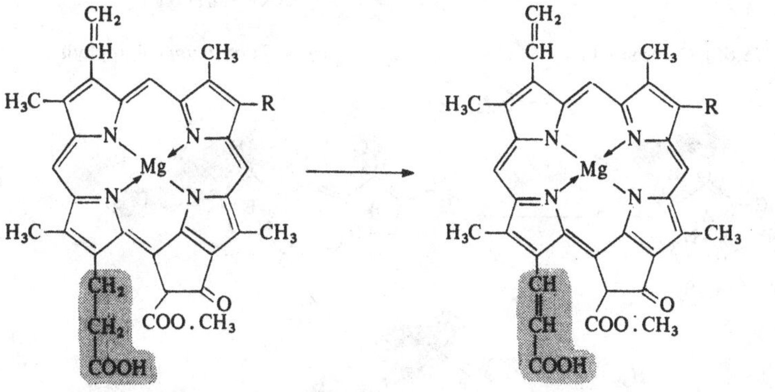

(5.64) (R = CH=CH$_2$) Methyl magnesium 3,8-divinyl phaeoporphyrin $a_5$

(R = CH$_2$.CH$_3$) Protochlorophyllide *a*

(5.72) Chlorophyll $c_1$ (R = CH$_2$.CH$_3$)

(5.73) Chlorophyll $c_2$ (R = CH=CH$_2$)

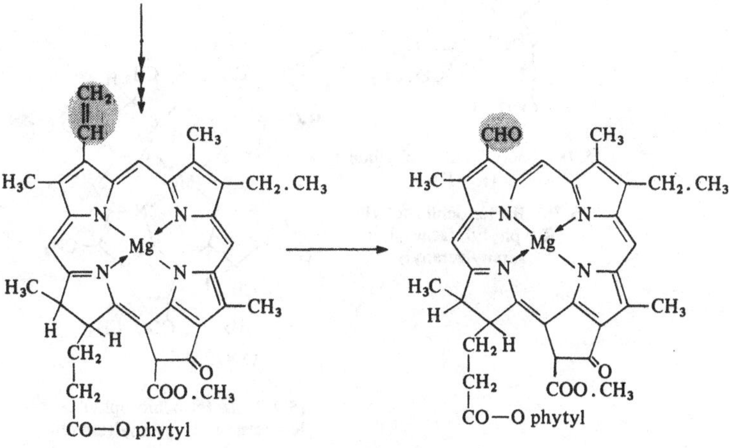

(5.67) Chlorophyll *a*

(5.74) Chlorophyll *d*

Fig. 5.26. Formation of bacteriochlorophylls and *Chlorobium* chlorophylls.

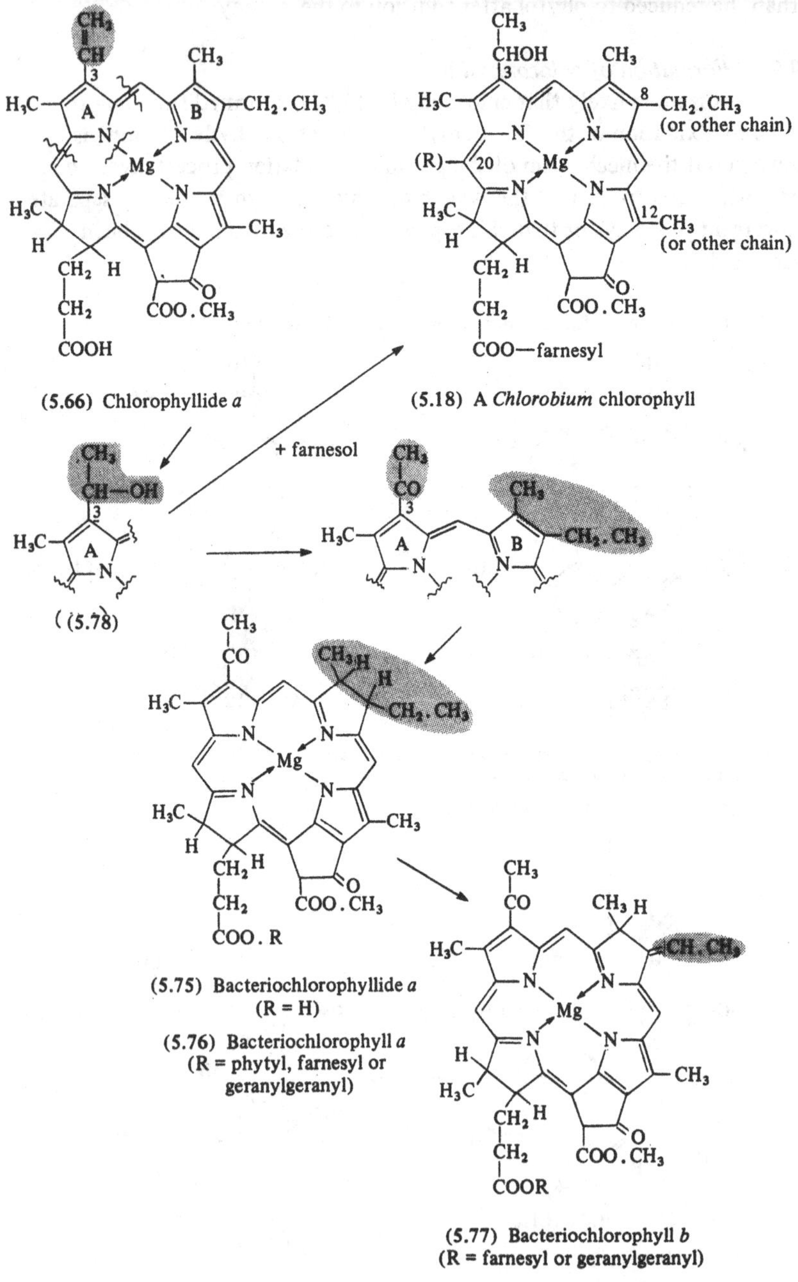

(5.66) Chlorophyllide *a*

(5.18) A *Chlorobium* chlorophyll

((5.78)

+ farnesol

(5.75) Bacteriochlorophyllide *a*
(R = H)

(5.76) Bacteriochlorophyll *a*
(R = phytyl, farnesyl or
geranylgeranyl)

(5.77) Bacteriochlorophyll *b*
(R = farnesyl or geranylgeranyl)

### 5.9.10 Formation of chlorophylls c and d

Nothing is known about the biosynthesis of these chlorophylls. Consideration of their structures suggests that chlorophylls $c_1$ (**5.72**) and $c_2$ (**5.73**) may be formed from intermediates on the pathway to chlorophyll $a$, and chlorophyll $d$ (**5.74**) perhaps from chlorophyll $a$ itself (fig. 5.25).

### 5.9.11 Formation of bacteriochlorophylls

Mutant strains of bacteria (*Rhodopseudomonas* spp.) have been produced which accumulate protochlorophyllide $a$, chlorophyllide $a$, and other related compounds in which the ring A vinyl group is modified into a hydroxyethyl or acetyl substituent. This points to the biosynthetic scheme outlined in fig. 5.26, with modification of the C-3 sidechain taking place before ring B is hydrogenated to give bacteriochlorophyllide $a$ (**5.75**). Esterification with phytol or a related alcohol such as farnesol is the final stage. Bacteriochlorophyll $b$ (**5.77**) is presumably produced by dehydrogenation of bacteriochlorophyll $a$ (**5.76**).

### 5.9.12 Formation of Chlorobium chlorophylls (bacteriochlorophylls c and d)

Little is known concerning the biosynthesis of the several *Chlorobium* chlorophylls. It seems likely that they arise from the hydroxyethyl intermediate (**5.78**) in the bacteriochlorophyll pathway, and that the extra alkyl substituents at C-8, 12 and 20 are introduced late in the sequence, presumably by addition to vinyl groups from $S$-adenosylmethionine.

### 5.9.13 Biosynthesis of vitamin $B_{12}$

The type III arrangement of the sidechains indicates the basic similarity of the $B_{12}$ biosynthetic pathway to that of haem and chlorophyll. The corrin ring is derived from ALA and PBG, and the early stages in the pathway are identical to those in porphyrin synthesis. Indeed uroporphyrinogen III is incorporated into the corrin system without fragmentation or rearrangement of the four rings. Later stages in $B_{12}$ biosynthesis are currently being studied intensively.

### 5.9.14 Formation of bilins

*In animals.* Animal bilins are produced by catabolism of haem, mostly haem from haemoglobin, by oxidative elimination of one of the methine bridge carbon atoms and loss of the chelated iron atom. The breakdown occurs principally in the reticulo-endothelial cells of spleen, liver, bone-marrow and, to a lesser extent, kidney. Although the mechanism is not fully understood it is known that the two ketone or lactam oxygen atoms introduced are derived from two different $O_2$ molecules, and a cytochrome-linked haem oxygenase enzyme, $NADP^+$ and ascorbic acid have been implicated. The

Fig. 5.27. A scheme for the formation of mammalian bilins from haem. (*a*) Breakdown of haem to bilirubin in reticulo-endothelial tissues. (*b*) Conjugation of bilirubin and its transport to the intestine. (*c*) Microbial and autoxidative transformations of bilirubin in the intestine.

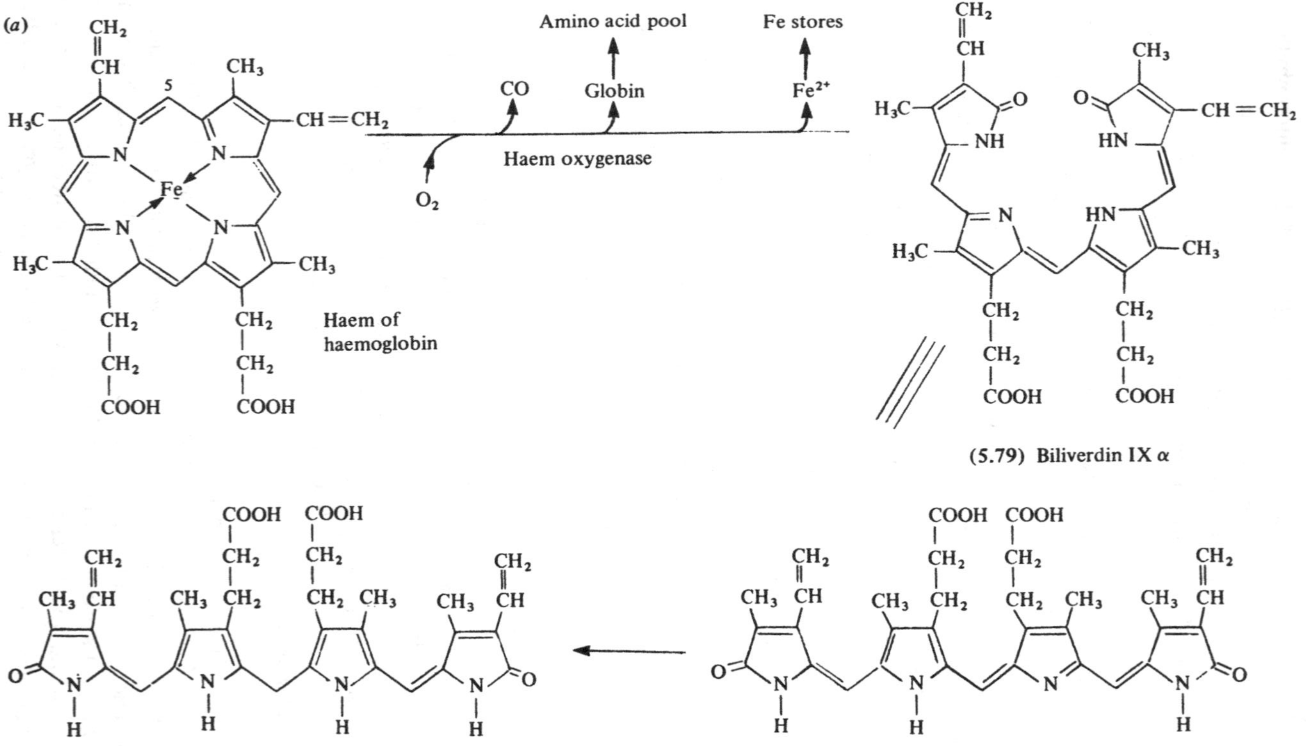

(5.79) Biliverdin IX α

(5.80) Bilirubin

methine carbon is released as carbon monoxide, the iron is transported to the body's general iron stores and the globin is broken down into amino acids which enter the general metabolic pool. In mammals the cleavage occurs exclusively at the α-methine bridge (C-5) to give biliverdin IXα (**5.79**). The central methine carbon of biliverdin is then reduced to $CH_2$ by a reductase which requires NAD(P)H as cofactor and is present in liver and spleen. The bilirubin (**5.80**) thus formed is transported as an albumin complex to the liver where it is conjugated with glucuronic acid and other sugars by a microsomal enzyme. The water-soluble conjugates pass into the bile, and thence to the

Fig. 5.27. *contd.*

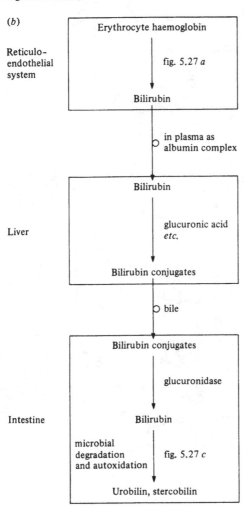

(*b*)

Reticulo-endothelial system

Erythrocyte haemoglobin

fig. 5.27 *a*

Bilirubin

in plasma as albumin complex

Bilirubin

Liver

glucuronic acid *etc.*

Bilirubin conjugates

bile

Bilirubin conjugates

glucuronidase

Intestine

Bilirubin

microbial degradation and autoxidation

fig. 5.27 *c*

Urobilin, stercobilin

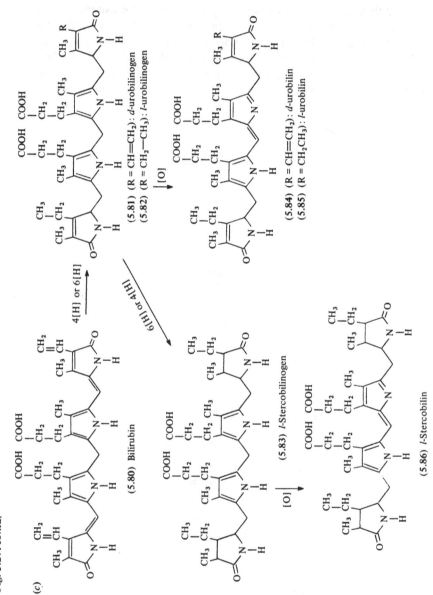

Fig. 5.27. contd.

intestine. In the intestine the bilirubin conjugates are hydrolysed by a
β-glucuronidase enzyme before further transformations take place which
are mainly reductive processes brought about by intestinal bacteria. The
steps involve hydrogenation of the vinyl sidechains, successive reduction of
the remaining two methine bridges and stepwise reduction of some (two?) of
the pyrrole ring double bonds to give *d*- and *l*-urobilinogen (**5.81, 5.82**) and
*l*-stercobilinogen (**5.83**). These bilinogens are oxidised in air to give *d*- and
*l*-urobilin (**5.84, 5.85**) and *l*-stercobilin (**5.86**), yellow-brown pigments that
colour the urine and faeces.

A possible overall scheme for bilin formation in mammals is given in
Fig. 5.27.

No evidence is available concerning the formation of bilins by lower
animals.

*In plants.* In algae, as in animals, oxidative porphyrin cleavage occurs virtually
exclusively at the α-methine bridge carbon, which is lost as carbon monoxide.
The source of the algal bilins appears to be haem rather than chlorophyll,
and the mechanism of cleavage seems to be similar to that in animal systems.
Details of the later transformations which lead to the characteristic structural
features of the phycobilins have not been worked out. Concomitant protein
synthesis, *de novo*, is necessary for algal biliprotein formation and incorporation into phycobilisomes. It is not yet known if the bilin prosthetic group of
phytochrome is biosynthesised from haem in the same way.

### 5.9.15 Factors controlling tetrapyrrole synthesis and accumulation
*External factors – environmental and nutritional control.* Many
external (*e.g.* environmental or nutritional) factors may influence tetrapyrrole
synthesis. The most familiar of these effects is the requirement for light for
chlorophyll or bacteriochlorophyll biosynthesis in most photosynthetic
systems. For example, in *Rhodopseudomonas sphaeroides*, bacteriochlorophyll is synthesised in light, anaerobic conditions but not in the dark in air,
although haem synthesis (cytochromes) does take place under the latter
conditions. ALA synthetase activity is much higher in the light. It is well
known that most higher plants become etiolated when maintained in the
dark, *i.e.* they are unable to synthesise their normal chloroplasts or chloroplast pigments, notably chlorophylls. On transfer into the light, however,
etiolated seedlings produce large amounts of chlorophyll over a period of
about 48 h, and take on the normal green appearance. This chlorophyll
synthesis takes place in stages. Very rapidly after even a brief period of
illumination the small amount of protochlorophyllide *a* (associated with the
protochlorophyllide holochrome protein) that is present in the etiolated
tissues is converted into chlorophyllide *a* which is subsequently and more
slowly esterified to give chlorophyll *a*. The small amount of chlorophyll

produced in the first two hours or so of greening is formed in this way. After this there is a lag phase of a few hours, followed by rapid and much more extensive chlorophyll synthesis, *de novo*. During the lag phase, substantial enzyme synthesis occurs, and ALA synthetase activity increases markedly. However, the formation of chlorophyll is just a part of the greater, complex process of chloroplast development, and the factors which regulate the formation of the various chloroplast components are closely coordinated (see chapter **10**).

Chlorosis, *i.e.* chlorophyll deficiency, is a common sympton of many plant diseases, and nutritional deficiencies, especially of nitrogen, iron and magnesium, also lead to decreased chlorophyll levels. It is difficult to assess whether these effects are on chlorophyll synthesis alone or on chloroplast development as a whole.

The intensity and spectral quality of the incident light influences biliprotein synthesis and levels in most red and blue-green algae, and determines whether phycocyanin or phycoerythrin synthesis predominates. In red incident light the red-absorbing phycocyanin is formed preferentially, whereas green light conditions favour formation of the green-absorbing phycoerythrin.

In animals, the syntheses of the two components, haem and globin, of haemoglobin appear to be very closely coordinated. Increased production of haemoglobin occurs (up to 20% in man) in response to lowered oxygen levels, *e.g.* at high altitude. Synthesis of the allosteric effector 2,3-bisphosphoglycerate is also increased at high altitude to allow more efficient 'unloading' of oxygen from oxyhaemoglobin. These effects were brought to public attention during the 1968 Olympic Games held in Mexico City. Kenyan and Ethiopian athletes who were fully acclimatised to high-altitude conditions won most of the medals in the middle- and long-distance events. Athletes from other, lower-lying, countries who had not been suitably acclimatised did not perform well, since their haemoglobin and 2,3-bisphosphoglycerate levels were not increased sufficiently to allow their muscles to work efficiently for long periods in the rarified atmosphere at high altitude.

It is likely that in many of the cases considered above the primary response to the external stimulus is a change in the amount or activity of ALA synthetase.

*Internal factors - biochemical control.* In the biosynthesis of all cyclic tetrapyrroles - haem, chlorophylls and corrins - in all living organisms - animals, plants and microorganisms - the main biochemical control centres around ALA and the key enzymes ALA synthetase and ALA dehydrase. The actual mechanism of this control in individual cases is less well understood; the large amount of experimental information available indicates that several different mechanisms may operate in different organisms and tissues. The fact that ALA synthetase is generally an insoluble, mitochondrial enzyme whereas almost all the other enzymes of the porphyrin biosynthetic pathway are cytoplasmic and soluble suggests that permeability, for example of the mitochondrial membrane to ALA, may be part of the controlling mechanism. In most cases, however, it is the activity or the amount of ALA synthetase, or of an ALA synthetase-activating enzyme that is regulated. End-product control of enzyme synthesis or activity appears to be the most important process and has been implicated in many examples.

If an organism or a tissue is synthesising different tetrapyrroles, *e.g.* bacteriochlorophyll and haem, the two pathways are likely to be under separate control.

## 5.10    Disorders of porphyrin metabolism
### 5.10.1    Porphyrias

In normal animals, porphyrin synthesis and breakdown are very accurately controlled. In man, a group of disorders has been recognised in which inborn or acquired derangements of enzymes involved in haem biosynthesis result in the presence of free porphyrins and/or their precursors in various body tissues. These conditions, porphyrias, are indicated clinically by increased excretion of these compounds.

In erythropoietic porphyria there is a deficiency in uroporphyrinogen III cosynthetase, and large amounts of uroporphyrinogen I accumulate, giving rise to wine-red urine and strong red fluorescence of teeth and other tissues under u.v. light. Acute intermittent porphyria affects the liver, and porphyrins and their precursors, especially ALA and porphobilinogen, accumulate there and in the urine. In the intermittent attacks, acute abdominal pain is experienced. Within the two main groups there are several types of porphyria which can be recognised from the nature of the porphyrins or precursors accumulated, the mode of inheritance, and the clinical features and symptoms.

It seems most probable that an acute intermittent porphyria – variegate porphyria – was responsible for the mania of King George III of England. Each bout of mental derangement was preceded by acute abdominal pain, and contemporary reports indicate that his urine was a characteristic red colour. Symptoms now known to be consistent with porphyria were also reported for various of his ancestors and descendants.

In the various types of porphyrias, uro- and coproporphyrins (**5.30, 5.31**), uro- and coproporphyrinogens (I and III isomers) (**5.51, 5.52, 5.57, 5.58**) protoporphyrin IX (**5.60**), ALA (**5.45**) and PBG (**5.50**) may accumulate in the urine and faeces, largely as a result of defects in the normal mechanisms controlling ALA synthetase activity. Similar experimental porphyrias can be produced by administration of a variety of drugs, including barbiturates.

Free porphyrin in body tissues can act as a sensitiser of harmful photo-oxidations, and many porphyria patients are photosensitive, suffering adverse effects, especially on exposure to sunlight. Interestingly, in some cases treatment with β-carotene affords protection against the harmful effects of photosensitisation. This resembles the protective effect of carotenoids against singlet oxygen photooxidation, often sensitised by porphyrins, in bacteria and plants (see chapters 10 and 11).

### 5.10.2   Disorders of bilin production
*Jaundice.* Pathological disorders involving excess bilins are called jaundice or icterus, and may be a result of either excess production, decreased destruction or faulty elimination of bilins. In jaundice in man, excess bilirubin (5.80) imparts a strong yellow colour to the plasma and other tissues. The bilirubin passes from the blood to tissue spaces and is absorbed strongly by connective tissues..

There are various causes of jaundice in the adult. Normally bilirubin passes from the the reticulo-endothelial system to the liver for conjugation with glucuronic acid and excretion into the bile. If, for some reason, haemoglobin breakdown is very greatly increased, the liver may not be able to remove bilirubin from the blood quickly enough ('retention jaundice'). More commonly, jaundice arises when disorders such as obstruction of the biliary passages or necrosis of the liver cause bile to be regurgitated into the blood stream ('regurgitation jaundice'). In this case it is the conjugated bilirubin which is found in increased levels in the blood and tissues.

Newborn infants not uncommonly exhibit a transitory and harmless jaundice. The liver of the newborn is deficient in the enzymes responsible for conjugating bilirubin, which is present in large amounts due to the destruction of the excess red cells present during life in the uterus. Free bilirubin thus accumulates. This harmless, mild jaundice must be distinguished from the persistent and dangerous jaundice, kernicterus, that is associated with haemolytic diseases of the newborn, and in which the bilirubin can be absorbed selectively by brain tissues, causing permanent brain damage.

Jaundice is not restricted to humans or mammals. A green jaundice, due to biliverdin (5.79), has been reported in the pike.

*Bruising.* The short-lived 'black-and-blue' and yellow colours associated with bruising in humans are caused when blood leaks into tissue spaces, and its haemoglobin is broken down into products such as biliverdin (green), urobilin (orange), and the violet-blue biliviolins.

### 5.11   Functions of tetrapyrrole pigments
Some functions which are not a consequence of the light-absorbing properties of tetrapyrroles have already been mentioned, especially the

oxygen-carrying role of haemoglobin. However, most of the biological functions of these pigments are related to their light-absorbing properties, *e.g.* as external colours, in photosynthesis, and as photoreceptors. These subjects will be dealt with in chapters **10** and **11**.

### 5.12    Conclusions and comments

As befits a group of compounds with such vital biological functions, tetrapyrrole pigments have been studied extremely intensively, so that probably more is known of the mechanisms of their formation and functioning than of any other group of pigments. The elucidation of the three-dimensional structures of myoglobin and haemoglobin, and the mechanism by which haemoglobin functions in oxygen transport is one of the classical stories of scientific investigation. Much is also understood about the way that chlorophyll is used as the main light-harvesting pigment in photosynthesis (see chapter **10**). The main aspects of porphyrin (and corrin) biosynthesis, including details of mechanism and stereochemistry, have now been worked out by most elegant studies with classical radioisotope methods and sophisticated development of $^{13}C$ and $^2H$ labelling techniques. The student wishing to learn about biosynthesis could do no better than to spend time reading about this work. Even so, the course of some of the biosynthetic transformations is still not fully established. There is much scope for study of the formation of bacteriochlorophylls, the unusual algal chlorophylls *c* and *d*, and the modified haems of some cytochromes, *etc.*; even some of the structural work is not yet complete. The free porphyrins found in lower animals are also worth detailed chemical study. The interactions between chlorophylls and proteins in specific complexes in the photosynthetic apparatus, and the orientation of the chlorophylls within these complexes, remain to be elucidated fully. The factors which regulate the formation of these complexes, and indeed the regulation of the biosynthesis of the chlorophyll and other porphyrin molecules themselves, should be a fruitful area for study.

The properties of the linear tetrapyrroles or bilins are quite different from those of the macrocyclic porphyrins, but again there are many fascinating problems. These compounds are familiar in animals, including ourselves, as 'bile pigments', formed by breakdown of haem. The mechanisms of breakdown and transformation into the products that are eventually excreted perhaps merit reinvestigation with modern physicochemical techniques. Similarly the formation of bilins as a result of bruising or jaundice, and in the pigmentation of birds' egg-shells deserves further study. The mechanism of deposition and association of bilins and free porphyrins with minerals in the shell, and the factors which regulate the intricate and distinctive patterning of shells of different species should be a fascinating field of investigation. Plant bilins (phycobilins) in blue-green and red algae occur in phycobilisomes and function as accessory pigments in photosynthesis. Many questions, such as

the manner of their uptake by and association with protein remain unanswered. The most intriguing plant bilin is phytochrome, the elusive substance which mediates many light responses in the plant world. The full
story of the phytochrome photocycles, when these can eventually be
described fully in molecular terms, should be as fascinating as that of the
mammalian visual pigment cycles.

## 5.13    Suggested further reading

Of the many books which deal with the tetrapyrrole pigments, the
monograph edited by Smith (1975) gives a comprehensive account of the
chemistry of these compounds, and also much information on their biosynthesis. The extensive seven-volume series edited by Dolphin (1978) will
also be a major source of information, but only two volumes have yet
appeared. An earlier book, reporting the proceedings of a Symposium on
porphyrins and related compounds (Goodwin, 1968) remains useful. Selected
topics are the subject of other articles and books. For example, Jackson
(1976) deals with the chemistry and distribution of chlorophylls, and the
book by Vernon and Seely (1966) is devoted to all aspects of chlorophyll
chemistry and biochemistry. For information on porphyrins in animals the
reader is referred to specialist articles by Kennedy (1969, 1976) and
Rimington and Kennedy (1962), but will also find sections in the general
animal pigment books of Fox (1976), Fox and Vevers (1960) and Needham
(1974) useful. Of many publications devoted to bilins in both animals and
plants the following are recommended: Bouchier and Billing (1967); Hudson
and Smith (1975); Lathe (1972); Ó Carra and Ó hEocha (1976); With (1968);
and Rüdiger (1980). The chemistry and biochemistry of phytochrome is
described by Smith and Kendrick (1976) and much information on cytochrome $P_{450}$ may be found in the book edited by Cooper, Rosenthal, Snyder
and Witmer (1975).

For the biosynthesis of porphyrins, the article by Akhtar and Jordan
(1979) gives the overall picture, with emphasis on reaction mechanisms and
stereochemistry, whereas Bogorad (1976) and Schneider (1980) adopt a more
biochemical or biological approach. A very recent article by Battersby,
Fookes, Matcham and McDonald (1980) clarifies a number of points
previously in doubt. Information on porphyrin and bilin biosynthesis can also
be obtained from the general books cited above, especially Smith (1975).
Finally, two specialist articles which make interesting reading are the review
by Perutz (1970), which explains the mechanism of cooperativity in haemoglobin–oxygen binding, and an article in which Tschudy and Schmid (1972)
discuss porphyria diseases.

## 5.14 Selected bibliography

Akhtar, M. and Jordan, P. M. (1979) Porphyrin, chlorophyll and corrin biosynthesis, in *Comprehensive organic chemistry*, vol. 5, ed. E. Haslam, p. 1121. Oxford: Pergamon.

Battersby, A. R., Fookes, C. J. R., Matcham, G. W. J. and McDonald, E. (1980) Biosynthesis of the pigments of life: formation of the macrocycle, *Nature*, 285, 17.

Bogorad, L. (1976) Chlorophyll biosynthesis, in *Chemistry and biochemistry of plant pigments*, 2nd edition, vol. 1, ed. T. W. Goodwin, p. 64. London, New York and San Francisco: Academic Press.

Bouchier, I. A. D. and Billing, B. H. (eds) (1967) *Bilirubin metabolism*. Oxford: Blackwell.

Cooper, D. Y., Rosenthal, O., Snyder, R. and Witmer, C. (eds) (1975) *Cytochromes* $P_{450}$ *and* $b_5$. New York: Plenum.

Dolphin, D. (ed.) (1978) *The porphyrins*, vols 1 and 2 (7 vols projected). New York: Academic Press.

Fox, D. L. (1976) *Animal biochromes and structural colors*, 2nd edition. Berkeley, Los Angeles and London: University of California Press.

Fox, H. M. and Vevers, G. (1960) *The nature of animal colours*. London: Sidgwick and Jackson.

Goodwin, T. W. (ed.) (1968) *Porphyrins and related compounds*. London and New York: Academic Press.

Hudson, M. F. and Smith, K. M. (1975) Bile pigments, *Chem. Soc. Revs.*, 4, 363.

Jackson, A. H. (1976). Structure, properties and distribution of chlorophylls, in *Chemistry and biochemistry of plant pigments*, 2nd edition, vol. 1, ed. T. W. Goodwin, p. 1. London, New York and San Francisco: Academic Press.

Kennedy, G. Y. (1969) Pigments of Annelida, Echiuroidea, Sipunculoidea, Priapuloidea and Phoronidea, *Chem. Zool.*, 4, 311.

Kennedy, G. Y. (1976) Survey of avian eggshell pigments, *Comp. Biochem. Physiol.*, B55, 117.

Lathe, G. H. (1972) The degradation of haem by mammals and its excretion as conjugated bilirubins, *Essays in biochemistry*, 8, p. 107. London and New York: The Biochemical Society – Academic Press.

Needham, A. E. (1974) *The significance of zoochromes*. Berlin, Heidelberg and New York: Springer-Verlag.

O Carra, P. and Ó hEocha, C. (1976) Algal biliproteins and phycobilins, in *Chemistry and biochemistry of plant pigments*, 2nd edition, vol. 1, ed. T. W. Goodwin, p. 328. London, New York and San Francisco: Academic Press.

Perutz, M. F. (1970) Stereochemistry of cooperative effects of haemoglobin, *Nature*, 228, 726.

Rimington, C. and Kennedy, G. Y. (1962) Porphyrins, in *Comparative biochemistry*, vol. 4, eds M. Florkin and H. S. Mason, p. 557. New York and London: Academic Press.

Rüdiger, W. (1980) Plant biliproteins, in *Pigments in plants*, 2nd edition, ed. F.-C. Czygan, p. 314. Stuttgart and New York: Gustav Fischer.

Schneider, H. (1980) Chlorophyll biosynthesis. Enzymes and regulation of enzyme activities, in *Pigments in plants*, 2nd edition, ed. F.-C. Czygan, p. 237. Stuttgart and New York: Gustav Fischer.

Smith, H. and Kendrick, R. E. (1976) The structure and properties of phytochrome, in *Chemistry and biochemistry of plant pigments*, 2nd edition, vol. 1, ed. T. W. Goodwin, p. 378. London, New York and San Francisco: Academic Press.

Smith, K. M. (ed.) (1975) *Porphyrins and metalloporphyrins*. Amsterdam: Elsevier.

Tschudy, D. P. and Schmid, R. (1972) The porphyrias, in *The metabolic basis of inherited diseases,* 3rd edition, eds J. B. Stanbury, J. B. Wyngaarden and D. S. Fredrickson, p. 1087. New York: McGraw-Hill.

Vernon, L. P. and Seely, G. R. (eds) (1966) *The chlorophylls.* New York and London: Academic Press.

With, T. K. (1968) *Bile pigments.* (Trans. J. P. Kennedy.) New York and London: Academic Press.

# 6      Other non-polymeric *N*-heterocyclic pigments

## 6.1      Introduction

As described in the previous chapter, the simple nitrogenous hetero-aromatic ring system of pyrrole forms the basis of many extremely important natural pigments. In this chapter it will be seen that several other nitrogen heterocyclic systems also give rise to pigment classes. The basic skeletons of most of these pigments are condensed bi-, tri- or oligocyclic heteroaromatic ring systems and their partially reduced derivatives. In these compounds electronic excitation is usually relatively easy, especially in ring systems with extended conjugation or when several substituents are present, and yellow, red, purple or blue colours may be produced. The $\pi \rightarrow \pi^*$ absorption bands usually occur at similar wavelengths to those of the corresponding carbocycle, but in addition $n \rightarrow \pi^*$ transitions give rise to important, though less intense absorption (forbidden transitions) at longer wavelengths.

Of the several groups of these pigments to be described, the purines and pteridines are extremely important substances, synthesised by all living organisms, but they are used as pigments only by a small number of animals. The closely related riboflavin is produced only by plants and microbes but is also extremely important in animals as a vitamin, though it rarely serves as a pigment. The ommochromes are exclusively animal products (arthropods). The phenazine group is produced only by bacteria, which also elaborate a miscellany of other nitrogenous pigments. The betalains, which do not have condensed ring systems, are exclusively plant products, and are not present in or used by animals.

In this chapter the very different main characteristics, *e.g.* structures, distribution, properties, functions and biosynthesis of each of these pigment groups will be described in turn.

## 6.2      Purines, pterins and flavins
### 6.2.1      Introduction

After the tetrapyrroles, probably the most important simple (*i.e.* not polymeric) *N*-heterocyclic pigments in Nature, certainly in the animal kingdom, are the pterins. It is convenient to consider at the same time as these the closely related purines and flavins. The purines, although they do not absorb light of visible wavelengths, nevertheless provide the basis of many structural whites and structural colours in animals (chapter 1). The flavins make little or no contribution to external colours but riboflavin is an important photoreceptor molecule which will be considered in chapter 11.

In the following sections the basic ring skeletons will be described first, then the occurrence, distribution, properties and biological significance will be considered of each pigment group in turn. Finally the biosynthetic pathway which gives rise to all three groups will be described.

### 6.2.2   Basic ring skeletons

The purine, pteridine and flavin ring skeletons all have four heterocyclic nitrogen atoms and are illustrated in fig. 6.1. The purine skeleton (6.1) has a six-membered pyrimidine ring and a five-membered imidazole system. Pteridine (6.2) is rather similar but has a six-membered pyrazine system in place of the imidazole. The isoalloxazine skeleton (6.3), on which flavin is based, is essentially a substituted (benzo)pteridine. As also illustrated in fig. 6.1, different numbering systems are used for the three groups, so great care is needed to avoid confusion, for example when comparing substitution patterns in members of the different series.

Fig. 6.1. Basic ring skeletons of purine, pteridine and isoalloxazine, illustrating the different numbering systems used for the different pigment classes.

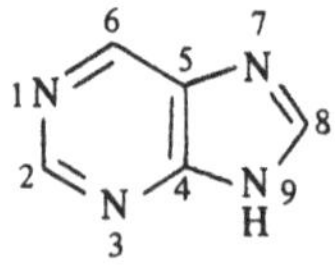

Purine (6.1) 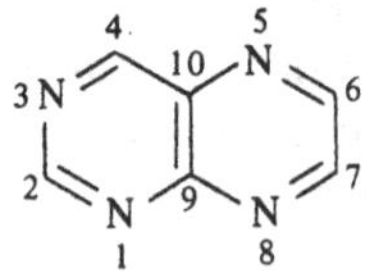          Pteridine (6.2) 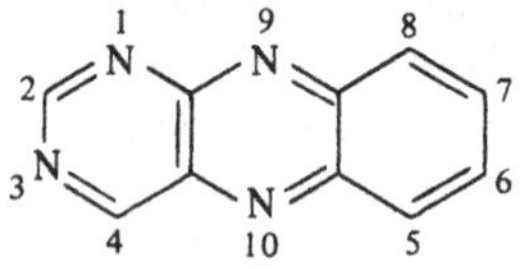          Isoalloxazine (6.3)

### 6.2.3   Purines

*Occurrence and distribution.* The purines adenine (6.4) and guanine (6.5) occur universally as components of nucleic acids and nucleotides. Guanine is also one of the purines that contribute most frequently to external colours and patterns in animals; uric acid (6.6) is also quite widespread, and xanthine (6.7) and isoguanine (6.8) are sometimes found. These purines do not absorb light of visible wavelengths, although they do absorb strongly in the ultraviolet and can therefore be seen by some animals, especially insects. Their important contribution to animal coloration is as the basis of structural colours, especially white and silver. Guanine in particular accumulates abundantly in certain tissues in microcrystalline or granular form. These crystals or

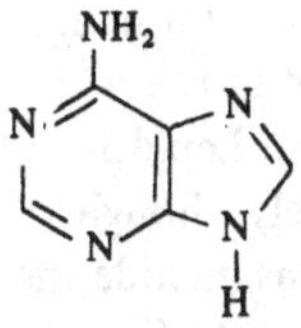

(6.4) Adenine 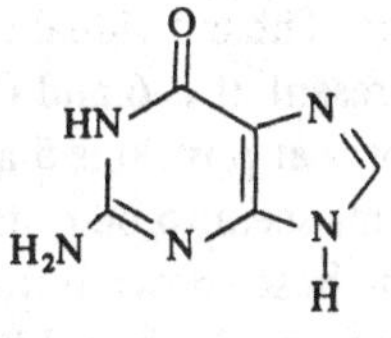         (6.5) Guanine 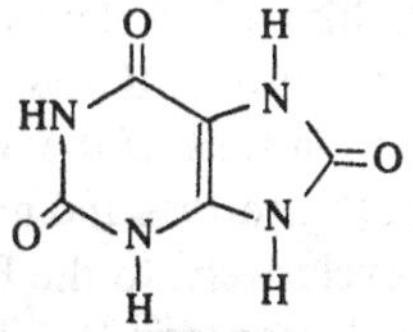         (6.6) Uric acid

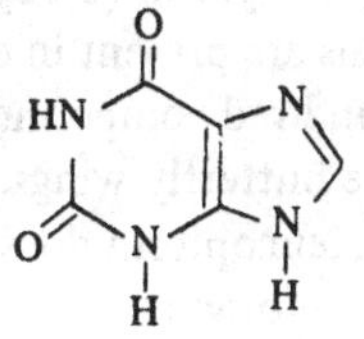

(6.7) Xanthine 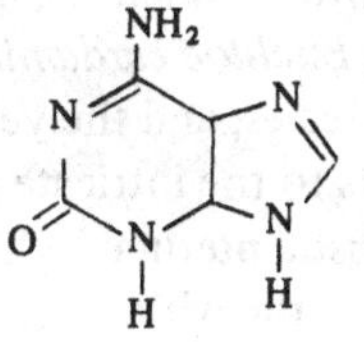              (6.8) Isoguanine

particles, when suitably orientated in the tissues, will reflect all incident light and so give a white or silvery appearance, such as that familiar in the skin of fishes. In fish skin and scales, guanine is contained in special cells known as guanophores or iridophores, and the reflective properties of the crystals can be altered in response to changes in background light levels (see chapter 8). In some examples, such as the goldfish, guanophores are overlain with carotenoid to give a shiny golden appearance. Structural whites or silvers are also produced by guanine and other purines in some amphibians and invertebrates. Tyndall scattering by minute guanine particles, in conjunction with a black melanin background, produces a blue colour. Purines make no contribution to the colour or external appearance of any plants or microorganisms.

### 6.2.4  Pterins

Pterins were first encountered as butterfly wing pigments, but have now become just as familiar as coloured constituents of many other insects, and also of crustaceans, amphibians and reptiles. A pterin is defined as a 2-amino-4-hydroxypteridine (6.9). The number of known animal pigments

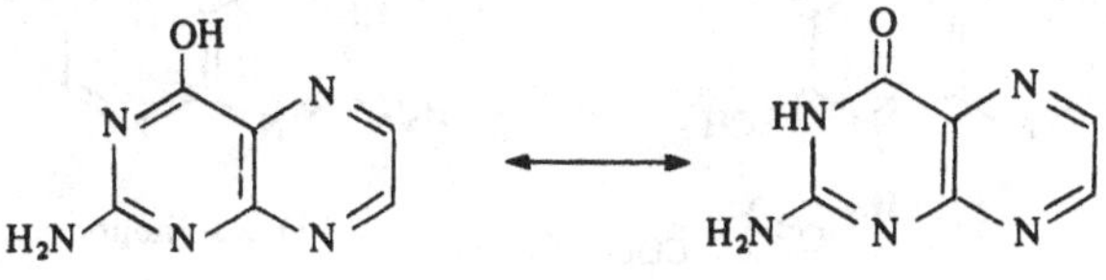

(6.9) Pterin = 2-amino-4-hydroxypteridine

with this basic skeleton is now large. The individual compounds differ mainly in the nature of the substituents present at C-6 and C-7 and also in the oxidation state of the nitrogen atoms at positions 5 and 8. In the Lepidoptera (butterflies and moths), pterins seem to be restricted largely, though not exclusively, to the Pieridae. The best known butterfly pterins include the white leucopterin (6.10) which is present in the cabbage white butterflies (*Pieris brassicae* and *P. rapae*), chrysopterin (6.11) the yellow pigment of the brimstone (*Gonepteryx rhamni*), and the red erythropterin (6.12) present in the orange tip, *Euchloë cardamines*. The pterins are present in crystalline form in the wing scales, and the yellow, orange and red compounds contribute those colours to the intricate patterns in the butterfly wings. Although it is the characteristic pterin of white butterflies, leucopterin does not serve as a white pigment; the white colour is structural in origin.

Although they were first recognised in the Lepidoptera, pterins are now also known to be common pigments in the Hymenoptera; xanthopterin (6.13) provides the yellow colour of the common wasp (*Vespula vulgaris*). Pterins are not normally present, however, in the integument of Diptera species – even dipteran syrphids (hoverflies) which mimic the external appearance of wasps do not employ pterins for their yellow colours – although these pigments may be present in the eyes. For example, eyes of the fruitfly *Drosophila melanogaster* contain a mixture of pterins, including the dimeric drosopterin (6.14). Pterins have been recognised as pigments in the integument and eyes of crustaceans and other arthropods, but these examples have been studied much less intensively than those of insects.

It is now known that many yellow, orange and red colours in fishes, and in amphibians and reptiles such as frogs, toads, salamanders and snakes, are

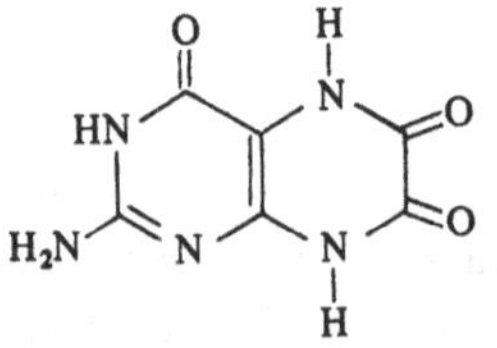

(6.10) Leucopterin

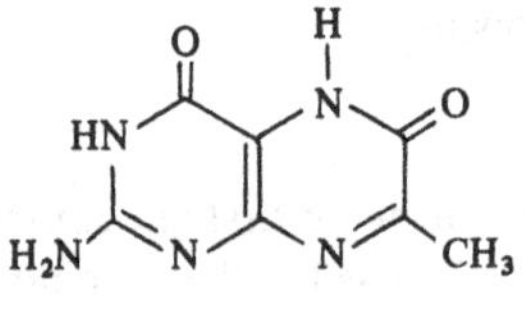

(6.11) Chrysopterin

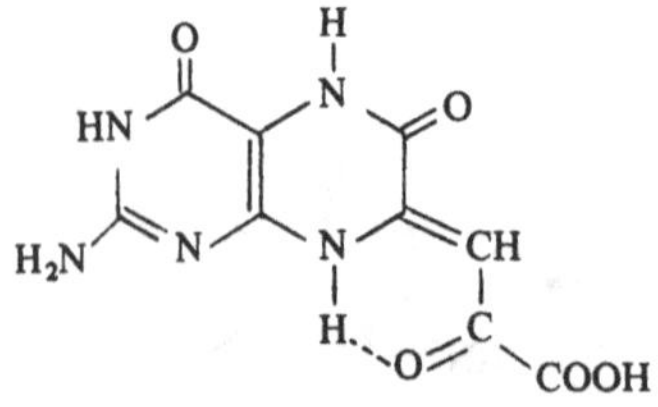

(6.12) Erythropterin

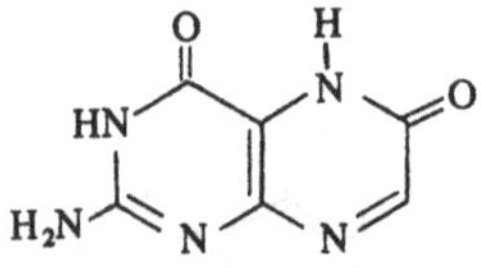

(6.13) Xanthopterin

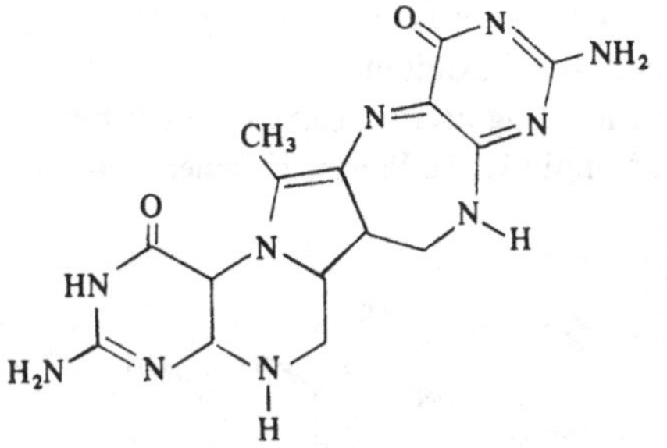

(6.14) Drosopterin

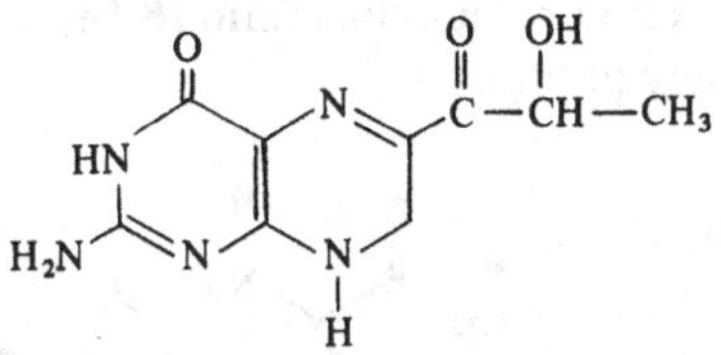

(6.15) Sepiapterin

due to pterins, which are mainly located, in granular form, in xanthophore and erythrophore cells (see chapter 8). The most common pterin pigments in these vertebrates are sepiapterin (6.15) and dimers of the drosopterin type.

Pterins sometimes serve in the eyes of vertebrates as light reflectors, a role which they very commonly fill in insects and other arthropods (see chapter 9).

Elsewhere in the animal kingdom, and in plants and microbes, coloration by pterins (or indeed by other pteridines) does not occur, though some are important in other ways. The pteridine derivative folic acid (6.16), which is synthesised by microorganisms, is an essential vitamin for man and most animals. Biopterin (6.17) is also of considerable biological importance, for example as a cofactor in the hydroxylation of phenylalanine to tyrosine.

*General properties.* The pterins (and the naturally occurring purines) are amphoteric molecules with weak acidic and basic groups. Thus although they are only sparingly soluble in water, they nevertheless dissolve in dilute acid or alkali. They may have moderate solubility in polar but not in non-polar

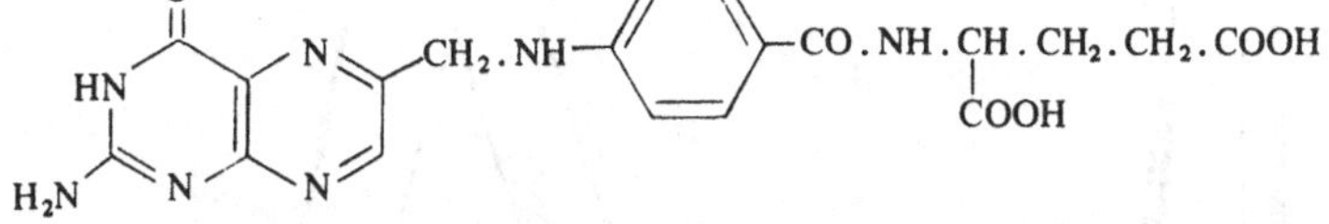

(6.16) Folic acid

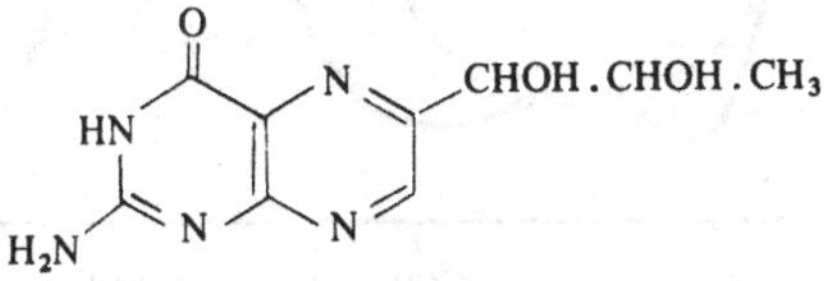

(6.17) Biopterin

organic solvents. Most pterins are potentially redox agents, but at physiological oxygen pressures they are normally fully oxidised.

Pteridines with oxygen substituents, including pterins, can exist in either a keto or quinonoid form **(6.18)** or as an enol **(6.19)**. Pterins in general are very photolabile.

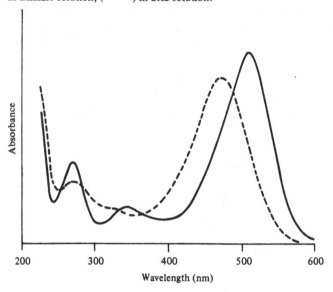

(6.18) Keto or quinonoid form          (6.19) Enol form

*Light-absorption properties.* Usually three (sometimes two) absorption peaks are seen in the absorption spectrum of a pterin but rarely is more than one of these in the visible region (fig. 6.2). The effect of substituent groups is seen from the comparison of the absorption maxima (in alkaline solution) of leucopterin **(6.10)** (240, 285, 340 nm), xanthopterin **(6.13)** (255, 391 nm), chrysopterin **(6.11)** (252, 385 nm), and erythropterin **(6.12)** (240, 310, 475 nm). The wavelengths of maximal light absorption by pterins are influenced by pH changes, as might be expected of amphoteric molecules. The

Fig. 6.2. Light absorption spectrum of a pterin, drosopterin (**6.14**); (———) in alkaline solution, (– – – –) in acid solution.

shifts in the maxima are usually not greater than *ca* 40 nm over the pH range 1–13.

The colours attributable to the various pterins in living tissues are usually more bathochromic than would be expected from the absorption maxima. For example xanthopterin with $\lambda_{max}$ at about 390 nm is only very faintly yellow *in vitro* but provides orange colour *in situ*. Conjugation with protein, or strong stabilisation in the keto (quinonoid) tautomeric form **(6.18)** may be the agency responsible for these bathochromic effects.

Most free pterins are strongly fluorescent in u.v. light, with emission bands usually at 150–200 nm longer wavelength than the corresponding absorption bands. This fluorescence is quenched *in situ* by protein conjugation.

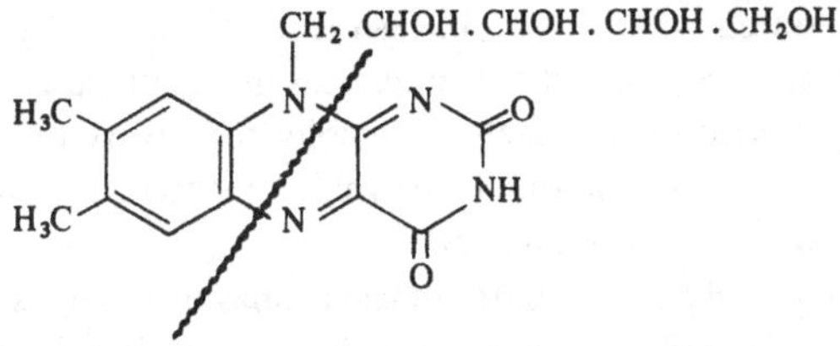

$$CH_2 . CHOH . CHOH . CHOH . CH_2OH$$

**(6.20)** Riboflavin

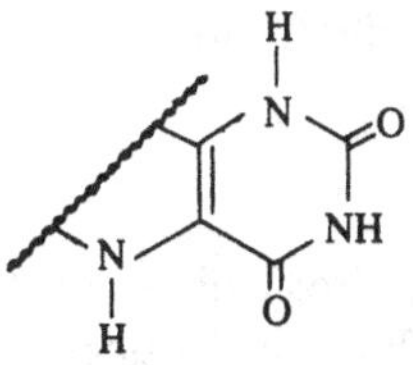

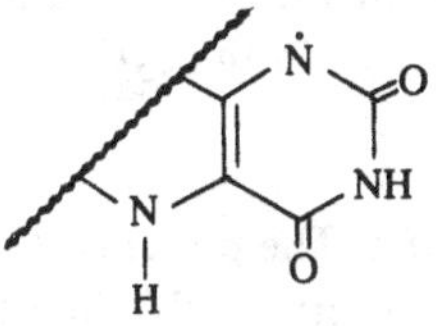

**(6.21)** Reduced riboflavin    **(6.22)** Semiquinone radical

*6.2.5    Flavins*

Flavins are based on the structure of isoalloxazine, which is essentially a pteridine (not a pterin) condensed with a benzene ring. The flavin structure has methyl substituents at positions 6 and 7 of this benzene ring and a further substituent attached to the heterocyclic nitrogen at position 9. In the only common biological example, riboflavin or ribitylflavin (6,7-dimethyl-9-ribitylisoalloxazine, **6.20**), this substituent is the sugar alcohol, ribitol. The structure of the reduced form **(6.21)** is also given.

*Occurrence and distribution.* Riboflavin is of extreme biological importance as the functional part of flavoproteins and the two coenzymes, flavin mononucleotide, FMN (= riboflavin phosphate) and flavin adenine dinucleotide, FAD, which are utilised in many redox reactions. The reduced forms of these coenzymes, $FMNH_2$ and $FADH_2$, have the reduced riboflavin **(6.21)**. Flavins and flavin coenzymes are common as firmly bound prosthetic groups of many enzymes, notably respiratory enzymes. Riboflavin occurs universally in all living organisms as an essential component of these coenzymes and enzymes. It is thought that it cannot be synthesised by animals and must be supplied in the diet, *i.e.* it is a vitamin ($B_2$). Plants and microorganisms synthesise the riboflavin that animals demand. The production of vitamin $B_2$ by yeasts has been exploited commercially for many years.

The endogenous synthesis of riboflavin in a few insects, such as the cockroach *Periplaneta*, has been reported, but it seems likely that microbial endosymbionts are responsible.

Although of ubiquitous occurrence in Nature, riboflavin rarely contributes to the external colour of living organisms – never to that of higher plants. Those microorganisms that are used for the commercial production of riboflavin may be coloured yellow by it, but generally these are artificially produced mutant strains and the yellow colour has no significance. Riboflavin also occasionally contributes to the yellow colour in invertebrates such as leeches and worms and may be the major yellow pigment of the integument of the sea-cucumber *Holothuria forskali*.

Riboflavin commonly accumulates in the retinas of the eyes of vertebrates. Good examples are provided by animals such as the bush baby, *Galago*, in which golden-yellow crystals of riboflavin constitute the reflective tapetum lucidum behind the retina. Unlike the pterins, riboflavin is not found in appreciable amounts in the eyes of arthropods.

*General properties.* The presence of the polyhydroxy ribitol substituent renders riboflavin very soluble in water. The flavin nucleotides containing additional phosphate and, in the case of FAD, sugar groups, have even greater water-solubility. Free riboflavin is soluble in polar organic solvents such as acetone and alcohols but not in chloroform. Riboflavin is readily reduced, and the reduced form readily reoxidised. This behaviour is utilised in the biological functioning of the flavin coenzymes and in the electron transport chain (see chapter **10**). The reduction occurs in two single-electron steps, *i.e.* it proceeds *via* a semiquinone radical (**6.22**).

*Light-absorption properties.* Riboflavin has light absorption maxima at 223, 267, 373, 445 and 475 nm (fig. 6.3) and is thus bright yellow in solution. The similarity of this spectrum to that of $\beta$-carotene (chapter 2) has led to much confusion over which of these pigments, occurring in minute amounts, is the true primary receptor for a number of photoresponses (see chapter **11**).

U.v. light induces strong fluorescence in the region 520–565 nm. This fluorescence is quenched in the flavin nucleotides and by other aromatic structures, including aromatic amino acid residues of proteins.

Fig. 6.3. Light absorption spectrum of riboflavin at pH 7.0.

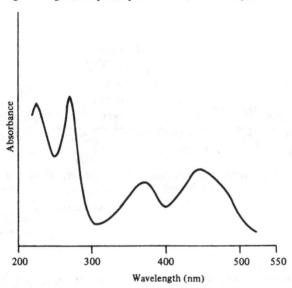

### 6.2.6 *Biosynthesis of purines, pterins and riboflavin*

The same basic pathway is used for the biosynthesis of all these groups of compounds; pterins and riboflavin are both biosynthesised *via* purine intermediates. In the description which follows, therefore, the biosynthetic pathway to the purine guanine will be presented first, and then the reactions by which guanine (as guanosine-9-triphosphate, GTP) can lead to either pterins or riboflavin will be outlined.

*Formation of guanine.* The purines guanine (**6.5**) and adenine (**6.4**) occur universally in the nucleic acids of all living organisms. Their biosynthesis has therefore been studied extensively and the pathway by which the purine ring system is built up from small fragments is very well understood. All living organisms synthesise these purines by essentially the same pathway (fig. 6.4).

The purine ring system is constructed on a molecule of ribose-5-phosphate (**6.23**). This is first activated by conversion into $\alpha$-5-phosphoribosyl-1-pyrophosphate (**6.24**) by a kinase enzyme and ATP. The pyrophosphate group of (**6.24**) is then replaced by an amino group to give 5-phosphoribosyl-1-amine (**6.26**). The amino group transferred originates as the amide $-NH_2$ of glutamine (**6.25**); this nitrogen constitutes the first part of the purine ring system. Inversion of configuration at C-1 of the ribose occurs during this reaction. The $\beta$-configuration thus produced is then retained throughout the remainder of the biosynthetic pathway.

Fig. 6.4. Main features of the pathway of biosynthesis of guanine and other purines.

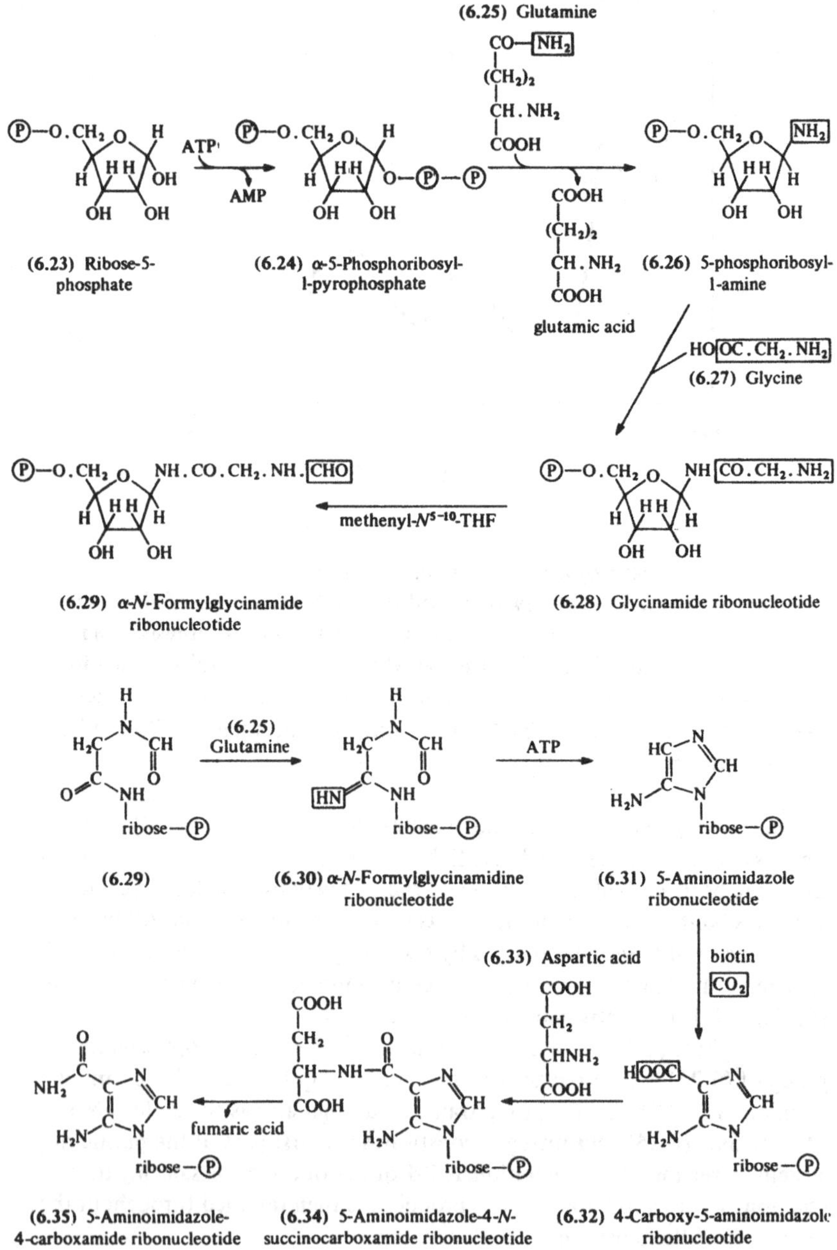
(6.25) Glutamine
ATP
AMP
(6.23) Ribose-5-phosphate
(6.24) α-5-Phosphoribosyl-1-pyrophosphate
glutamic acid
(6.26) 5-phosphoribosyl-1-amine
(6.27) Glycine
methenyl-$N^{5-10}$-THF
(6.29) α-N-Formylglycinamide ribonucleotide
(6.28) Glycinamide ribonucleotide
(6.29)
(6.30) α-N-Formylglycinamidine ribonucleotide
(6.31) 5-Aminoimidazole ribonucleotide
biotin
(6.33) Aspartic acid
fumaric acid
(6.35) 5-Aminoimidazole-4-carboxamide ribonucleotide
(6.34) 5-Aminoimidazole-4-N-succinocarboxamide ribonucleotide
(6.32) 4-Carboxy-5-aminoimidazole ribonucleotide

Fig. 6.4. *contd.*

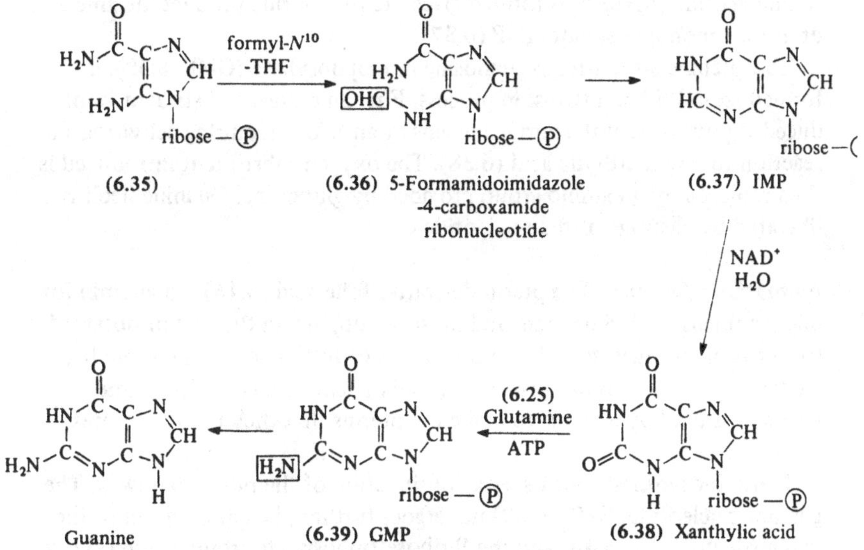

(6.35)  (6.36) 5-Formamidoimidazole -4-carboxamide ribonucleotide  (6.37) IMP

NAD⁺ H₂O

Guanine  (6.39) GMP  (6.38) Xanthylic acid

The amino acid glycine (6.27) is then added on to the C-1 amino group of the phosphoribosylamine, *via* an amide linkage. ATP is required, and the product is known as glycinamide ribonucleotide (6.28). A formyl (CHO) group is next transferred to the free amino group of (6.28) from the coenzyme methenyl-$N^{5-10}$-tetrahydrofolic acid (interestingly, itself a pterin formed by this pathway), in the presence of a transformylase enzyme, to produce formylglycinamide ribonucleotide (6.29).

At this point all the atoms that are to form the imidazole ring of the purine nucleus have been attached to the phosphoribose. Before the imidazole ring is closed, however, a further nitrogen transfer occurs, again from glutamine. The glycinamide (6.29) amide oxygen is replaced by =NH, which will eventually constitute one of the pyrimidine-ring nitrogens of the purine. It is the product of this reaction, α-$N$-formylglycinamidine ribonucleotide (6.30), which undergoes ring closure by an ATP-requiring dehydration process. The imidazole nucleus of the 5-aminoimidazole ribonucleotide (6.31) thus formed is then carboxylated with $CO_2$ (biotin required) to give 4-carboxy-5-amino-imidazole ribonucleotide (6.32).

Aspartic acid (6.33) is used to provide the second nitrogen atom of the pyrimidine ring. The aspartate is first attached by an amide link to the imidazole carboxy-group. The succinocarboxamide derivative (6.34) produced is then cleaved in an aspartase type of reaction to leave only the aspartate amino group as part of the imidazole carboxamide product (6.35). Formyl-$N^{10}$-tetrahydrofolic acid then provides the final carbon atom of the purine nucleus and ring closure of the 5-formamidoimidazole-4-carboxamide ribonucleotide (**6.36**) thus formed gives the first purine product inosinic acid or inosine monophosphate, IMP (**6.37**).

The guanine structure, as guanosine monophosphate (GMP, **6.39**), is formed from IMP in a two-step process. First an oxygen substituent is introduced at position 2 of the purine nucleus by an $NAD^+$-requiring dehydrogenase reaction to give xanthylic acid (**6.38**). The oxygen substituent introduced is then replaced by an amino group provided by glutamine. Guanine itself is liberated by cleavage of the nucleotide GMP.

*Formation of pterins.* The pterin derivative folic acid (**6.16**) is a vitamin for many animals, including man, and must be supplied in the diet or obtained from gut microorganisms. The biosynthesis of folic acid has therefore been studied extensively in bacteria, particularly *Escherichia coli*. The same pathway is used by animals to make the pterins that they use for coloration purposes.

Pterin biosynthesis occurs by a continuation of the purine pathway. The guanine nucleotide, GMP (**6.39**) undergoes further phosphorylation to the triphosphate, GTP (**6.40**) and the 9-ribosetriphosphate group of the GTP is used in the enlargement of the imidazole to a pyrazine ring. Details of the ring enlargement are given in fig. 6.5. (*N.B.* It is important to remember that the numbering schemes used in the purine and pteridine series are different, see §6.2.2.) The imidazole ring of the GTP is cleaved and carbon atom 8 is lost. Carbon atoms 1 and 2 of the triphosphorylated ribose then become the remaining two carbons of the pyrazine ring, *i.e.* C-6 and C-7 of the pterin. The 2-amino and 4-hydroxy (or keto) substituents are in place, so the first pteridine product is already a pterin, 7,8-dihydroneopterin triphosphate (**6.41**). This is dephosphorylated to give 7,8-dihydroneopterin itself (**6.42**).

7,8-Dihydroneopterin is the key compound which gives rise by various branching pathways to folic acid, biopterin and the many animal pterin pigments. There is still uncertainty about the subsequent branches and interconversions, but the interrelationships that seem likely on current evidence are outlined in fig. 6.6. The main features of this scheme are: (*i*) the C-6 sidechain undergoes progressive shortening and modification, and new substituents are introduced at C-7, and (*ii*) the 7,8-dihydropterin structure may be dehydrogenated to the more common fully aromatic pterins at various stages during the sidechain modifications. All the natural pterins and dihydropterins could be produced by these various transformations, but the biochemical details of most of the postulated reactions have not been fully worked out.

It is probable that the dimerisations to produce pterins such as drosopterin (**6.14**) occur late in the biosynthetic sequence, but again no detailed studies have been reported.

Fig. 6.5. Mechanism of ring enlargement by which pterins are formed from purines.

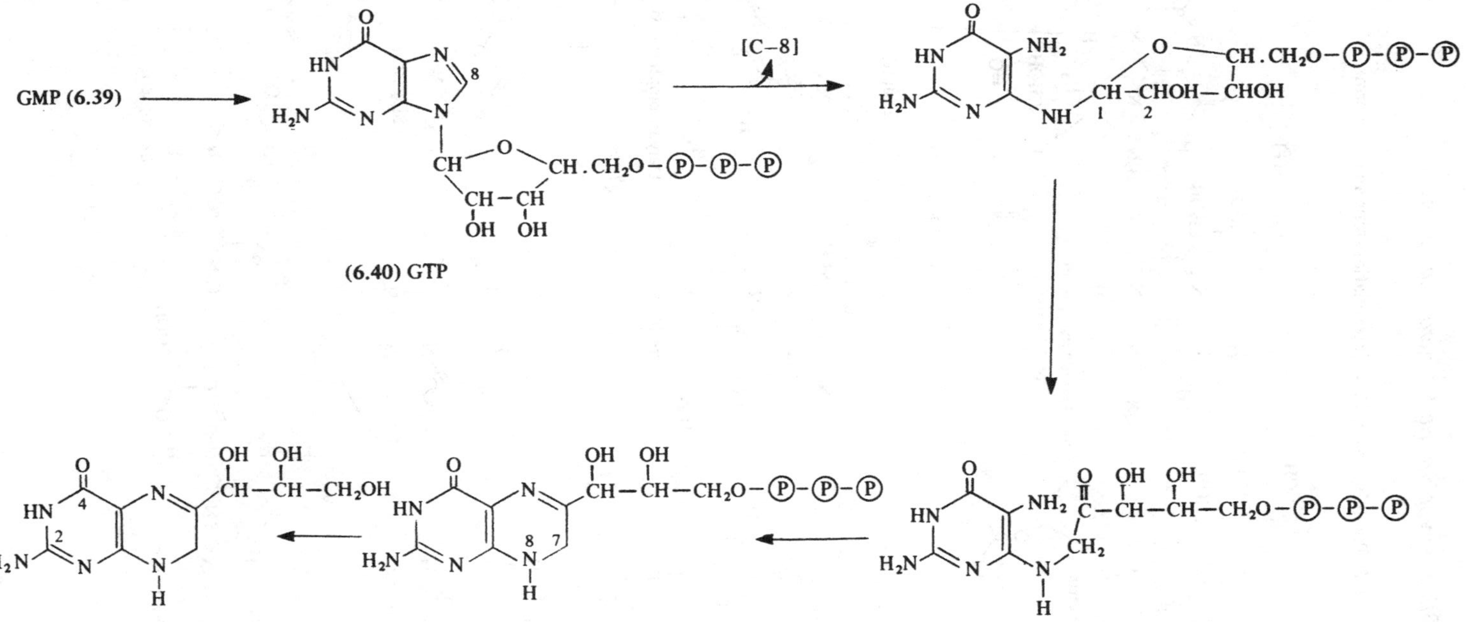

GMP (6.39)

(6.40) GTP

(6.41) 7, 8-Dihydroneopterin triphosphate

(6.42) Dihydroneopterin

Fig. 6.6. Possible biosynthetic interrelationships among pterins.

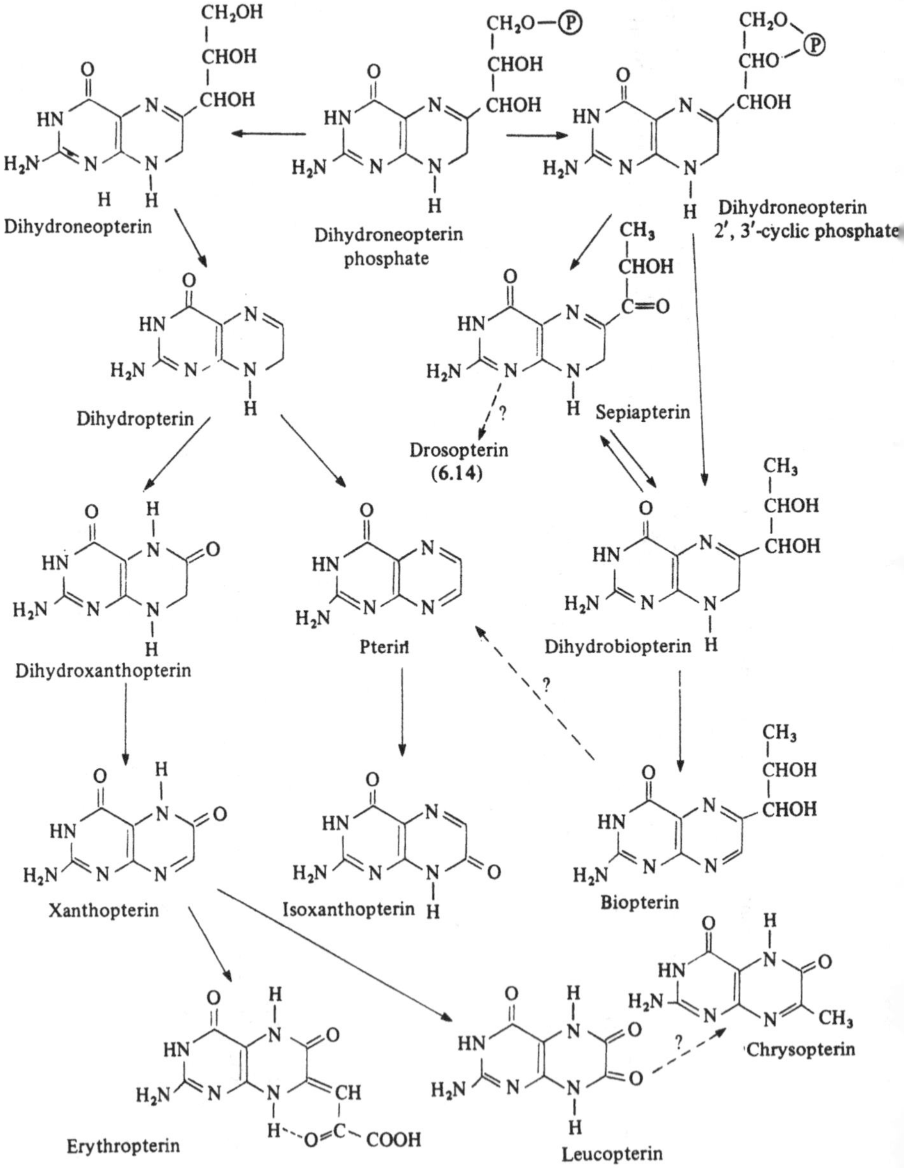

Fig. 6.7. Outline of the pathway of biosynthesis of riboflavin from GTP
(-ribityl = —CH.CHOH.CHOH.CHOH.CH$_2$OH).

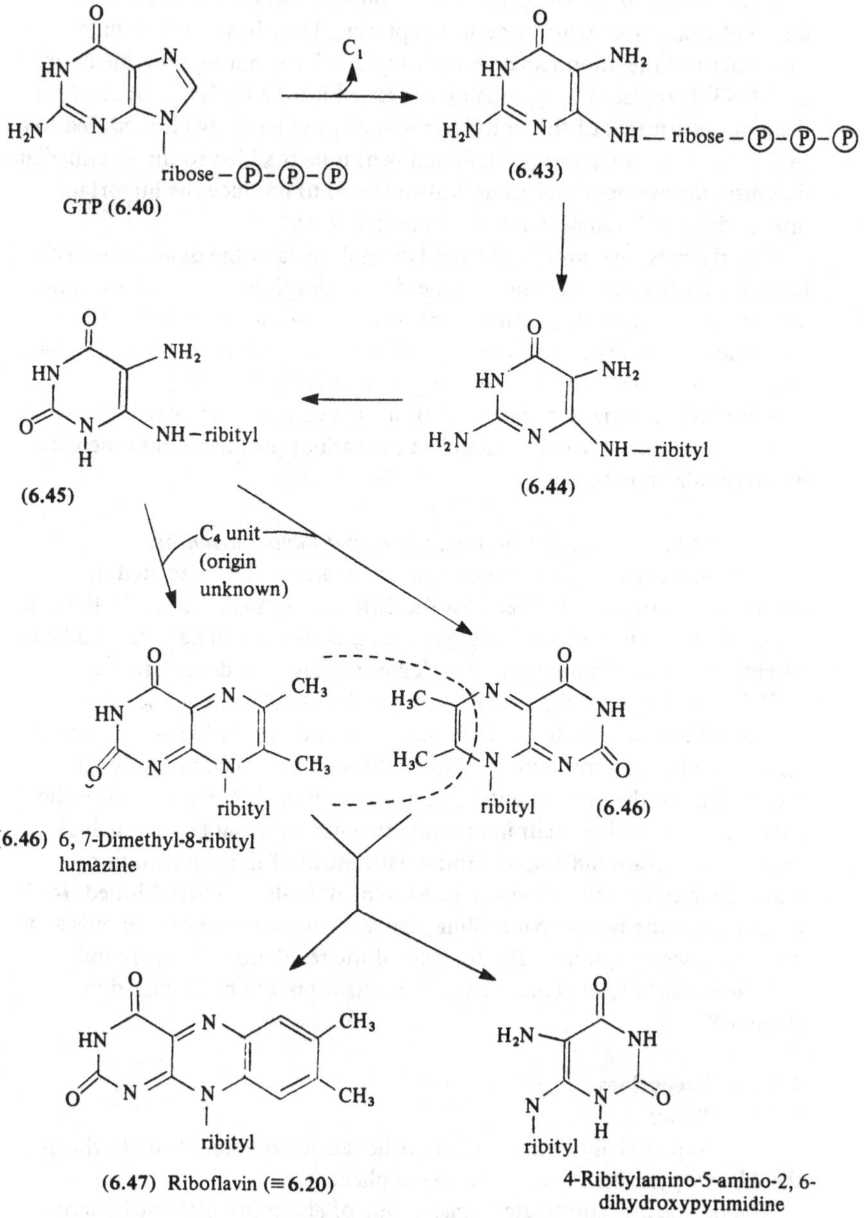

GTP (6.40)

(6.43)

(6.45)

(6.44)

C$_4$ unit
(origin
unknown)

(6.46) 6, 7-Dimethyl-8-ribityl
lumazine

ribityl (6.46)

(6.47) Riboflavin (≡ 6.20)

4-Ribitylamino-5-amino-2, 6-
dihydroxypyrimidine

*Biosynthesis of riboflavin (fig. 6.7).* The biosynthesis of riboflavin in plants and especially in microorganisms also involves cleavage of the imidazole ring of GTP and loss of the C-8 carbon atom. Subsequent events, though, are different from those which lead to the pterins. The ribose triphosphate substituent of the intermediate 6-hydroxy-2,4,5-triaminopyrimidine derivative (6.43) is replaced by (probably converted into) a ribityl group, and the 2-amino-substituent of the ribityl pyrimidine product (6.44) is modified to a hydroxy-group. A $C_4$ unit (origin unknown) is next added to the intermediate dihydroxydiaminopyrimidine derivative (6.45) to produce the important intermediate 6,7-dimethyl-8-ribityllumazine (6.46).

Riboflavin is then produced from two molecules of the dimethylribityl-lumazine by the action of the enzyme riboflavin synthetase. In this reaction a $C_4$ moiety, comprising carbon atoms 6 and 7 and their attached methyl substituents, is transferred from one molecule of the lumazine to the other to complete the dimethylbenzene ring of the riboflavin (6.47).

Riboflavin-deficient mutant strains of the yeast *Saccharomyces cerevisiae* have been produced which accumulate the various postulated intermediates in the riboflavin pathway.

### 6.2.7    *Factors controlling purine, pterin and flavin biosynthesis*

Although the biosynthetic pathways are so closely related, the controlling factors are different for the different pigment classes. In the case of riboflavin synthesis by microorganisms, genetic control has been studied extensively and end-product feedback control has been demonstrated.

Purine and especially pterin synthesis and deposition in animals are influenced by many factors. These substances are usually located in specific pigment cells, the xanthophores and erythrophores, which not only are responsible for the integumental colour of the animal, but also control the colour changes which occur in response to environmental factors such as changes in background colour. Hormonal control of the pigmentation, especially during maturation and development, is also well established. In all these aspects the factors controlling pigmentation embrace carotenoids and melanin as well as pterins. The subjects of the regulation of colour and pattern in animals, and colour change mechanisms will be discussed in chapter 8.

### 6.3    Phenazines
### 6.3.1    *Structures*

Superficially similar to the isoalloxazine skeleton of the flavins is the dibenzopyrazine structure (6.48) of phenazine.

The phenazines constitute a small group of about 30, often strikingly coloured, pigments produced exclusively by bacteria. The substituents most frequently present are hydroxy- and carboxy-groups at positions 1 and 6,

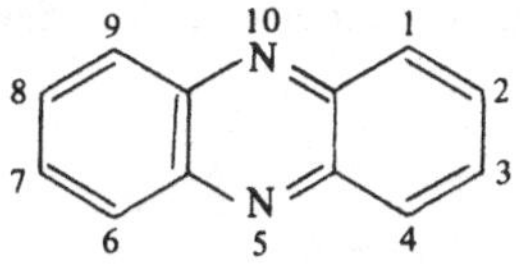

(6.48) Basic dibenzopyrazine
skeleton of phenazines

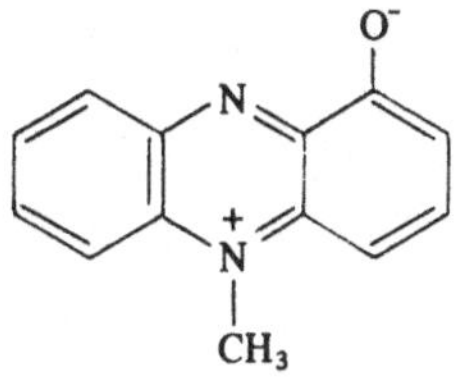

(6.49) Pyocyanine

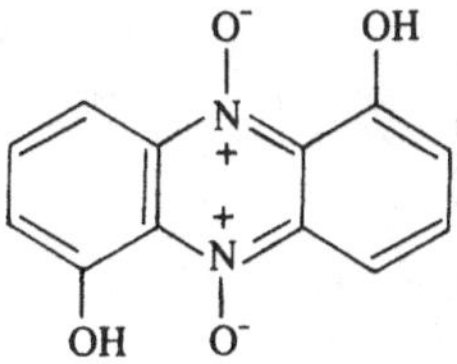

(6.50) Iodinin

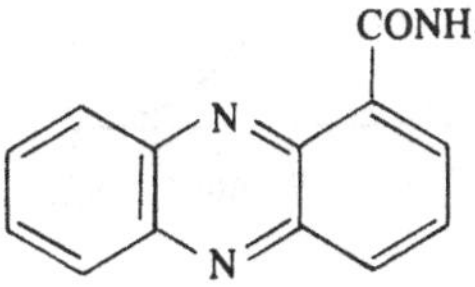

(6.51) Oxychlororaphine

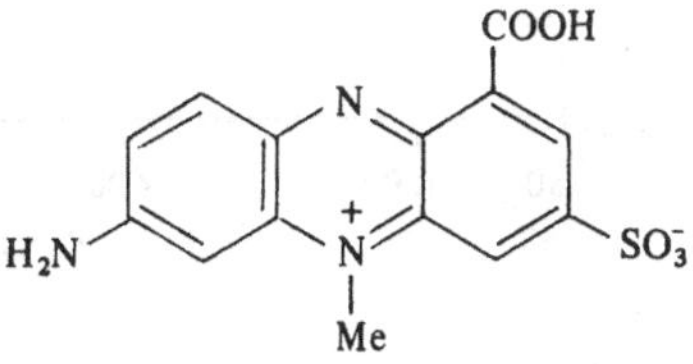

(6.52) Aeruginosin B

and substitution of the pyrazine ring nitrogens by oxygen or a methyl group is also common. The best-known natural phenazines are probably pyocyanine (6.49) and the phenazine-*N*-oxide iodinin (6.50). Other examples are oxychlororaphine (6.51) and aeruginosin B (6.52), which contain carboxamide and sulphonic acid substituents, respectively.

### 6.3.2 Properties

Many of the phenazines, especially the carboxylic acids, are appreciably water-soluble and accumulate in the culture medium. Often such high concentrations are produced that the pigments are then precipitated as solid deposits. Most naturally occurring phenazines have only limited solubility in organic solvents, although many may be extracted from acidified aqueous solution with chloroform.

Phenazines in general are intensely coloured and provide a wide diversity of hues. Usually there are several absorption bands in the ultraviolet and at least one main band in the visible region (400–600 nm) to which the phenazines owe their colours (fig. 6.8). Most phenazines are yellow ($\lambda_{max}$ 400–450 nm), but iodinin is purple ($\lambda_{max}$ 530 nm) and pyocyanine blue

Fig. 6.8. Light absorption spectrum of a phenazine, iodinin (6.50), in chloroform.

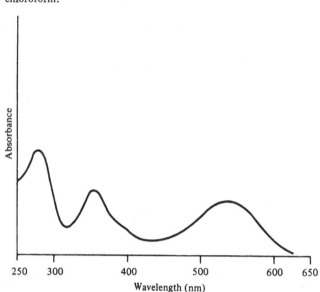

($\lambda_{max}$ 695 nm). Oxychlororaphine and its dihydro-derivative exist in the form of a green $\pi$-complex.

### 6.3.3   Distribution

The phenazines are of restricted distribution being limited almost entirely to some *Pseudomonas*, *Streptomyces*, *Brevibacterium* and *Nocardia* species. Occasionally, however, phenazine colour may be seen in animal tissues. *Pseudomonas aeruginosa* is a common microbial parasite of the skin of humans and other animals, and pyocyanine produced by this organism may on occasions colour blue the pus of infected wounds. A blue-green colour sometimes noticeable in sheep's wool has also been attributed to pyocyanine from *P. aeruginosa*.

### 6.3.4   Biosynthesis

Phenazine biosynthesis has been studied most intensively in *Pseudomonas aeruginosa* and *P. phenazinium* which produce mainly pyocyanine (6.49) and iodinin (6.50), respectively. The main features of the pathway are apparent and have been deduced mainly from the results of incorporation experiments with labelled postulated precursors, and from studies of the accumulation of possible intermediates in mutant strains. Details of most of the individual reactions and the enzymes which catalyse them have not yet been established.

The biosynthesis of phenazines is yet another branch of the great shikimic acid pathway of aromatic-compound biosynthesis (see chapters **3** and **4**). The postulated pathway for the formation of the phenazine nucleus is outlined in fig. 6.9. Shikimic acid **(6.53)** is efficiently incorporated into phenazines. The origin of the nitrogen atoms is not known, but a nitrogen-substituted shikimate intermediate (still unidentified) such as **6.54** has been suggested. An alternative route *via* chorismic acid **(6.55)** and a nitrogen-substituted derivative could be involved. Anthranilic acid **(6.56)** seems an obvious candidate for a precursor, but is now known not to be incorporated into phenazines.

Whatever the mechanism of its formation may be, it now seems clear that phenazine-1,6-dicarboxylic acid **(6.57)** is the first phenazine product, and precursor to the other phenazines by a series of branching or alternative pathways, some of which are illustrated in fig. 6.10. It is believed that the hydroxy-groups at C-1 and C-6 are introduced by direct oxidative displacement of the carboxy-groups, whereas hydroxy-groups in other positions arise by conventional aromatic hydroxylation. *N*-Oxidation to give the *N*-oxide structure in iodinin is an enzymic process; the enzyme involved can appar-

Fig. 6.9. Possible routes for formation of the basic phenazine structure from shikimic acid.

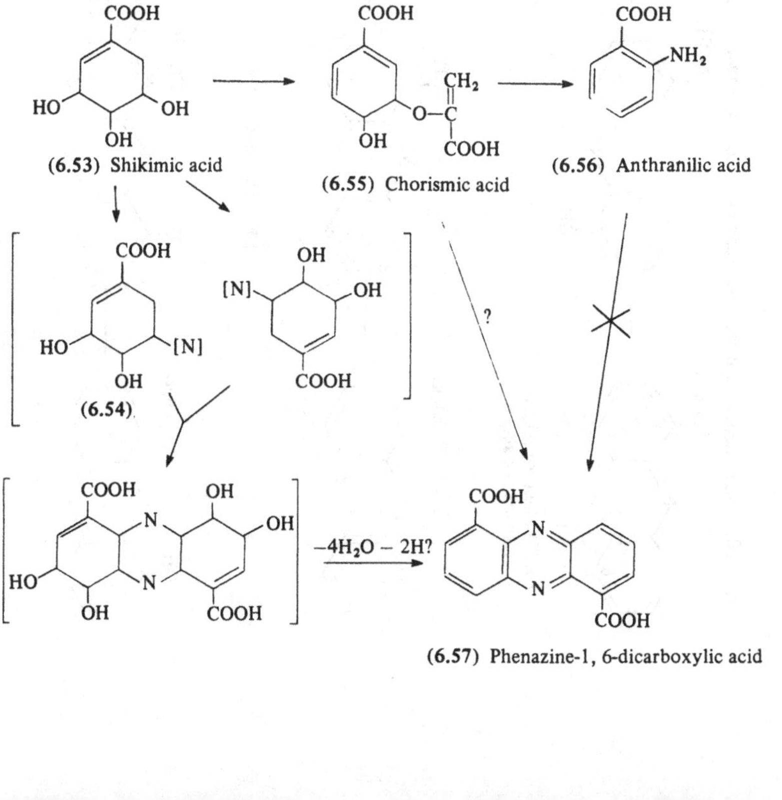

(6.53) Shikimic acid

(6.55) Chorismic acid

(6.56) Anthranilic acid

(6.54)

$-4H_2O - 2H$?

(6.57) Phenazine-1, 6-dicarboxylic acid

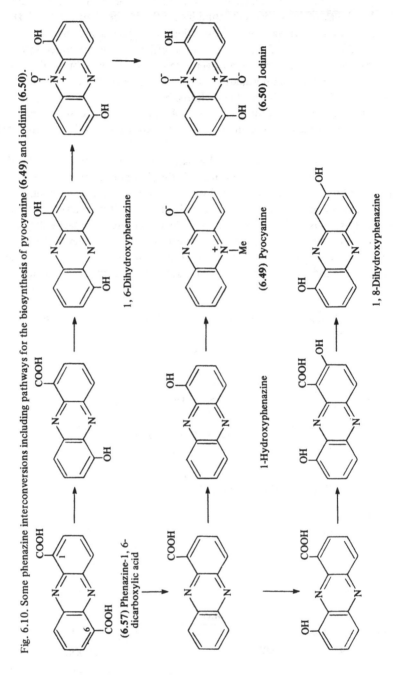

Fig. 6.10. Some phenazine interconversions including pathways for the biosynthesis of pyocyanine (6.49) and iodinin (6.50).

ently utilise a range of phenazines as substrates. The $N$-methyl substituent of pyocyanine arises conventionally from $S$-adenosylmethionine.

Phenazine production by bacteria is greatly dependent on the culture medium and growth conditions, but details of the controlling and regulating mechanisms remain to be elucidated.

### 6.3.5 Biological activity

There are various ways in which phenazines appear to affect other living organisms and tissues. They were, for example, the first bacterial products to be shown to exert antibiotic activity against other microbes. The bacteriostatic properties of iodinin and pyocyanine are now well known, and appear to be related to the interaction of the phenazines with DNA, presumably by intercalation of the planar aromatic ring system. There have also been reports that phenazine di-$N$-oxides may have carcinostatic activity.

The redox properties of phenazines may also have some biological significance. Pyocyanine inhibits brain succinate dehydrogenase although some other dehydrogenases are activated. The active absorption and concentration of phenazines by animal tissues has been reported, but it is not known whether this is fortuitous or has any physiological significance.

### 6.4 Phenoxazines

The phenoxazine ring system (**6.58**) is structurally very similar to phenazine. The best-known phenoxazine pigments are the ommochromes.

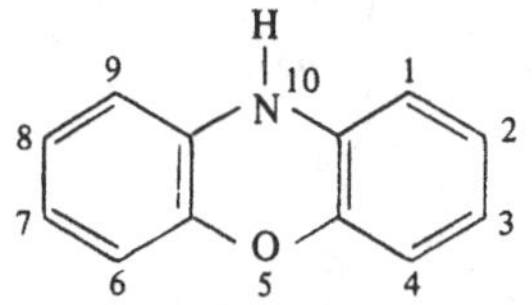

(**6.58**) Phenoxazine ring system

### 6.4.1 Ommochromes

*Structures.* These dark-coloured substances were first extracted from the ommatidia of arthropod compound eyes, and were originally thought to be melanins. There are two main sub-groups, the **ommatins** and **ommins**, which are dimers and oligomers, respectively, of kynurenine (**6.59**) derivatives; other sub-groups have been recognised but not fully characterised. The most common ommochrome is the yellow xanthommatin (**6.60**), and this readily undergoes reduction to the red dihydroxanthommatin (**6.61**). These compounds are examples of the basic ommatin structure, which consists of the phenoxazine ring system with two aspartic acid sidechains, one of which is cyclised to form the fourth ring. The ommins, on the other hand, are larger,

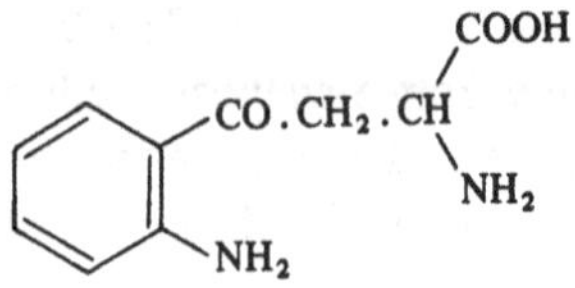

(6.59) Kynurenine

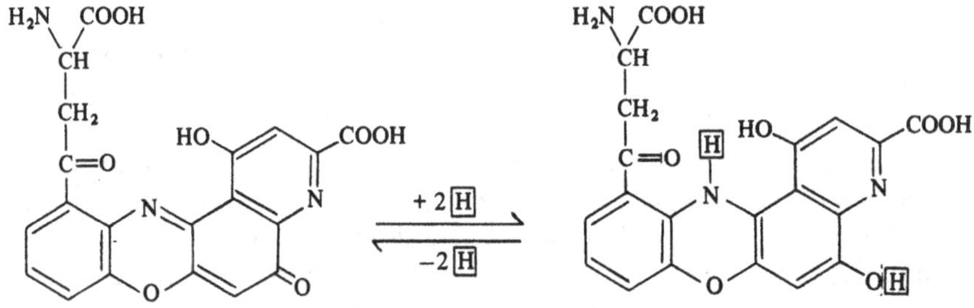

(6.60) Xanthommatin        (6.61) Dihydroxanthommatin

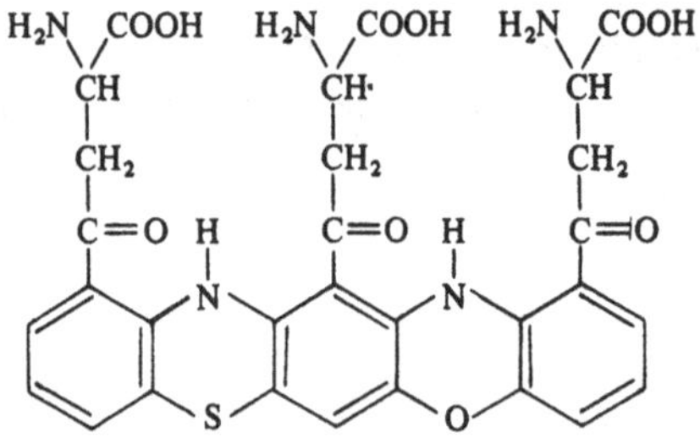

(6.62) Ommin A

non-dialysable molecules, and contain sulphur. The structure of a relatively simple, trimeric ommin, ommin A, is illustrated (6.62).

In the invertebrate eye, the ommochromes exist in granules as complexes with protein.

*Light-absorption properties.* The ommochromes show strong absorption in the ultraviolet as well as characteristic absorption in the visible region at around 440–500 nm (fig. 6.11), the wavelength varying with pH. *In vivo*, both ommatins and ommins appear dark-coloured and may be yellow-brown, purple or almost black, hence the confusion that has arisen between them and melanins.

Fig. 6.11. Light absorption spectra (in weakly acidic methanol) of xanthommatin (——) and dihydroxanthommatin (– – – –).

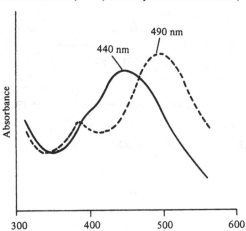

*Occurrence and distribution.* Ommochromes are characteristic pigments in the eyes of insects and other arthropods. They function in the eyes not as photoreceptors but as screening pigments which prevent the photoreceptors from being affected by stray light (see chapter 9). They are also widespread in the integuments of arthropods and other invertebrates, notably cephalopods, and have been found in the eggs and various tissues of some worms. Their occurrence in some vertebrates has been suggested but never proved.

*Biosynthesis.* Much information about the biosynthesis of the ommochromes has been obtained from detailed genetic studies with the fruitfly *Drosophila melanogaster*, the wild-type of which has xanthommatin as the brown pigment in its eyes. Many mutants of *Drosophila* have been produced which have abnormal eye colours due to lesions in the ommochrome biosynthetic pathway, so that intermediate compounds accumulate.

The pathway for the biosynthesis of xanthommatin is outlined in fig. 6.12. Although the essential features of the pathway have been established, details of the individual reactions are in most cases not well understood. The amino acid tryptophan (6.63) undergoes ring opening to give formylkynurenine (6.64) which has an aspartate sidechain. Spontaneous deformylation gives kynurenine itself (6.59), and this is hydroxylated to 3-hydroxykynurenine (6.65). This compound undergoes oxidative dimerisation to generate the oxazine bridge of the ommatin structure (6.66) and the fourth ring is then formed by cyclisation of an aspartate sidechain onto the amino substituent of the second hydroxykynurenine molecule. The ommins are also synthesised from 3-hydroxykynurenine, but not *via* an intermediate ommatin structure.

Fig. 6.12. A pathway for xanthommatin biosynthesis.

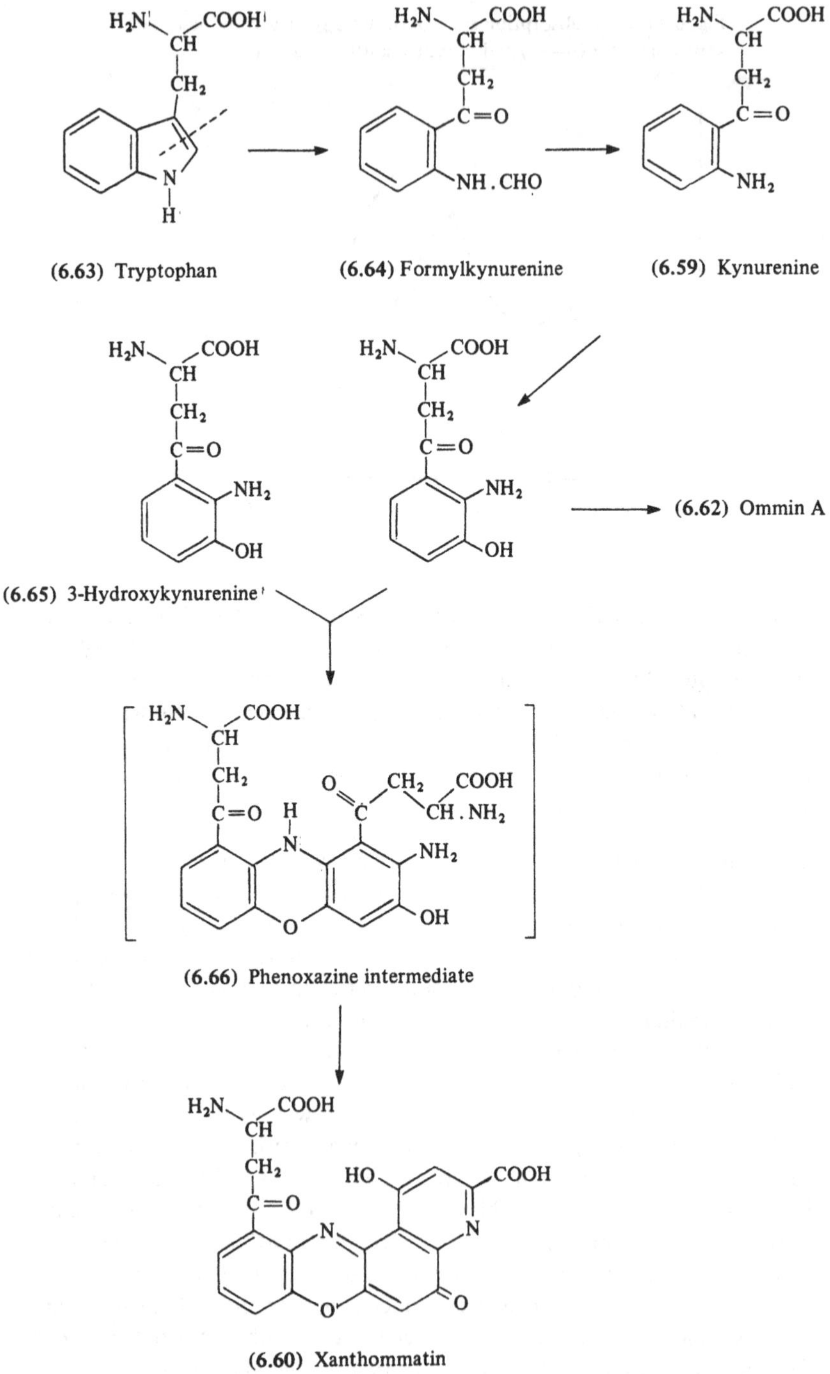

(6.63) Tryptophan      (6.64) Formylkynurenine      (6.59) Kynurenine

(6.65) 3-Hydroxykynurenine

(6.62) Ommin A

(6.66) Phenoxazine intermediate

(6.60) Xanthommatin

Related to the ommochromes are the papiliochromes, white and yellow compounds found only in the wings of butterflies of the Papilionidae. These pigments are also derivatives of kynurenine, probably complexed with non-nitrogenous quinones. Kynurenine and hydroxykynurenine themselves sometimes serve as yellow pigments.

*Functions.* No functions are known for ommochromes except their roles in external coloration, and as screening pigments in the invertebrate eye.

### 6.4.2 Microbial phenoxazines. Actinomycins

The phenoxazinone chromophore (6.67) is also present in several microbial pigments, many of which have important antibiotic properties. Best known amongst these are the actinomycins (6.68), red antibiotics produced by the mould *Streptomyces*. It is highly likely that the biosynthesis of these microbial pigments involves coupling of 3-hydroxyanthranilic acid (6.69) or its derivatives by a process similar to that which produces xanthommatin from 3-hydroxykynurenine.

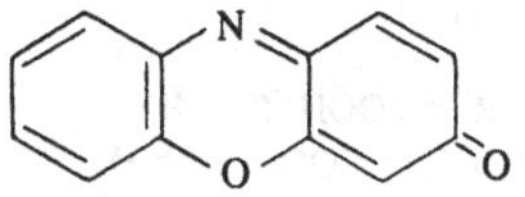

(6.67) Phenoxazinone chromophore

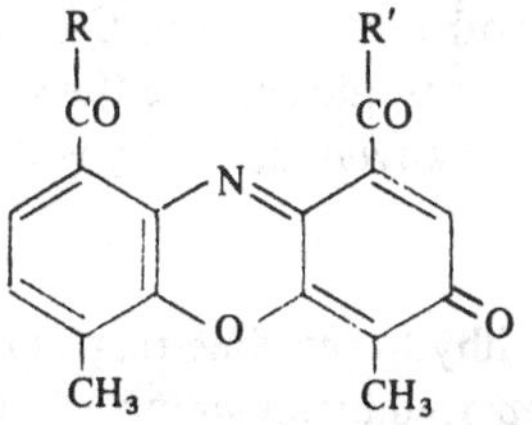

(6.68) Actinomycins (R, R' are pentapeptide chains) 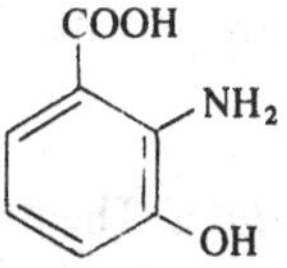     (6.69) 3-Hydroxyanthranilic acid

## 6.5 Betalains
### 6.5.1 Introduction

The betalains are exclusively plant pigments. They have a very limited distribution, yet are extremely familiar in the deep red-purple colour of beetroot. They are water-soluble and show a superficial similarity to the anthocyanins; the occurrence of the two groups may be mutually exclusive. Before about 1960 virtually nothing was known of their chemistry or bio-chemistry, but it is now known that there are two main groups, the red-violet betacyanins and the yellow betaxanthins.

### 6.5.2    Structures

*Betacyanins.* With one exception, all the naturally occurring beta-cyanins are based on only two aglycones, betanidin (**6.70**) and its C-15 epimer isobetanidin (**6.71**). The single exception is 2-decarboxybetanidin (**6.72**), so far found in only one flower species (*Carpobrotus acinaciformis*). The basic structure thus consists of two *N*-heterocyclic systems, a dihydro-indole and a dihydropyridine, linked by a conjugated $C_2$ unit.

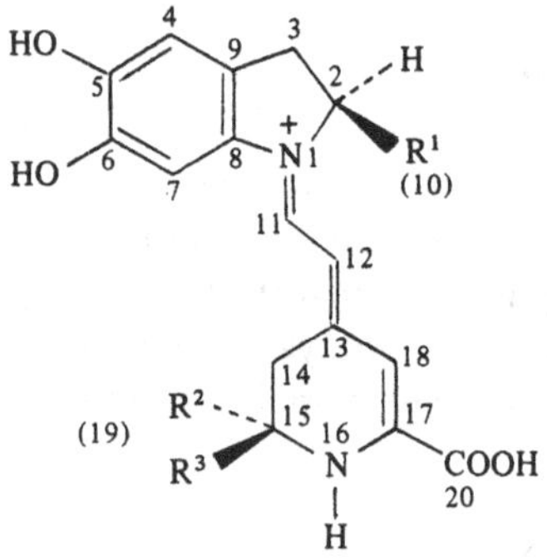

(**6.70**) Betanidin ($R^1 = R^2 = COOH; R^3 = H$)
(**6.71**) Isobetanidin ($R^1 = R^3 = COOH; R^2 = H$)
(**6.72**) Carboxybetanidin ($R^1 = R^3 = H; R^2 = COOH$)

In most cases the natural betacyanins are glycosides with a mono- or disaccharide attached to either (but never both) of the C-5 or C-6 hydroxy-groups of the dihydroindole. Glucose and glucuronic acid are the sugars most commonly present. The sugar residues may be acylated, usually with cinnamic acids.

*Betaxanthins.* The betaxanthins retain the dihydropyridine ring and the linking $C_2$ unit, but the dihydroindole is replaced by an amino acid or amine group, *e.g.* proline in indicaxanthin (**6.73**). Other examples have hydroxy-proline, aspartic acid, glutamic acid or glutamine, methionine, dihydroxy-phenylalanine or mono- or dihydroxyphenylethylamine in place of proline. In all cases, linking to the common unit occurs through the *N*-atom of the amino acid. The numbering schemes cause much confusion. Different schemes are used for the different betaxanthin aglycones, since numbering begins with the amine nitrogen of the amino acid or amine moiety. All beta-xanthin schemes differ from the betacyanin numbering.

### 6.5.3    Properties

Both the aglycones and the glycosides are water-soluble, ionic (acidic) compounds. They are optically active and the absolute configurations

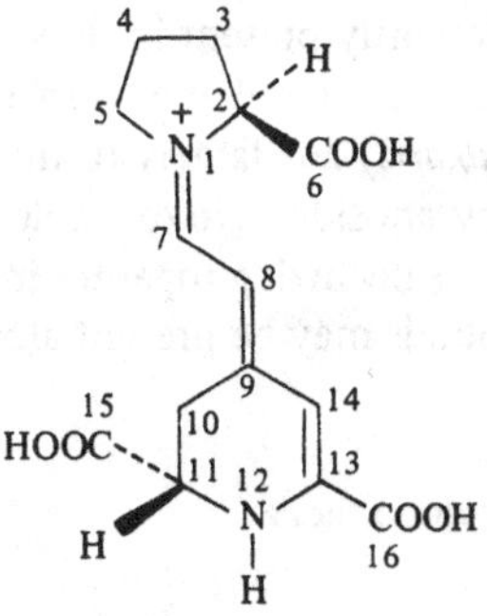

(6.73) Indicaxanthin

at C-2 and C-15 of betanidin have been established. Epimerisation at C-15 occurs very readily, perhaps spontaneously. The betaxanthins are yellow, with light absorption maxima around 480 nm, whereas the betacyanins have an extended conjugated system and absorb at longer wavelength, 534–554 nm, and are thus red-violet in colour. The chemistry of the betalains has not been studied extensively, but one property worth noting is the relatively facile interconversion of different betalains. For example, in the presence of L-proline and ammonia, betanin (betanidin-5-*O*-β-D-glucoside) is readily converted into the betaxanthin indicaxanthin (6.73).

### 6.5.4    Distribution

The poisonous toadstool *Amanita muscaria* (fly agaric) contains one violet and several yellow pigments that have been identified as betalains, *e.g.* musca-aurin I (6.74). Elsewhere betalains are restricted to higher plants and are found only in certain families of the Centrospermae, including many

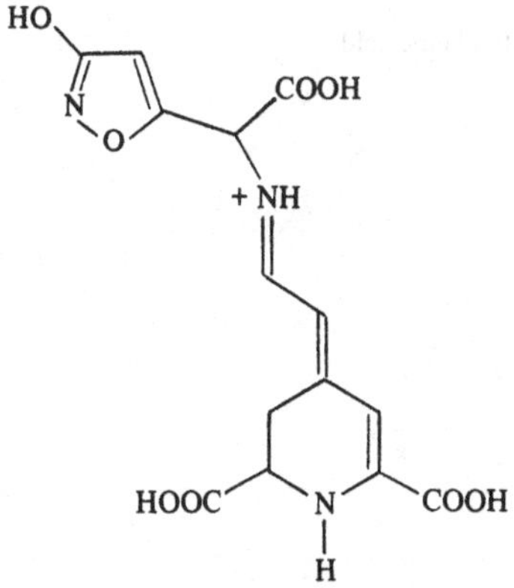

(6.74) Musca-aurin I

exotic plants, such as cacti. They are most frequently present in flowers, but
may also be found in leaves, fruits or roots; the most familiar example is the
intense red-purple colour of beetroot (*Beta vulgaris*) due largely to the
presence of betanin (betanidin-5-*O*-β-D-glucopyranoside). No example has
yet been found of betacyanins and anthocyanins occurring together in the
same species, although other flavonoid compounds may be present along
with betacyanins.

Fig. 6.13. Proposed pathway for betanidin biosynthesis.

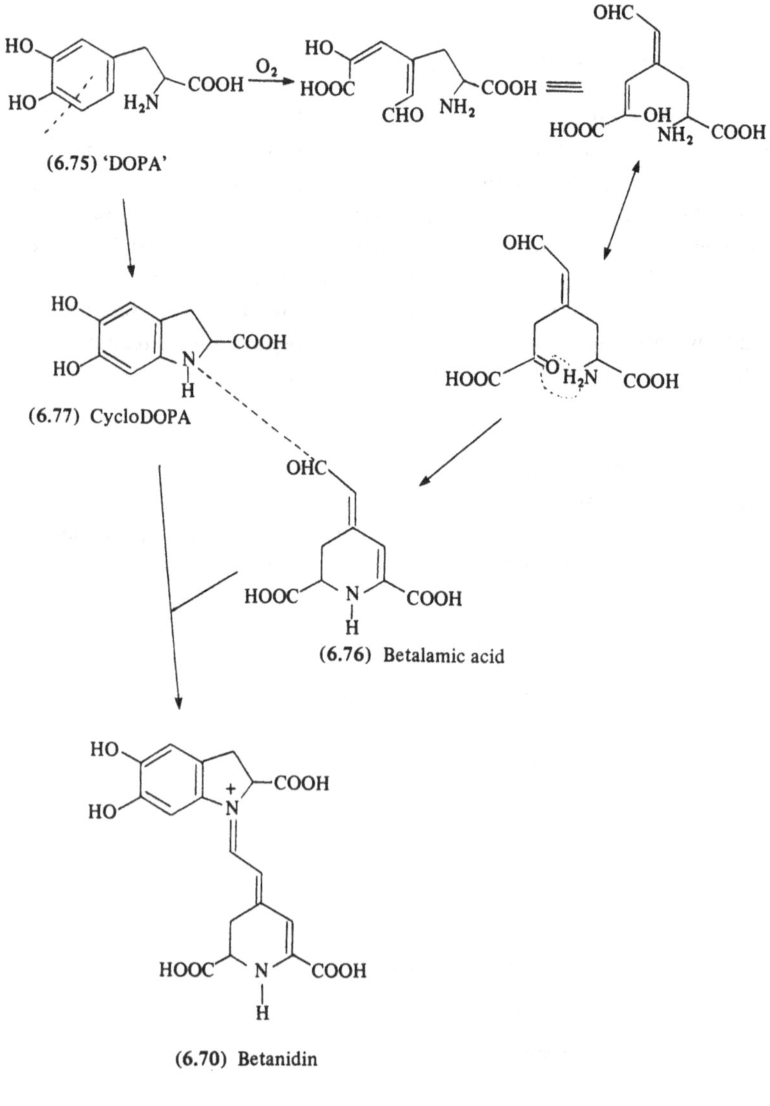

## 6.5.5    Biosynthesis

Not much definite information is available concerning betalain biosynthesis, though it is known that aromatic amino acids such as tyrosine and dihydroxyphenylalanine (DOPA, **6.75**) are incorporated into both the dihydroindole and the dihydropyridine rings of betanidin. The pathway (fig. 6.13) is thought to involve cleavage of the dihydroxyphenyl ring of DOPA followed by recyclisation to betalamic acid (**6.76**). The dihydro-indole ring ('cycloDOPA' (**6.77**) formed by cyclisation of DOPA) is then attached through the nitrogen to the aldehyde group of betalamic acid to give betanidin. Similar attachment of other amino acids or amines to betalamic acid gives the betaxanthins.

There is conflicting evidence as to whether glycosylation occurs at an early stage (*e.g.* cycloDOPA) or a late stage (*e.g.* betanidin) in the bio-synthesis. Isobetanidin is formed by epimerisation of betanidin. This epi-merisation is known to occur very readily, even spontaneously, *in vitro*, but *in vivo* it is presumably a controlled, enzyme-catalysed process. In general, virtually nothing is known of the enzymology or reaction mechanisms involved in the biosynthesis.

Light appears to be an absolute requirement for betalain synthesis in some species. In others, pigment formation takes place in darkness but is increased by irradiation with white light. The biosynthesis appears to be under direct nuclear genetic control.

## 6.5.6    Functions and uses

The physiological function of betalains in the plant is not known. It is assumed that they fulfil the same roles as the anthocyanins which they replace, so that in fruits and flowers they may serve to attract insects, birds, *etc.*, as an aid to seed dispersal and pollination (see chapter 8). Familiarity with betanin as a natural pigment in beetroot has led to increased interest in the possible use of this compound, or extracts containing it, as commercial food colorants.

## 6.6    Miscellaneous *N*-heterocyclic pigments

Nitrogenous heterocyclic pigments that do not belong to any of the main groups already described are only likely to be encountered with any frequency in microorganisms, particularly bacteria, although some higher-plant alkaloids, *e.g.* berberine (**6.78**) are coloured and therefore could be considered as plant pigments.

Intensely coloured nitrogenous bacterial products include indole deriva-tives, *e.g.* violacein (**6.79**) from *Chromobacterium violaceum*, pyrroles, *e.g.* the tripyrrole prodigiosin (**6.80**) from *Serratia* spp., and dimeric pyridine

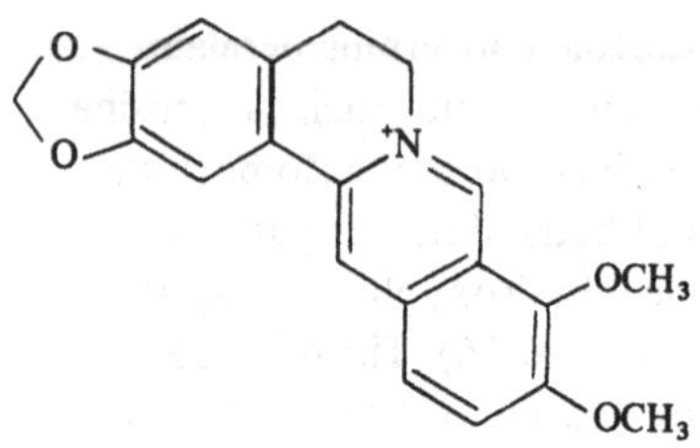

(6.78) Berberine

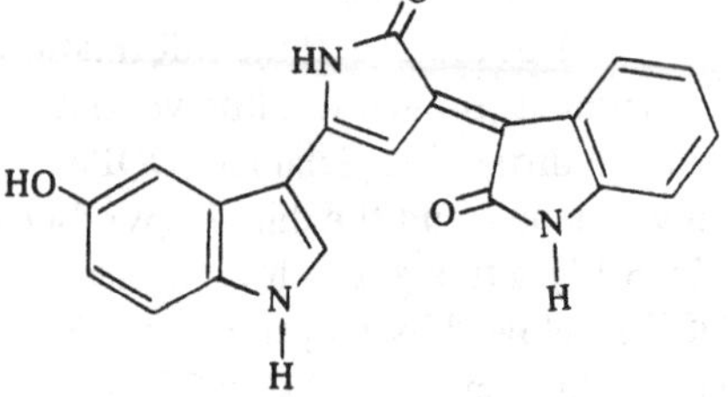

(6.79) Violacein

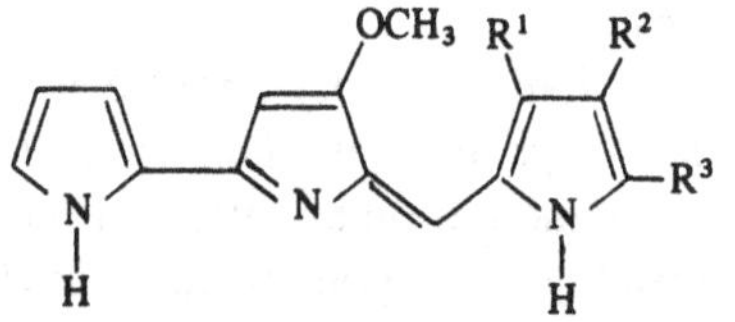

$(R^1, R^2, R^3$ are H or alkyl)

(6.80) Prodigiosin

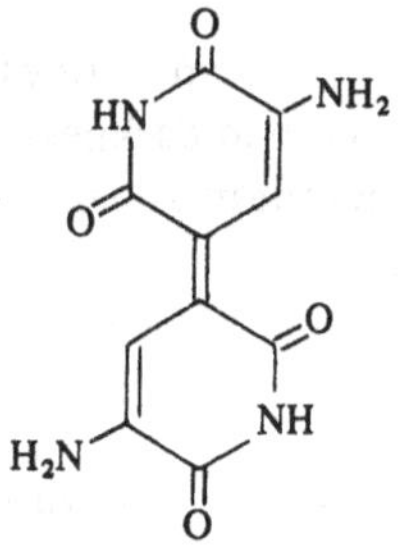

(6.81) Indigoidine

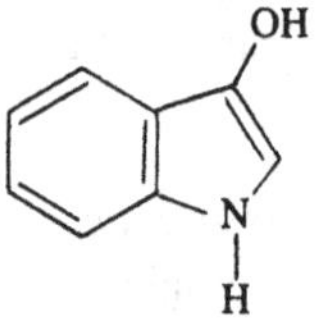

(6.82) 3-Hydroxyindole

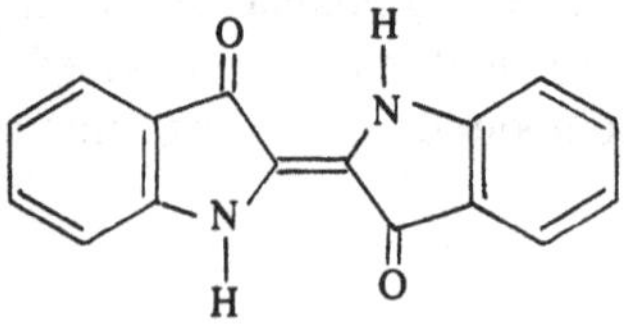

(6.83) Indigo

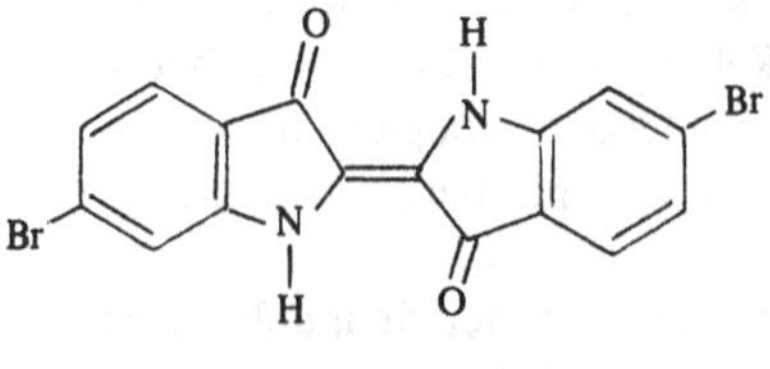

(6.84)  6, 6′-Dibromoindigo
(Tyrian purple)

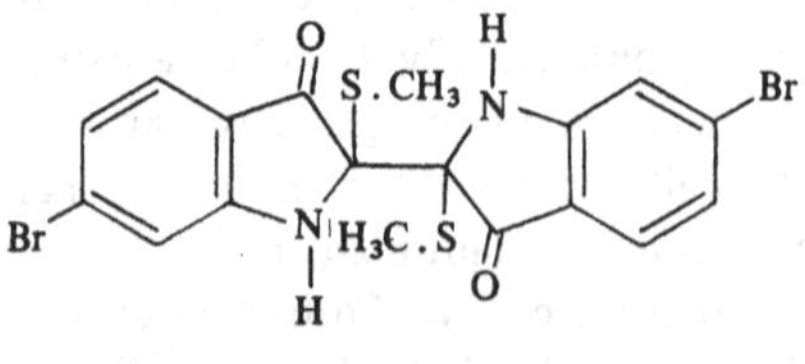

(6.85) Tyriverdin

derivatives, *e.g.* indigoidine ('bacterial indigo') from *Pseudomonas indigofera* (**6.81**) which owes its light-absorption properties to a diaza-*o*-diphenoquinone structure. The biosynthesis and functional significance of these compounds in most cases has not been investigated.

Indigo is usually thought of as a plant pigment, but the plant indigo used as a fabric dye and body colour (woad) by the ancients is an artefact. The juices of the plants concerned, *Indigofera tinctoria* and *Isatis tinctoria*, contain indican, the colourless glucoside of 3-hydroxyindole (**6.82**). The indigo pigment (**6.83**) is only obtained after hydrolysis and oxidation.

A similar situation arises with the 6,6'-dibromoindigo(tin) (**6.84**) used by the Romans and others as the dye Tyrian purple. This is also an artefact, derived in this case by photochemical oxidation of an animal product. The natural precursor, produced by various marine molluscs such as *Murex* and *Nucella*, has recently been identified as tyriverdin (**6.85**).

## 6.7    Conclusions and comments

The pigments described in this chapter do not belong all to one group, but have been collected together for convenience, on the basis of their common structural feature, the presence of at least one nitrogen-containing ring. There are, however, structural and functional similarities between some classes; purines, pterins and flavins, for example, do have somewhat similar heteroaromatic ring systems and properties. There is much scope for further chemical work. No comprehensive systematic surveys have been undertaken and novel *N*-heterocyclic pigments surely remain to be discovered, perhaps even entirely new groups. There may be structures of considerable complexity, with much interesting stereochemistry, and difficult challenges in the field of synthetic methods. There is also a great deal of work for the biochemist. The broad outlines of the biosynthetic pathways for most groups are recognised, but details of the mechanisms of individual reactions, and of the enzymes that bring them about, remain to be elucidated. Also, little is known about the regulation of any of these pathways. The *N*-heterocyclic pigments in general have not been exploited commercially as natural food colours, although the highly coloured, water-soluble betalains are currently arousing interest in this area.

Overall it seems that the study of many of these *N*-heterocyclic groups will remain one of the more neglected areas of pigment research, despite the variety of interesting work which could be done.

## 6.8    Suggested further reading

The miscellaneous *N*-heterocyclic pigments described in this chapter are such a heterogeneous collection that, not surprisingly, there is no book or major review which covers all the different groups. Once again, the general books on animal pigments (Fox and Vevers, 1960; Needham, 1974; Fox,

1976) and plant pigments (Goodwin, 1976) between them contain information on most of these nitrogenous pigments. The student really needs to be directed towards more specialised books or articles which concentrate on specific pigment groups.

In the case of purines and pterins, articles by Brown (1971), Ziegler (1961, 1965) and Ziegler and Harmsen (1969), and the proceedings of the two latest in a series of Symposia (Pfleiderer, 1975; Kisliuk and Brown, 1979) are especially useful. Several articles, *e.g.* Plaut, Smith and Alworth (1974), deal with riboflavin and its biosynthesis. Recent advances in methodology in the pteridine and flavin fields are included in two volumes of *Methods in enzymology* (McCormick and Wright, 1971, 1980). Ommochromes are less well served. A review by Linzen (1974) gives details of biosynthetic work and provides references to earlier work on the chemistry and biochemistry of these compounds.

For the microbial phenazines, articles by Leisinger and Margraff (1979) and by Ingram and Blackwood (1970) give a general account, and a survey of the relevant applied microbiology, respectively. A chapter by Piattelli (1976) in the Goodwin (1976) plant pigments book describes the chemistry and biochemistry of the betalains. A brief description of some other miscellaneous nitrogenous pigments is included in an article by Thomson (1976) in the same book.

## 6.9    Selected bibliography

Brown, G. M. (1971) The biosynthesis of pteridines, *Adv. Enzymol.*, **35**, 35.
Fox, D. L. (1976) *Animal biochromes and structural colors*, 2nd edition. Berkeley, Los Angeles and London: University of California Press.
Fox, H. M. and Vevers, G. (1960) *The nature of animal colours*. London: Sidgwick and Jackson.
Goodwin, T. W. (ed.) (1976) *Chemistry and biochemistry of plant pigments*, 2nd edition, vol 1. London, New York and San Francisco: Academic Press.
Ingram, J. M. and Blackwood, A. C. (1970) Microbial production of phenazines, *Adv. Appl. Microbiol.*, **13**, 267.
Kisliuk, R. L. and Brown, G. M. (eds) (1979) *Chemistry and biology of pteridines*. New York: Elsevier–North Holland.
Leisinger, T. and Margraff, R. (1979) Secondary metabolites of the fluorescent Pseudomonads, *Microbiol. Rev.*, **43**, 422.
Linzen, B. (1974) The tryptophan-ommochrome pathway in insects, *Adv. Insect Physiol.*, **10**, 117.
McCormick, D. B. and Wright, L. D. (eds) (1971) *Methods in enzymology*, vol. 18B. New York, London and San Francisco: Academic Press.
McCormick, D. B. and Wright, L. D. (eds) (1980) *Methods in enzymology*, vol. 66. New York, London and San Francisco: Academic Press.
Needham, A. E. (1974) *The significance of zoochromes*. Berlin, Heidelberg and New York: Springer-Verlag.
Pfleiderer, W. (ed.) (1975) *Chemistry and biology of pteridines*. Berlin: De Gruyter.

Piattelli, M. (1976) Betalains, in *Chemistry and biochemistry of plant pigments*, 2nd edition, vol. 1, ed. T. W. Goodwin, p. 560. London, New York and San Francisco: Academic Press.

Plaut, G. W. E., Smith, C. M. and Alworth, W. L. (1974) Biosynthesis of water-soluble vitamins, *Ann. Rev. Biochem.*, **43**, 899.

Thomson, R. H. (1976) Miscellaneous pigments, in *Chemistry and biochemistry of plant pigments*, 2nd edition, vol. 1, ed. T. W. Goodwin, p. 597. London, New York and San Francisco: Academic Press.

Ziegler, I. (1961) Genetic aspects of ommochrome and pterin pigments, *Adv. Genet.*, **10**, 349.

Ziegler, I. (1965) Pterine als Wirkstoffe und Pigmente, *Ergebn. Physiol.*, **56**, 1.

Ziegler, I. and Harmsen, R. (1969) The biology of pterins in insects, *Adv. Insect Physiol.*, **6**, 139.

# 7    Melanins

## 7.1    Introduction

Black is a very common colour in living organisms. The term 'melanin' was coined to describe the insoluble polymeric pigments responsible for most of these natural black colours. The expression is now also used to describe similar polymeric materials that provide some natural brown, red and yellow colours, especially in feathers and hair. Melanins are the only pigments to be widely used for coloration by mammals, including man.

The description 'melanin' conveys no information about the chemical structures of the pigments, except in so far as it is usually understood that these are high-relative-molecular-mass polymers. Attempts have been made to recognise and define certain classes of melanin pigments. Thus the black animal pigments are generally called **eumelanins**, whereas the yellow-brown variety are known as **phaeomelanins**. Somewhat similar black plant pigments which lack nitrogen are frequently termed **allomelanins**.

Knowledge of the chemistry and biochemistry of melanins is seriously limited because of the extreme technical difficulties which studies of these pigments present. Melanins are usually insoluble in almost all solvents, and therefore difficult to isolate and purify. Indeed it is almost impossible to know whether a melanin is pure, if it ever can be pure in the sense that only one molecular species is present. There are no useful diagnostic tests for melanins and no general methods for specific degradation of the melanin molecule into recognisable fragments or subunit structures. Nor is it possible to prove whether two melanin samples are identical. Association with protein, which is normal at least in the case of animal melanins, makes the isolation and study of the pigments even more difficult.

It is against this background of experimental difficulty that the present state of knowledge of the biochemistry of melanins must be viewed.

## 7.2    Chemistry

### 7.2.1    General characteristics

In the study of natural products the elucidation of a chemical structure frequently gives a first indication of the possible mode of biosynthesis of the compound under investigation. With the melanins this situation has, to a considerable extent, been reversed; it was largely biochemical studies that gave the first clues to the chemical structures.

The melanins are now known to be polymers of quinonoid substances. The classical picture of the structure of melanin was that of a long-chain polymer of indole-5,6-quinone **(7.1)** units (fig. 7.1). The situation is, however, not nearly so simple as this. In the first place, any melanin sample will almost certainly consist of a mixture of macromolecular species. Secondly, melanins are rarely homopolymers; usually a variety of monomeric units will be present. Also, different melanins, for example from different living organisms, are built up from different groups of monomeric units. The situation is complicated further by the possibilities of branching of the polymer chain and cross-linking between chains.

Despite the complexity of the problems, biochemical studies, together with drastic chemical degradation, have allowed a picture to be built up of the main structural features of the various classes of melanins. It must be stressed, however, that in no case has it been possible even to approach a full elucidation of the structure. In the discussion that follows, the main structural features and properties will be described, first of the most widely studied group, the eumelanins, and then of the other main melanin groups.

### 7.2.2    Eumelanins

*Occurrence.* The name 'eumelanins' is applied to the black, nitrogen-containing polymeric pigments, which are usually of animal origin but are also

Fig. 7.1. Classical idea of the structure of melanin as a long-chain polymer of indole-5,6-quinone (7.1).

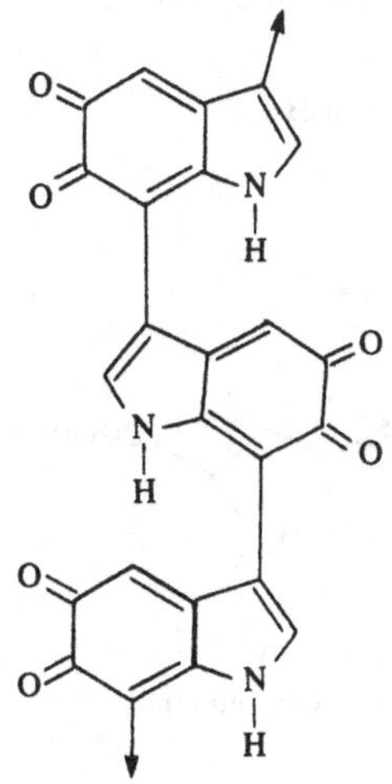

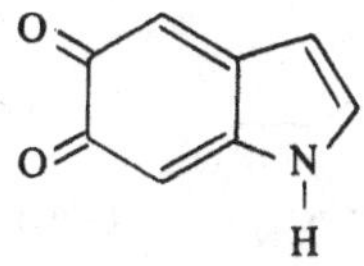

(7.1) Indole-5, 6-quinone

produced by some plant tissues. Several different kinds of eumelanin have
been recognised from a range of sources. These different eumelanins each
give characteristic elemental analyses and yield characteristic degradation
products under drastic conditions, *e.g.* boiling in 6M hydrochloric acid or alkali
fusion at 300°C. The examples that have been studied most intensively are
sepiomelanin, the very dark brown pigment of the defensive ink of the
cuttlefish (*Sepia officinalis*) and the melanin from mammalian melanomas.
Other familiar, though less well-characterised, examples occur in black hair,
feathers and skin, and the choroid of the eye, as well as in some internal
tissues. Some black plant pigments, including those produced when potatoes
and bananas are damaged, may also be eumelanin in character. Similar
pigments (DOPA-melanins) can be produced from dihydroxyphenylalanine
(DOPA, **7.2**) by plant and fungal enzymes.

*Structures.* The classical idea (fig. 7.1) that eumelanin is a linear polymer of
indole-5,6-quinone (**7.1**) is correct in essence but is an oversimplification. The
main monomeric units are indole-5,6-quinone and its reduced derivatives, but
small amounts are also present of pyrrolic monomers (**7.3**), presumably
formed from indole precursors. DOPA and DOPAquinone (**7.4**) units may
also be present in melanoma melanin. Electron spin resonance (e.s.r.) studies
indicate the presence of small amounts of free radicals, perhaps semiquinone
structures. Heavy metals, especially copper, zinc or iron, may be associated
with eumelanin *in vivo*. The extent of cross-linking has not been determined
but is likely to be generally small.

   A suggested partial structure for sepiomelanin is illustrated in fig. 7.2. The
general structures of the other eumelanins are probably related to this, but
the range of monomeric subunits and their frequency of occurrence will vary.
The natural polymers are very long chain macromolecules. *In vivo* they are
linked to protein, probably *via* the sulphydryl group of cysteine.

(7.2)  Dihydroxyphenylalanine
('DOPA')

(7.3)

(7.4)  DOPAquinone

*Properties.* Eumelanins are virtually insoluble in water and all organic solvents. They are extremely inert and stable, except under the most forcing chemical conditions, but can be bleached by prolonged exposure to air and bright sunlight or especially by prolonged oxidation with hydrogen peroxide. The bleaching of human hair to give the 'peroxide blonde' is well known.

As might be expected of black substances, eumelanins absorb light over the entire visible region. The absorption is more intense at the shorter wavelengths, and is probably not simply a consequence of the highly conjugated system.

Fig. 7.2. Suggested partial structure for sepiomelanin. The frequency of occurrence of the pyrrolic monomer units has deliberately been greatly exaggerated to illustrate the several possible modes of binding. In reality these units represent only a very small part of the sepiomelanin macromolecule.

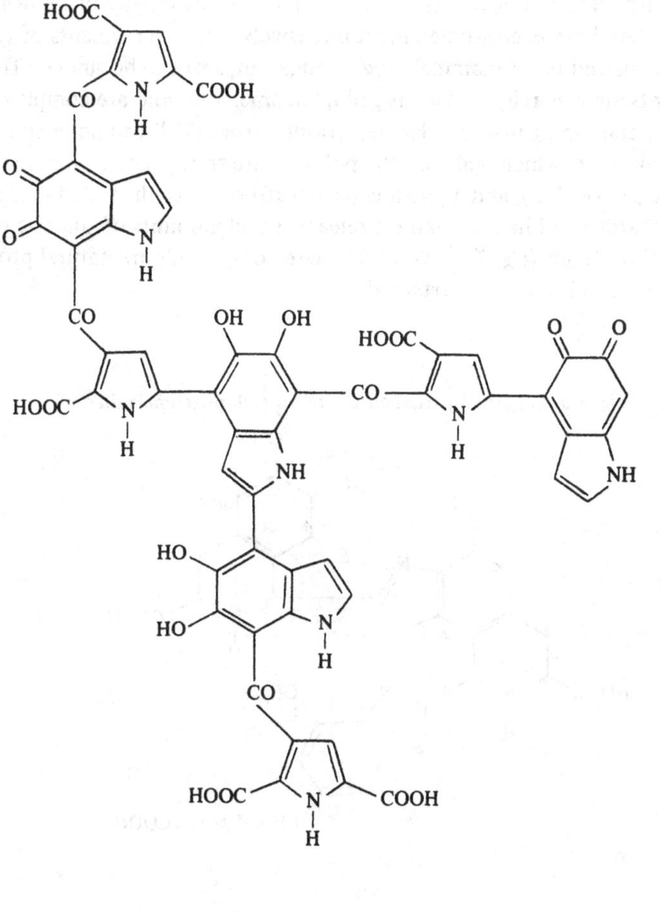

X-ray diffraction studies of a solid, animal eumelanin sample provide evidence of stacking of the indole and other aromatic units to form a π-complex. The stacked units may be members of different polymer chains or distant members of the same chain.

### 7.2.3    Phaeomelanins

*Occurrence.* The phaeomelanins are yellow, red or brown in colour. Some authors restrict the use of the name 'phaeomelanin' to the yellow or blond pigments, and call the red materials 'erythromelainin'. In this book the entire range of these pigments will be called 'phaeomelanins'. The occurrence of phaeomelanins appears to be restricted to hair and feathers and perhaps also freckles, which, in humans, are often associated with red hair. The best-studied examples are the pigments of human red hair and of the red feathers of some breeds of domestic chickens. The phaeomelanin content of red feathers is generally much higher than that of red hair.

*Structures.* The red pigments of chicken feathers, especially of the Rhode Island Red, have been studied most intensively, but the pigments of red hair of humans and other mammals, *e.g.* orang-utan, seem to be similar. These red pigments are generally known as gallophaeomelanins and are complex sulphur-containing macromolecules, formed from DOPAquinone and cysteine. The main units which make up the polymer are thought to be derived from benzothiazole (7.5), and a possible partial structure is illustrated in fig. 7.3. Acid treatment of hair or feathers releases small amounts of dimeric pigments, the trichosiderins (*e.g.* 7.6). Whether these compounds are natural products or artefacts remains to be ascertained.

Fig. 7.3. Suggested partial structure for gallophaeomelanin.

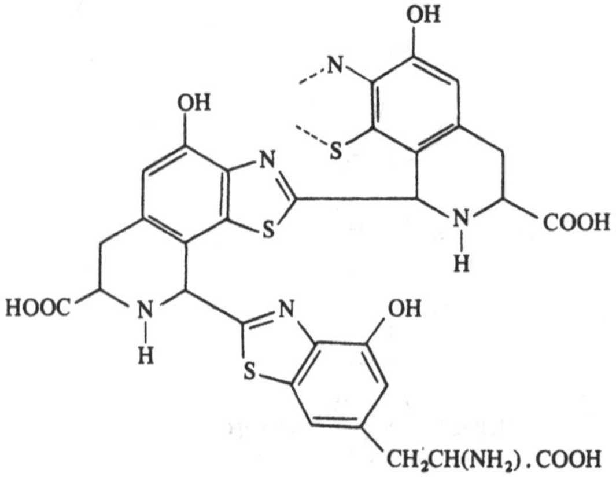

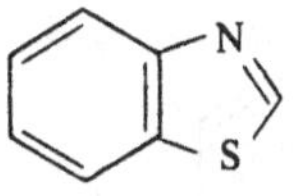

(7.5) Benzothiazole 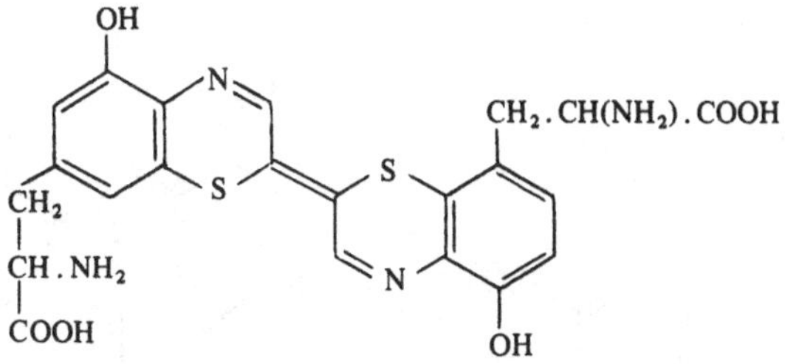                    (7.6) Trichosiderin

*Properties.* Phaeomelanins differ from most eumelanins by dissolving in dilute alkali. Their general insolubility in other solvents, and their chemical stability are otherwise very similar to those of eumelanin.

Phaeomelanins and trichosiderins exhibit a strong light absorption band in the visible region at around 500–550 nm, which is responsible for their reddish colour.

## 7.2.4    Allomelanins

*Occurrence.* The black pigments known as allomelanins appear to be exclusively products of higher plants where, for example, they constitute an important part of the black protective coating of many ripe seeds, and of fungi, where they are found primarily in the gills and spores, as in the mushroom or in the black hyphae of moulds such as *Phycomyces*.

*Structures.* The allomelanins of fungi have been studied most intensively, but the structures are still far from clear. The characteristic feature of allomelanins is that they contain little or no nitrogen, and thus cannot be indolic polymers. Many appear to be polymers of simple phenols, such as catechol (**7.7**), and their quinones, and are considered as catechol melanins. The available evidence indicates that extensive condensations may occur between catechol units in various stages of oxidation to give polymeric structures in which the monomeric units are joined by C—C and C—O—C linkages, *e.g.* partial structures such as **7.8**. Catechol also appears to be the main degradation product of plant allomelanins. The allomelanins of the fungi *Aspergillus niger* and *Daldinia concentrica* appear to contain perylene units such as (**7.9**) derived from 1,8-dihydroxynaphthalene (**7.10**).

*Properties.* The general insolubility and stability of allomelanins are similar to those of the other melanin groups. The allomelanins generally absorb light over the whole of the visible region, but absorption peaks are sometimes observed, notably at around 450 nm.

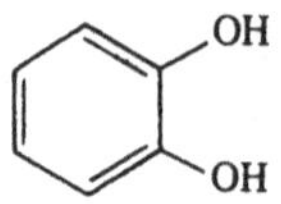

(7.7) Catechol

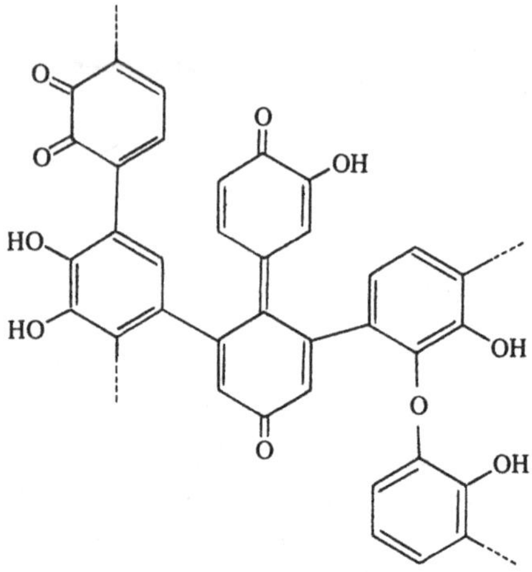

(7.8)

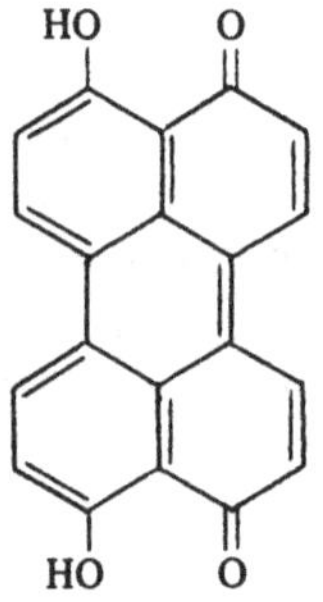

(7.9)

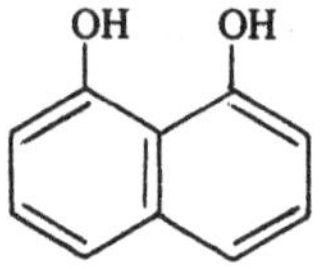

(7.10) 1, 8-Dihydroxynaphthalene

### 7.2.5    Sclerotins

Sclerotins are polymeric materials most characteristic of the exo-skeleton of arthropods. They are responsible for the hardening and the associated darkening of the cuticle of many insect species.

The chemistry of sclerotin and the biochemistry of the sclerotisation process are by no means fully understood, but some features are apparent. Sclerotin is not a simple quinone polymerisation product, like melanin. It is a protein, but one which can undergo copolymerisation with enzyme-generated quinones such as DOPAquinone and other molecules related to tyrosine. The copolymerising units may be, at least in part, tyrosine residues of the protein. Extensive melanin synthesis usually follows after the hardening process of sclerotisation is complete.

## 7.3 Distribution

### 7.3.1 In animals

Apart from haemoglobin in the blood, melanins are the only pigments to be synthesised and used extensively by mammals. They are also the most common feather pigments in birds. Black skin, hair, fur and feathers are coloured by eumelanin. Yellow, red and brown hair and fur owe their colours to phaeomelanins, which also pigment brown and some yellow and red feathers. (Carotenoids are also frequently the pigments of yellow and red feathers, see chapter 2.) Phaeomelanins are found only in mammals and birds, but eumelanins also occur widely in fishes and invertebrate animals. The best-known examples are the black colours of many insects and other arthropods (along with sclerotin). However, several dark insect colours previously thought to be melanin in nature are now known to be due to ommochromes (see chapter 6).

In addition to their value as black, brown and red pigments, melanin particles or layers commonly constitute the structural basis of the physical or structural colours (chapter 1).

Melanins are usually regarded as pigments of the integument, but they are also found in other tissues. Eyes contain melanin, which is responsible for the eye colour, either directly in the case of brown or black eyes, or indirectly as the colloidal particles responsible for the light scattering which produces blue eyes. The functioning of melanin as a light-screening pigment in the eyes of some species will be outlined in chapter 9. Internal tissues too, such as the substantia nigra of the mammalian brain frequently contain considerable quantities of melanin pigment. Excessive melanin production is characteristic of certain human melanoma tumours.

Sepiomelanin is secreted and ejected as a defensive ink cloud by many cephalopod molluscs, such as octopus and squid. This ink used to be collected and used by artists as the dark-brown 'sepia'. The inert nature of the melanin made sepia an ideal long-lasting material with good colour retention.

*Intracellular location.* In the animal integument, melanins are usually present in specific cells known as melanocytes or melanophores. Within these cells the melanin may be located in specific organelles, melanosomes, and, in association with protein, as granules, usually about 1 $\mu$m long, which have different forms in different species. These various structures and the factors that affect their morphology and physiology are extremely important in animal colour change mechanisms and will be described in more detail in chapter 8.

### 7.3.2 In plants and microorganisms

Black spots, streaks and other markings are frequently seen on the leaves, flowers *etc.* of higher plants, where they may serve as directional guides for pollinating insects. Black seed-coats and seed-pods are also very

common, though the inertness of the material may be more important than
the black colour. Usually the pigments responsible for these black colours
have not been investigated, but are assumed to be of the melanin type. Indolic
eumelanins do occur in the plant kingdom and may be produced when some
plant tissues are damaged, *e.g.* banana, potato. The characteristic plant
melanins are, however, allomelanins, containing little or no nitrogen. Allomelanins occur in many dark seed-coats, where they are located in pigmented
epidermal cells. Allomelanins are also common as spore or gill pigments in
moulds and other fungi. Some bacteria produce black pigments which appear
to be of the melanin type but have not been characterised. Phaeomelanins do
not appear in the plant kingdom.

## 7.4    Biosynthesis

### 7.4.1    *Introduction*

Tyrosine is a precursor of eumelanin and phaeomelanin. Incorporation of tyrosine into eumelanin in hair, feathers and sepia ink, and of tyrosine
and cysteine into phaeomelanin has been achieved. Other phenols, notably
catechol (**7.7**), are used in the biosynthesis of allomelanins.The enzymic
conversion of tyrosine into indole derivatives has been studied thoroughly,
and it is clear that indolic monomers polymerise to produce melanins.
Plausible schemes have been put forward for the mechanisms of these poly-
merisations but have not been verified experimentally.

### 7.4.2    *Biosynthesis of eumelanins*

*Enzymic oxidation of tyrosine.* The enzymic production of melanin-
type pigments from tyrosine was first demonstrated with a fungal enzyme
preparation which was given the name tyrosinase. Similar oxidation of tyrosine
by an enzyme from mealworms (*Tenebrio molitor*) and by similar preparations
from potato and mushroom was studied intensively by Raper. This work,
later extended by Mason, led to the proposal of the classic Raper–Mason
scheme of melanogenesis (fig. 7.4). According to this proposal, melanin
formation occurs essentially in three stages. First, tyrosine (**7.11**) is oxidised
successively to DOPA (**7.12**) and DOPAquinone (**7.13**) which then cyclises
to give the red DOPAchrome (**7.14**). In the second stage this is converted
into the colourless 5,6-dihydroxyindole-2-carboxylic acid (**7.15**) and 5,6-
dihydroxyindole (**7.16**) itself. Finally, oxidation to indole-5,6-quinone (**7.17**)
and polymerisation of this gives the melanin macromolecule. A similar process
is envisaged for the formation of eumelanin by mammals and birds.

The enzyme, tyrosinase, is a polyphenol oxidase (*o*-diphenol:$O_2$ oxidoreductase) with $Cu^{2+}$ as prosthetic group. It seems to have two distinct
activities, the aerobic oxidation of tyrosine to DOPA and that of DOPA
itself. Other steps in melanin formation may occur non-enzymically.

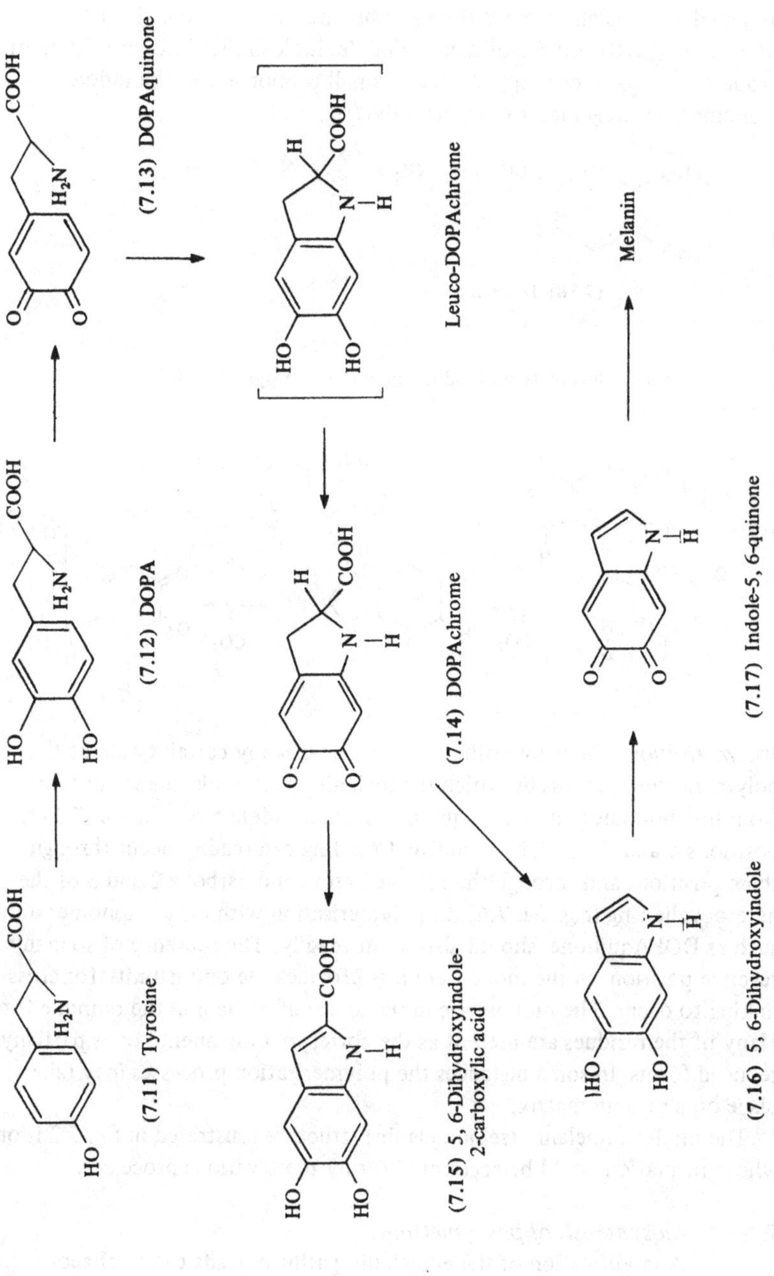

Fig. 7.4. The Raper-Mason scheme of melanogenesis.

(7.11) Tyrosine

(7.12) DOPA

(7.13) DOPAquinone

Leuco-DOPAchrome

(7.14) DOPAchrome

(7.15) 5, 6-Dihydroxyindole-2-carboxylic acid

(7.16) 5, 6-Dihydroxyindole

(7.17) Indole-5, 6-quinone

Melanin

The Raper–Mason scheme appears to be essentially correct, but more recent work has inevitably led to its extension and modification. Melanins prepared enzymically from different substrates, *e.g.* tyrosine, DOPA, DOPamine (7.18) and 5,6-dihydroxyindole, look similar but have different properties. Also, it now appears that a small proportion of the indole monomers are degraded to pyrrole units (fig. 7.5).

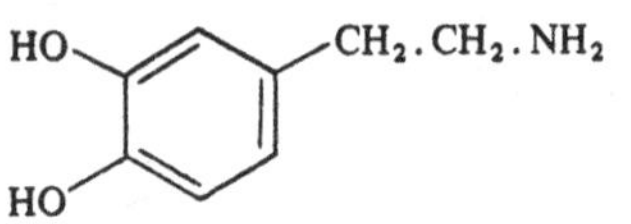

(7.18) DOPamine

Fig. 7.5. Degradation of indolic monomers to pyrrole units.

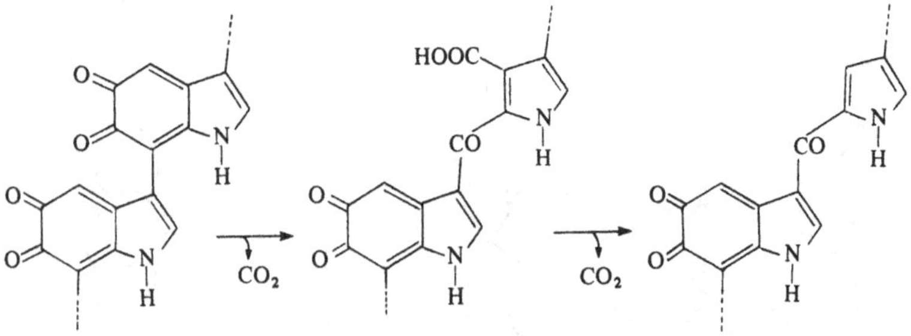

*Polymerisation.* Virtually nothing is known with any certainty about the polymerisation process by which the melanin macromolecule is built up from the monomer units. In a quinone such as indole-5,6-quinone (7.17), positions 4 and 7 are highly reactive. Coupling can readily occur through these positions and through the nitrogen atom and carbons 2 and 3 of the heterocyclic ring (*e.g.* fig. 7.6). Copolymerisation with other monomer units such as DOPAquinone, should also occur readily. The presence of so many reactive positions in the monomer units provides the opportunity for cross-linking to occur. The monomeric units do not all remain in the quinone form. Many of the residues are present as the corresponding phenol or as partially reduced forms. In some instances the polymerisation processes may take place on a protein matrix.

The model eumelanin (sepiomelanin) structure illustrated in fig. 7.2 is one whose formation could be accounted for by these various processes.

### 7.4.3  Biosynthesis of phaeomelanins

A modification of the eumelanin pathway leads to the phaeomelanins. Enzymic oxidation of tyrosine again produces DOPAquinone

Fig. 7.6. Mechanism of oxidative polymerisation of indole-5,6-quinone.

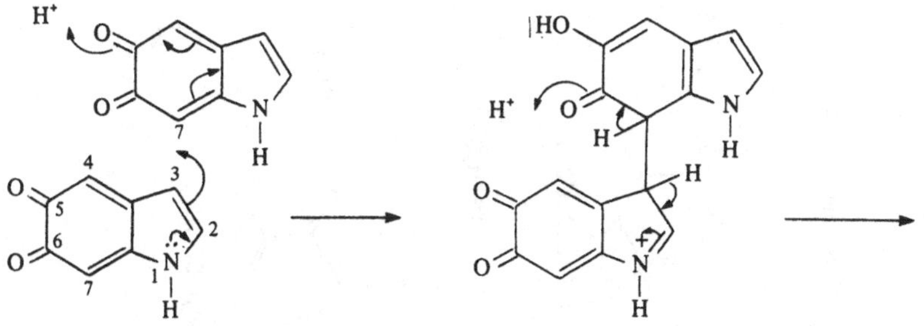

(7.17)

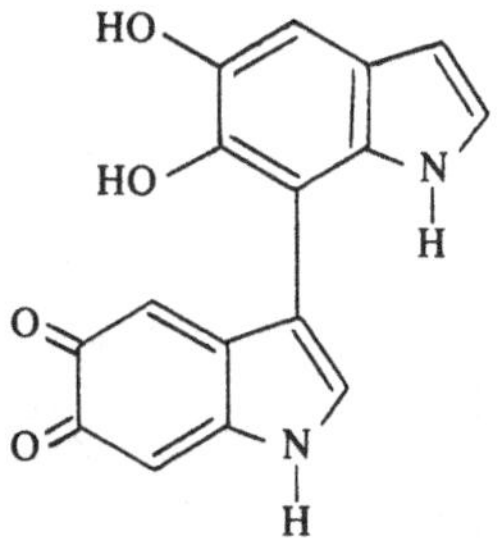

(7.13), but this interacts with the sulphur-containing amino acid cysteine (7.19) to give probable benzothiazole intermediates such as (7.20). Dimerisation or polymerisation, respectively, of the monomeric units can then lead to molecules of the trichosiderin (7.6) type or to phaeomelanins. The incorporation of labelled tyrosine and cysteine into phaeomelanin has been demonstrated, and a red-brown phaeomelanin-like pigment has been produced by oxidation of DOPA (7.2) with mushroom polyphenol oxidase in the presence of cysteine. The proposed mechanism illustrated in fig. 7.7 is, however, purely speculative and has not been verified experimentally.

### 7.4.4    *Biosynthesis of allomelanins*

Plants and fungi elaborate many phenolic substances. Oxidation of these phenols by polyphenol oxidase gives, in many cases, quinones which readily polymerise to black, nitrogen-free materials of the allomelanin type. Enzymic oxidation of catechol (7.7) gives not only *o*-benzoquinone but also products such as hydroxy-*p*-benzoquinone (7.21). Natural catechol allomelanins are considered to be formed by polymerisation of quinone molecules such as these to give branching structures like that illustrated (7.8).

The naphthol allomelanins produced by *Aspergillus niger* and other fungi are presumably formed in a similar way from 1,8-dihydroxynaphthalene

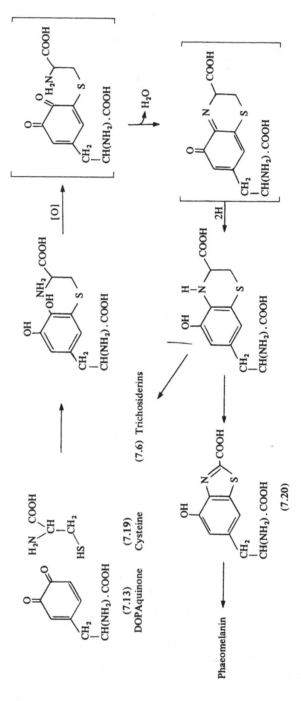

Fig. 7.7. Proposed mechanism for the biosynthesis of phaeomelanins and trichosiderins from DOPAquinone and cysteine.

(**7.10**) (a polyketide product) *via* quinone and perylene intermediates such as (**7.9**).

(7.21) Hydroxy-*p*-benzoquinone

### 7.4.5 General conclusions

Two general features of the biosynthesis of melanins are clear: (i) the early stages involve enzymic oxidation of phenolic or polyphenolic precursors to give quinones, and (ii) the highly reactive quinones readily undergo polymerisation or copolymerisation to produce coloured melanin macromolecules.

It seems likely that the attempts so far made to classify melanin material into eumelanins, phaeomelanins and allomelanins and to identify biosynthetic pathways for the three groups present a greatly oversimplified picture. Most, if not all, living tissues contain enzymes of the polyphenol oxidase type. All living tissues contain phenolic substrates, often (especially in plants) in wide variety. Polyphenol oxidase oxidation of any of these phenols, either alone or in combination with each other or with other molecules (*e.g.* cysteine), will give quinones that are capable of polymerisation to yield macromolecular products with the properties of melanins.

The same basic biosynthetic process is thus able to lead to an enormous number of products which may look similar but have widely differing structures. The differences arise from the variety and state of oxidation of the monomer units and the sequence and mechanism of polymerisation, copolymerisation and cross-linking. Details of these mechanisms have not yet been elucidated for any individual natural melanin. Indeed it is likely that new structural types of melanin, based upon hitherto unrecognised monomeric units, remain to be identified.

### 7.5 Factors affecting melanogenesis

### 7.5.1 In plants and fungi

The control of melanogenesis in plants and fungi has not been studied in detail. In plants, the synthesis of melanin in the dark spots, stripes and patterns that contribute to the overall appearance of the leaves and flowers is under genetic control. In other circumstances melanins are produced in response to mechanical damage to the tissues, *e.g.* the brown and black colours produced when fruit is bruised. In a number of fungi there is a correlation between melanin synthesis and spore formation, but the significance of this is unknown. It may be related to the production of a strong, inert spore-coat rather than coloration, as with plant seed-coats.

### 7.5.2    *In animals*

In animals, melanin is usually present in specific organelles in specific cells, so any regulatory mechanism may be controlling the overall process of differentiation and formation of these cells and organelles rather than affecting the melanin biosynthetic pathway directly. In general the main influence on synthesis and deposition of the melanin that is responsible for all or part of the external colour or pattern of an animal is genetic. For example, the negroid races of man are black, whatever their environment. However, seasonal and environmental factors may also be extremely important. Thus some examples are known of the stimulation of melanogenesis by low temperatures, and factors such as humidity and swarming have been implicated in some isolated examples.

*Invertebrates and poikilothermic vertebrates.* In these animals, colour changes, either rapid or longer-term, are common and usually involve melanins and melanin-containing cells. However, to deal with melanin in isolation at this stage would be unsatisfactory because other pigments and other cell or organelle types are also involved. The subject of colour changes and colour-change mechanisms is deferred until chapter **8**.

*Mammals and birds.* Melanin pigmentation in mammals and birds is under direct genetic control. Colour changes of the type common among invertebrates, amphibia, reptiles and fishes are not observed, although seasonal variations do occur.

The pigmentation of human skin has been studied extensively. Considerable turnover takes place. Melanin is lost with the old skin, and resynthesised in the new. Not surprisingly, the dark races synthesise and accumulate very much more melanin than do the so-called white races. Melanocytes can be recognised in the skin of the Negro foetus as early as the eleventh week of development.

Particularly noticeable in the white races of man is the synthesis of increased amounts of melanin in the skin on exposure to sunlight – the process of suntanning. The tanning takes place in two stages. First, immediately on exposure to sunlight, some melanin is made by photooxidation of a precursor substance, and forms a rapidly erected screen to reduce sunburn. The action spectrum for this rapid tanning shows a broad maximum at about 350 nm. The main and persistent tanning develops about two days after sunburn and reaches a peak after about seven days. Melanin granules are generated in the epidermal melanocytes and dispersed into the keratinocytes (skin cells). Light of about 300 nm (the same as for painful sunburn) is most effective for this longer-term tanning. An enormous holiday industry has grown on the basis of this biochemical process, but the mechanism is still not fully understood.

The synthesis of melanin in human skin is under the influence of a pituitary hormone, melanocyte-stimulating hormone, MSH. In white races, excessive melanin synthesis often occurs in some parts of the skin during pregnancy, a reflection of the generally greater activity of the pituitary. Melanin-stimulating hormones also promote melanogenesis and melanin dispersal in other vertebrate and invertebrate animals. Light is also known to stimulate melanogenesis in many lower animals, but apparently does not increase melanin synthesis in dark feathers or mammalian hair.

*Seasonal variations.* It is not uncommon for the plumage of birds in the breeding season to be quite different from that in winter. These differences are frequently a result of variations in melanin synthesis (eumelanin and phaeomelanin) in the new feathers that grow after a moult, and they are probably under hormonal as well as genetic control. Some mammals and birds which live in Arctic climates are dark-coloured in the summer season but completely or largely white in the winter, thus enjoying the maximum benefit from camouflage at all times of the year. Temperature or daylength may be the factors which trigger the seasonal moult and control melanogenesis in the new hair or feathers.

### 7.5.3   Abnormal melanogenesis
*Albinism.* Albino mammals and birds are recessive homozygotes that lack melanin in the skin, eyes, and hair or feathers and are thus white, in contrast to the dark, dominant phenotype. Such animals do have melanocytes but these contain only colourless ghosts of melanin granules because they lack the enzyme tyrosinase (polyphenol oxidase) and are thus unable to bring about the conversion of tyrosine into melanin. Albino animals may be abnormally susceptible to disease, though it is not known whether there is any direct relationship between this and the inability to make melanin.

*Human disorders affecting melanin metabolism.* In man and other animals many tumours are dark coloured. Such tumours (melanomas) synthesise and accumulate excessive amounts of melanin. Moles and naevi (birthmarks) also contain large amounts of melanin.

Increased or decreased melanogenesis are associated with a number of diseases, including Parkinson's disease, Addison's disease and vitiligo. The effects on melanin are probably secondary, however. In some cases adrenaline (**7.22**) or biosynthetic precursors that would normally be used to produce adrenaline may serve as the source of monomeric units for the melanin molecules.

$$HO \diagdown \quad \diagup CH(OH).CH_2.NH.CH_3$$

(7.22) Adrenaline

## 7.6    Functions

No functions are known for plant melanins except as contributors to the overall plant colour or pattern. The significance of the apparent correlation between melanin synthesis and sporulation in various fungi is also unknown.

The two main functions of melanins in animals, namely their roles as integumental pigments, and as light-screening pigments in eyes and skin will be dealt with in chapters **8** and **11**.

## 7.7    Conclusion and comments

Of all the groups of natural pigments considered in this book, the melanins are by far the most difficult to study. Their isolation, purification and chemical degradation present such problems that they have been characterised not on the basis of any chemical property (except perhaps inertness) but simply as a group of dark, extremely stable polymers. There are, therefore, opportunities for any scientist with some inspiration and a great deal of perseverance to introduce refinements to the structural analysis and to biosynthetic studies. We are a long way from being able to define the structure of any melanin or from understanding fully its biosynthesis; the melanins from many sources have not been studied at all. The proteinaceous environment of the melanins *in vivo* is another area where detailed work is needed.

There seem to be no interesting functions of melanins, save as dark pigments, though in this respect they are noteworthy as the only pigments responsible for colour in humans. An enormous holiday industry has been built around the desire of millions of members of the 'white' races to increase their skin melanin content by exposing themselves to sunlight. On a more serious note, it must be remembered that it is simply the obvious difference in skin melanin content that lies behind the evils of racial conflict and apartheid. For this reason no group of natural pigments could have such far-reaching social consequences. The chemical problems with melanins, although difficult enough, will probably be easier to resolve than the social ones.

## 7.8    Suggested further reading

For further reading about melanins, especially their chemistry, distribution and biosynthesis, the reader is referred to four books and review articles by Thomson (1962), Nicolaus (1968), Swan (1974) and Blois (1978). Although these were written at intervals over the past 20 years so that an increasing number of melanin samples have been examined and more and more information has been obtained, there have been no major fundamental changes in the melanin field over this time. Each of these articles has an extensive list of original references which can be used to obtain detailed information on any particular aspect. For material about special melanin-containing cells and their functioning in animal colour changes, and also about hormonal control of melanin synthesis and dispersal, the monograph by Bagnara and Hadley (1973) is most useful.

## 7.9    Selected bibliography

Bagnara, J. T. and Hadley, M. E. (1973) *Chromatophores and color change: the comparative physiology of animal pigmentation*. Englewood Cliffs, New Jersey: Prentice Hall.

Blois, M. S. (1978) Melanins, in *Photochemistry and photobiology reviews*, vol. 3, ed. K. C. Smith, p. 115. New York: Plenum.

Nicolaus, R. A. (1968). *Melanins*. Paris: Hermann.

Swan, G. A. (1974) Structure, chemistry and biosynthesis of the melanins, *Fortschr. Chem. Org. Naturst.*, 31, 521.

Thomson, R. H. (1962) Melanins, in *Comparative biochemistry*, vol. 3, eds M. Florkin and H. S. Mason, p. 727. New York and London: Academic Press.

**SECTION II**
**FUNCTIONAL ASPECTS**

# 8 The importance of colour in Nature

## 8.1 Introduction

The most obvious and fundamental function of pigments in living organisms is to bestow colour upon the tissues that contain them. This chapter will attempt to explain why this is important and how it has been exploited by animals and plants.

It is primarily by visual means that we recognise animals or plants as belonging to a particular species. Probably the first visual impressions registered are overall size and shape, but also of extreme importance are colour and pattern. These are usually used as a kind of 'fine tuning', a means of distinguishing between species of similar size and shape, but in some cases they may be the most immediately obvious features. Most animal species respond to such visual signals and use their powers of sight as a means of recognising food, enemies or mates. In parallel with this, most animals and many plants have developed visual effects of pattern and/or colour that are used to advertise or to conceal their presence. Colours and patterns may change. These changes may be developmental or seasonal or they may occur rapidly in response to variations in environmental conditions.

All these aspects will be considered in this chapter. The main aim will be to emphasise the significance and importance of coloration in the animal kingdom, but for many plants too it is important to be coloured or to produce coloured organs or tissues such as flowers and fruits to attract those animal vectors that help the spread of the plant species through their unwitting pollinating and seed-dispersing activities. The discussion will concentrate on today's species, which are the result of millions of years of evolution and selection for those characteristics that gave a slight increase in the chances of survival of the individual or of the species. No attempt will be made to speculate about the evolution of pigmentation. The aim is simply to outline the ways in which colour and pattern are used by and are useful to those animals and plants which exist today.

## 8.2     Colour and pattern in animals

Integumental colours in animals may be used for concealment or camouflage (**crypsis**) or for advertisement or warning (**semasis**). All classes of pigments, as well as structural colours, may be used for these purposes.

### 8.2.1     *Crypsis*

The purpose of cryptic colours is to conceal, *i.e.* to make it as diffi-cult as possible for an animal to be seen in its natural habitat. Cryptic colours are therefore most crucial when they are used by animals to escape detection by predators, but they are also used by predators to escape detection by prey and thus to facilitate approach. Crypsis may involve simply matching the background colour or it may extend to countershading, shape-disruption and shadow-elimination. Some animals do little more than become lighter or darker, matching various shades of grey, whereas others are capable of matching accurately a particular colour of background. Some may even copy a patchwork of colours. In some cases what appears at a distance to be a nondescript monotone may in reality be a mixture of bright colours, which are only revealed when viewed closely. It must also be remembered that the appearance of an animal must be considered in the context of its natural habitat. Thus a tiger or a zebra may be very conspicuous in a zoo, but their shape-disrupting patterns may be very effective means of concealment in the animals' natural surroundings.

Animals which move around on a variable background cannot be well camouflaged all the time. Such species, *e.g.* the crab *Carcinus*, may exhibit colour polymorphism, with animals of each of the various colour forms closely matching one type of background which they will deliberately seek out and spend most time against.

The value of cryptic coloration to survival of the species has received ample experimental support. If insects of the same species but two different colour forms are maintained against a single-coloured background, those individuals whose colour contrasts with the background are very much more likely to fall victim to predators than are background-matching specimens.

### 8.2.2     *Semasis*

Advertising or sematic coloration involves bright colours, usually in conspicuous large patches with sharp boundaries and contrasts. These are recognised and interpreted by a viewing animal as either warning or welcom-ing, depending on previous vivid experience.

*Aposemasis.* Colours and patterns are said to be aposematic when they warn the viewer that the creature possessing them is either dangerous or unpalatable. A well-known example of such a warning pattern is provided by wasps (*e.g. Vespula vulgaris*), whose yellow and black contrasting bands cause terror even in some members of the human race. Other conspicuous patterns in insects, such as the red and black of ladybird beetles (*Coccinella* spp.), serve a warning to would-be predators that the owners are decidedly unpalatable. Such animals do not need to hide or try to escape; predators, recalling previous experience, are likely to give them a wide berth.

*Episemasis.* Conspicuous colours and patterns that serve to attract other animals are known as episematic. In most cases they are directed towards the opposite sex of the same species as part of sexually attracting displays. Commonly only the male is brightly coloured and only he displays, but in some cases the sexes are alike and both display. In addition to its advertising function, sexual display probably accelerates gonad maturation and readiness to mate, and increases fertility. Development of sexual display colours is commonly seasonal, and likely to be under hormonal control.

It is frequently necessary for the sexual display to compromise with camouflage requirements. The bright colours and bold patterns must then remain hidden except during the process of courtship, or during threat-displays directed towards a rival.

The use of attracting colours and patterns for other than sexual purposes is less common, but in some cases serves to attract prey. Some deep-sea angler fish, which otherwise are well camouflaged, use a brightly coloured and often luminous lure to attract their prey. The tentacles of sea anemones serve a similar purpose.

*Pseudosemasis.* Pseudosematic patterns are those which mimic the sematic appearance of another species. In most cases pseudoaposemasis is employed, with the result that a harmless and 'tasty' creature escapes being attacked and eaten because it is disguised to look like an unrelated dangerous or unpalatable species. Many hoverflies, for example, have adopted a wasp-like yellow-and-black pattern, and some roaches closely mimic members of the ladybird family.

## 8.3     Animal pigment cells – chromatophores

In the integuments of animals, especially invertebrates and poikilo-thermic ('cold-blooded') vertebrates, pigments are normally located in special cells. In a most useful monograph on the subject of animal pigment cells and colour changes, Bagnara and Hadley (1973) have recommended the use of the term **'chromatophore'** to describe these colour-bearing cells, and their nomenclature will be adopted in this book, although other authors have used alternative names such as 'chromatocytes'. Several kinds of chromatophore are recognised; these are distinguished primarily by the colour effect that they produce.

### 8.3.1    Melanophores

Black pigment cells containing melanin (chapter 7) are known as melanophores, though in the case of mammals and birds such cells have usually, perhaps preferentially, been called melanocytes. In the vertebrates at least two distinct types of melanophore cell are recognised, differing in their location, general appearance and response to hormones. Dermal melanophores are prevalent, almost ubiquitous, in poikilotherms and are involved in rapid colour changes. They may be very large cells, up to 0.5 mm in diameter. Epidermal melanophores are generally thin, elongated cells and are not of great importance in rapid colour changes. The distribution of epidermal melanophores is variable; they are usual in reptiles and amphibia but rare in fishes. Similar epidermal cells (melanocytes) are responsible for melanic skin coloration in mammals and also give rise to pigmentation by melanin, including phaeomelanin, in hair and feathers.

It is a fundamental property of all melanophores that they produce their own melanin pigment. At least in vertebrates, melanin synthesis within the melanophore involves the deposition of melanin polymers upon a matrix of coiled protein fibres called a premelanosome. The premelanosome contains the enzyme tyrosinase (polyphenol oxidase) which catalyses melanin synthesis (chapter 7). The completed sub-cellular organelles are called melanosomes, or melanin granules, and the aggregation and dispersal of these particles are of extreme importance in animal colour changes (see §*8.4.2*).

There have been few studies of melanin-containing cells and organelles in invertebrate animals, but it seems likely that they resemble those of the vertebrates.

### 8.3.2    Xanthophores and erythrophores

Melanophores are usually black, though some, containing phaeomelanin, may be yellow or orange-red in colour. Most yellow, orange and red integumental cells are a different cell type known as xanthophores (yellow) or erythrophores (red). The predominant pigments in these cells are carotenoids (chapter 2), present in lipid droplets or carotenoid vesicles, but it is now known that pterins (chapter 6), either alone or in combination with carotenoid, are responsible for the bright colours of many poikilothermic vertebrates such as frogs and toads. The pterins are located in special small organelles called pterinosomes, distributed throughout the cytoplasm. Even in species which are coloured primarily by carotenoids, pterins are usually synthesised in developing xanthophores and erythrophores, and are the first pigments to be seen. Carotenoids, which must be obtained from the diet, do not appear until later.

The pterin eye-pigments of *Drosophila* are located in organelles remarkably similar to vertebrate pterinosomes.

### 8.3.3 Iridophores

Very common in invertebrates and poikilothermic vertebrates are integumental cells known as iridophores (sometimes guanophores). Although not strictly pigment-containing cells like melanophores, xanthophores and erythrophores, they nevertheless play an important part in the coloration and external appearance of the animal. Iridophores contain organelles that are orientated in such a way as to reflect light efficiently and thus form the basis of many iridescent or metallic structural colours. The principal 'pigments' are purines – most commonly guanine (chapter 6) – which are white or colourless but are arranged in stacks or platelets which reflect all or some wavelengths of visible light and are characteristically responsible for the silvery and golden metallic sheens perhaps most familiar in fish scales.

### 8.3.4 Chromatophore cell associations

*The dermal chromatophore unit.* In the dermis of those vertebrates that can complete rapid colour changes there is usually a specified localisation of the three types of chromatophore. Xanthophores (or erythrophores) are uppermost, iridophores just below them and dermal melanophores form a basal layer. The three cell types are not usually present in equal numbers. This association of the three cell types (fig. 8.1) constitutes a functioning dermal chromatophore unit that is responsible for rapid colour changes. The light-absorbing melanophores and the light-reflecting iridophores contribute darkness and lightness, respectively, whereas the xanthophores or erythrophores behave largely as a yellow filter. Iridophores may also contribute structural blue colours. Colour changes are brought about rapidly by variations in the contributions made by the different cell types in the functional unit (see §8.4).

*The epidermal melanin unit.* The vertebrate epidermal melanophore carries out its pigmentary function in association with other cells. Melanin pigment is

Fig. 8.1. Schematic representation of a dermal chromatophore unit.

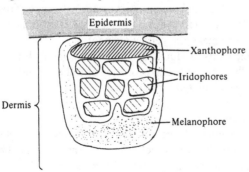

transferred from its site of formation, the melanophore, to an adjacent or surrounding pool of cells. The combined collection of melanised cells is considered as an epidermal melanin unit. This is especially important in mammals and birds, for it allows melanin to be transferred from the melano-phores themselves and deposited in those epidermal cells that give rise to the specialised structures of hair, feathers and bill.

### 8.3.5    Developmental origin of chromatophores

Pigment-cell development and pigment pattern formation in verte-brates have been studied extensively. The site of origin of chromatophores of all vertebrate classes is the neural crest. In amphibians, for example, all three basic pigment-cell types, melanophores, iridophores and xanthophores, are derived in this way. The neural crest can give rise to many different kinds of cells besides these different chromatophore types, but the factors that determine the various cell types are not known. Chromatophore determina-tion occurs very early, even before the neural folds appear, and the various chromatophore types appear in a definite sequence. Differentiation of dermal melanophores occurs first, followed by differentiation of xanthophores and iridophores. Epidermal melanophores appear relatively late, in some instances only shortly before metamorphosis.

It is clear that, during embryogenesis, pigment cells must migrate long distances from the neural crest region to the areas of the integument that they will eventually colour. The migration of chromatoblasts (incipient chromatophores) has been demonstrated experimentally in, for example, the frog *Rana pipiens*. The majority of pigment-cell movements are movements of chromatoblasts and take place before much pigment synthesis has occurred, but some limited movement of differentiated melanophores has been observed.

Pattern formation results from differential migration of chromatoblasts into genetically predetermined areas. White spots or patches occur in some species or albino mutant strains, usually because the chromatoblasts have failed to differentiate into chromatophores even though they have migrated into the white areas. Similarly, in blue mutants or normally green frogs, xanthoblasts (partly differentiated xanthophores) are present in their proper locations but are unable to make the yellow pigment.

The nature of the pigments produced by the differentiated chromatophores is determined genetically, but the actual processes of cell differentiation and pigment synthesis may be regulated by hormonal factors.

### 8.4    Animal colour changes

Some animals maintain the same colour and pattern throughout their lives. Apart from minor modifications such as suntanning and loss of hair pigments, this is true of man and many other mammals. The only pigment that has to be synthesised is that required to replace losses due to wear and tear or following a moult.

Many animals, however, do change their appearance, either at certain stages of their development or at different seasons. Obviously, these changes will usually be due, at least in part, to alterations in the colour or pattern of the integument. A further group of animals, mainly poikilothermic vertebrates and invertebrates, are able to change their colour and pattern, often very rapidly, in response to environmental changes or stress.

These various colour changes are brought about by either (i) alterations in the amount of pigment present, *i.e.* pigment synthesis or destruction, or (ii) alterations in the effectiveness with which the pigment is displayed.

### 8.4.1 *Colour changes due to pigment synthesis or destruction*

Colour changes of this type have variously been called morphological, morphogenetic or chromogenic. The special kinds of colour change associated with moulting of plumage or pelage and their replacement by feather or hair of different colour and/or pattern are obviously of this type, as are the seasonal changes related to camouflage (*e.g.* winter whitening) or to reproduction (development of breeding coloration).

Some colour changes that occur in response to environmental factors are also of this type. The usual cue for such colour changes is background adaptation, *e.g.* animals maintained on a dark background synthesise more melanin and develop more melanosomes, whereas animals transferred to a lighter background lose melanin. The total number of pigment cells in the skin of the adapting animal may increase or decrease markedly; either proliferation of existing melanophores or melanisation of undifferentiated melanoblasts may be involved. Other types of chromatophores also respond.With fishes, for example, a light background leads to increased production of the iridophore 'pigment' guanine, concomitant with melanin destruction; conversely the production of melanin as a response to a dark background is accompanied by disappearance of guanine. Yellow pigment in xanthophores tends to behave in a similar way to melanin, *i.e.* pigment synthesis occurs in response to background darkening.

### 8.4.2 *Physiological colour changes*

Physiological or chromomotor colour changes generally occur much more rapidly than the kind just described because they do not involve pigment synthesis or destruction. In this case the mobilisation of existing pigmented organelles within the cell determines how effectively the various pigments are displayed, and hence the overall colour that the animal adopts. Normally these responses are freely and rapidly reversible. They are characteristic of reptiles, amphibians, fishes and many invertebrates, but not of mammals or birds.

*Usual mechanism.* In all cases except for cephalopod and some pteropod molluscs, the colour change is brought about by the movement of pigmented organelles within the chromatophore (fig. 8.2). For example, in a melanophore the melanin granules or melanosomes may be dispersed around the peripheral regions of the cell where they will efficiently absorb light and thus produce a dark colour. Alternatively they may be aggregated in a small volume towards the centre of the cell, so that little of the incident light will be absorbed. Dispersal and aggregation of melanosomes thus provides a mechanism for rapid darkening and lightening of the skin. The dermal chromatophore unit (see §*8.3.4*) contains not only melanophores but also iridophore and xanthophore or erythrophore cells, and the pigment particles or granules within these may also be dispersed or aggregated so that they make a greater or lesser contribution to light absorption and colour. A wide range of colours may thus be produced, rapidly and reversibly, and the animal is able to match the colour of its background quite closely. The celebrated range of colours displayed by the chameleon is produced in this way.

Fig. 8.2. The usual mechanism for physiological colour changes (lightening–darkening) by aggregation and dispersion of melanin granules (melanosomes) within the melanophore cells.

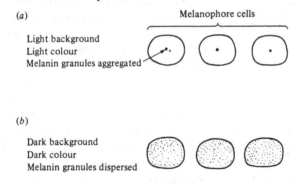

(a)                                    Melanophore cells

Light background
Light colour
Melanin granules aggregated

(b)

Dark background
Dark colour
Melanin granules dispersed

*Cephalopod mechanism.* Cephalopod molluscs, *e.g.* squid and octopus, and some pteropod molluscs use a different mechanism (fig. 8.3). These animals have developed structures that are, in effect, tiny organs which consist of five different types of cell, including a central chromatophore. Contraction of radial muscle fibres causes expansion of the chromatophore (up to seven times its original diameter) and consequent dispersal of its pigment granules. There are dark-brown, red and yellow chromatophores (the dark ones contain ommochrome (chapter 6) not melanin) in rather similar chromatophore organs and the colour of the skin at any particular moment depends upon the relative extents to which the three types are expanded. Colour changes in these animals occur extremely rapidly (less than 1 s).

Fig. 8.3. The cephalopod mechanism for physiological colour change. (*a*) Light colour. Melanin granules are aggregated by contraction of the central chromatophore into a small volume. (*b*) Dark colour. Melanin granules are dispersed as contraction of the radial muscle fibres causes great expansion of the central chromatophore.

(*a*)                                          (*b*)

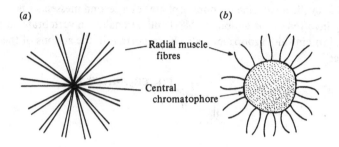

Radial muscle fibres

Central chromatophore

### 8.4.3   *Factors controlling colour changes*

The same factors appear to be instrumental in the regulation of all background adaptation colour changes, whether these involve quantitative variations in pigment concentrations or aggregation–dispersion phenomena. The main regulating factor appears to be the albedo, *i.e.* the ratio between the amount of light reflected from the background and the intensity of the direct incident light. Other environmental factors may also have an effect Thus low temperatures or increasing humidity frequently result in increased pigmentation, either by synthesis (melanin, pterin, ommochrome) or by increased uptake, transport and deposition (exogenous carotenoids). Diurnal cycles of colour variation have been recognised in several species of both vertebrate and invertebrate animals in either aquatic or terrestrial habitats.

Pigment synthesis during prolonged background adaptation is frequently preceded by, and may be a necessary consequence of, a more rapid physiological colour change brought about by aggregation or dispersal of pigmented organelles.

### 8.4.4   *Mechanisms of colour-change regulation*

Hormonal or neural mechanisms may regulate colour changes of both kinds. A pituitary hormone, known as melanophore- (or melanocyte-) stimulating hormone (MSH) or intermedin, from the pars intermedia, regulates melanin synthesis and melanosome dispersal. The more brightly coloured pigment cells, iridophores, xanthophores and erythrophores, are also influenced by MSH. MSH is a peptide hormone and some variations in amino acid composition have been detected in MSH samples from different animal species. All samples of vertebrate MSH so far characterised contain an active heptapeptide sequence, —Met-Glu-His-Phe-Arg-Trp-Gly—.

Another hormonal substance, melatonin (8.1), from the pineal area, may mediate the response to light or albedo in many species by inhibiting darkening processes. Adrenaline (8.2) and noradrenaline (8.3) acting as neurohormones, and thyroxine (8.4) acting as an initiator of morphological differentiation may also influence pigmentation and colour change. Cyclic AMP (3′,5′-cyclic adenosine monophosphate) as a second messenger is probably involved in the actions of MSH and adrenaline in vertebrate colour changes. Protein thiol groups may also be important in the actions of these hormones.

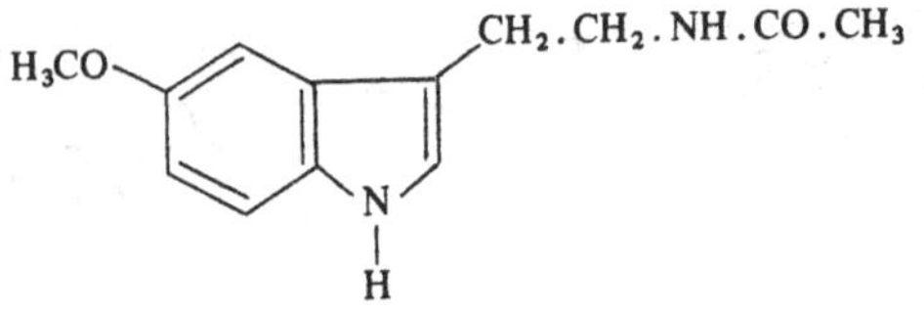

(8.1) Melatonin

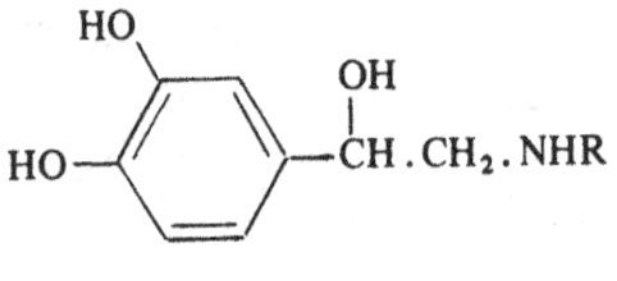

(8.2) Adrenaline (R = $CH_3$)
(8.3) Nor-adrenaline (R = H)

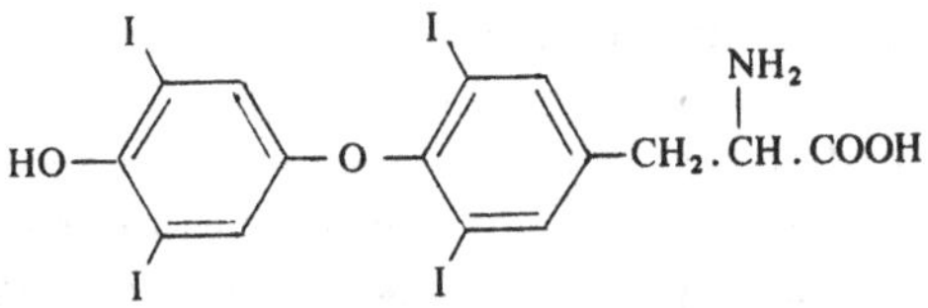

(8.4) Thyroxine

It is thought that several different hormones mediate physiological colour changes in crustaceans. Separate red-, black- and white-pigment-dispersing hormones and -pigment-concentrating hormones have been recognised. These are thought to be peptides and are produced by the eyestalks where the receptor pigments that detect background colour changes are located.

The study of the biochemistry of hormonal involvement in colour changes in animals is still at a very early stage.

## 8.5    Colour in plants

The plant kingdom is predominantly green. This green colour, in its variety of shades, is relaxing and pleasing to the human eye, but would surely become monotonous if not relieved by splashes of other bright, contrasting colours. The brilliant colours of many flowers and fruits provide a focal point to which the eye is drawn. However, the possession of brightly coloured flowers and fruits is obviously of much greater fundamental importance than this. The background green of plants and the chlorophyll that provides it are essential to the process of photosynthesis which keeps each individual plant alive. The importance of the contrasting hues of the flowers and fruits lies in their facilitating the propagation and survival of the species.

Flowers attract not only the human eye but also the eyes of many other smaller animals which render tremendous service by transferring pollen from one plant to another. Bees have probably been studied more in this respect than any other of the pollen-transferring animals. Bees can discriminate four basic 'colours', including the ultraviolet region around 340–380 nm. They are blind to red wavelengths. The natural preference of bees is for blue, yellow or u.v.-absorbing (white) flowers. Other animals that serve as pollinating vectors have different colour preferences. Humming birds are significant pollinators in tropical climates, and prefer red and orange colours. Several pollinating agents show a preference for pale colours; moths, beetles and (occasionally) bats prefer white (ultraviolet), and butterflies prefer pale colours such as pink and mauve.

In a similar way, the bright and obvious colours of fruits may be visible from afar and will attract fruit-eating animals, especially birds, which then disperse the indigestible seeds over large distances.

The evolutionary tendency in plants has thus been to select for the pigments which give flowers and fruits of the colour range most appreciated by the predominant pollinating and seed-dispersing animals. Thus plant species pollinated predominantly by bees will have either colourless flavonoids or blue anthocyanins in their flowers, whereas tropical species pollinated to a large extent by humming birds have red flowers, containing red anthocyanins.

The importance as plant pigments of compounds which absorb strongly in the near ultraviolet must be stressed. Although the human eye does not detect u.v. light, the eyes of many other animals, especially insects, are u.v.-sensitive. U.v.-absorbing compounds are therefore seen by these species.

Both visible- and u.v.-light-absorbing compounds are important in the production of flower patterns. Some very elaborate patterns have been evolved, many of which are related to the performance of the insect pollinators. For example many flowers, *e.g.* foxglove (*Digitalis purpurea*), have 'honey lines' or similar markings which serve to direct bees *etc.* towards the nectar. High concentrations of u.v.-absorbing 'pigment' may be used in the pattern, particularly at the base of the petals. The intense u.v. absorption serves to guide the insects to their target, the nectar- and pollen-containing centre of the flower. Patterns in flowers are usually brought about either by a local increase in pigment production in some parts of the petals or by the superimposition of a second pigment on the main colour. The patterning is nomally under strict and complex genetic control.

Two groups of pigments are responsible for the colours and patterns of almost all flowers, the carotenoids (chapter 2) and the flavonoids (chapter 4). Red, purple and blue colours are virtually always due to anthocyanins. White and cream flowers contain large amounts of flavones and flavonols which are the u.v.-absorbing pigments seen by insects. Yellow and orange are usually due to carotenoids but occasionally may be produced entirely or in part by flavonoids of the chalkone and aurone classes, or by betaxanthins (chapter 6). Mixtures of water-soluble flavonoid pigments give the expected colours, *e.g.* yellow plus red gives orange. On the other hand a purple water-soluble anthocyanin on a yellow lipid-soluble carotenoid background will usually result in a brown flower colour, *e.g.* in the wallflower *Cheiranthus cheirii*.

Fruit colours, especially reds, purples and blues, are usually anthocyanin in origin, though in some notable cases (*e.g.* tomato, orange) carotenoids are responsible.

In all cases these plant pigments serve to advertise the presence of the flower or fruit to animals. The purpose of this advertisement is virtually always to attract beneficial creatures rather than to warn off possible adverse influences. The role of pigmentary colours is thus much simpler in plants than in animals. There are, however, some individual cases in which the ecological factors involved may be extremely complex.

## 8.6    Conclusions and comments

Colour and pattern, especially in animals, is extremely important in relation to behaviour and ecology, but there may appear to be very little of biochemical interest, except perhaps the identification of the pigments concerned. There are, however, some most fascinating biochemical questions, which as yet remain largely unanswered. Of especial interest are problems concerning the mechanisms of regulation and control of pigment synthesis and deposition in tissues and of colour-change phenomena. The mechanisms by which environmental factors, such as light and temperature, perhaps mediated by hormones, bring about slow or rapid colour changes, and the ways in which genetic control of pattern formation is expressed are just a few examples. Biochemical aspects of these phenomena are likely to be extremely difficult to study, but will undoubtedly attract increasing attention in the near future.

In the plant kingdom, pigment synthesis is only one part of the enormous morphological changes which take place as flowers and fruits develop and mature, and is part of the chloroplast to chromoplast transformations that occur during these processes. Again the mechanisms of regulation and control

of pigment synthesis in relation to these massive changes present a real
challenge to the biochemist.

## 8.7    Suggested further reading

Many examples of the importance of colour and pattern in the animal
kingdom are given in the general books by Broughton (1964), Fox (1976,
1979), Fox and Vevers (1960), and Needham (1974), and the subject is dealt
with in more detail in the extensive and still useful work by Cott (1940).
There are many books and articles dealing with animal pigment cells and
colour changes. Among these, the monograph by Bagnara and Hadley (1973)
provides an extremely useful general account of the field, and the more
specialised publications of Della Porta and Mühlbock (1966) on melanocytes,
Fingerman (1970) on chromatophores, and Novales (1969) and Riley (1972)
on control of colour changes supplement the older book by Parker (1948).
An even more specialised volume by Searle (1968) presents a genetic survey
of mammalian coat colours.

An overall view of the importance of colour in plants is given by Harborne
(1976). Much interesting information about the ecology of flower colours and
pollination may be obtained from the book by Faegri and Van der Pijl (1971).

## 8.8    Selected bibliography

Bagnara, J. T. and Hadley, M. E. (1973) *Chromatophores and color change: the com-
    parative physiology of animal pigmentation.* Englewood Cliffs, New Jersey: Prentice-
    Hall.
Broughton, W. B. (ed.) (1964) *Colour and life.* London: Institute of Biology.
Cott, H. B. (1940) *Adaptive coloration in animals.* London and New York: Oxford
    University Press.
Della Porta, G. and Mühlbock, O. (eds) (1966) *Structure and control of the melanocyte.*
    New York: Springer-Verlag.
Faegri, K. and Van der Pijl, L. (1971) *The principles of pollination ecology*, 2nd edition.
    Oxford: Pergamon.
Fingerman, M. (1970) Comparative physiology: chromatophores, *Ann. Rev. Physiol.*, **32,**
    345.
Fox, D. L. (1976) *Animal biochromes and structural colors*, 2nd edition. Berkeley, Los
    Angeles and London: University of California Press.
Fox, D. L. (1979) *Biochromy: natural coloration of living things.* Berkeley, Los Angeles
    and London: University of California Press.
Fox, H. M. and Vevers, G. (1960) *The nature of animal colours.* London: Sidgwick and
    Jackson.
Harborne, J. B. (1976). Functions of flavonoids in plants, in *Chemistry and biochemistry
    of plant pigments*, 2nd edition, vol. 1, ed. T. W. Goodwin, p. 736. London, New
    York and San Francisco: Academic Press.
Needham, A. E. (1974) *The significance of zoochromes.* Berlin, Heidelberg and New
    York: Springer-Verlag.
Novales, R. R. (ed.) (1969) Cellular aspects of the control of color changes, *Amer. Zool.*,
    **9,** 427.

Parker, G. H. (1948) *Animal color changes and their neurohumors.* Cambridge University Press.
Riley, V. (ed.) (1972) *Pigmentation: its genesis and control.* New York: Appleton-Century-Crofts.
Searle, A. G. (1968) *Comparative genetics of coat colour in mammals.* New York: Academic Press.

# 9     Pigments in vision

## 9.1     Introduction

Section I of this book dealt with the biochemistry of the various classes of natural pigments which give colour to the tissues that contain them. Then in chapter **8**, the importance of the property of being coloured, both to the survival of the individual and to the propagation of the species, was stressed. Clearly this can only be the case if colours and patterns can be seen and distinguished by various animals. In other words, animals must be able to detect light and also to discriminate between different light wavelengths. For this purpose they have developed photoreceptor organs – eyes – and within these eyes light-absorbing photoreceptor molecules or visual pigments play the central role. In addition to the photoreceptors themselves, other pigments are frequently used in an accessory role. It is thus essential that a book on natural pigments should not overlook the processes of photoreception and vision, not only because of the intrinsic interest of the photoreceptor molecules themselves, but also because most of the other natural pigments would be unnecessary and might never have been evolved if no such colour-discrimination mechanisms existed.

## 9.2     The eye

Most living organisms are able to respond in a general way to light because they possess some kind of photoreceptor cells, organelles or molecules. It is in the animal kingdom that this ability to respond to light is exploited most effectively in the process of vision. The term 'vision' implies not simply the detection of light, but also the perception of position, shape and movement in the environment, and in many cases colour discrimination. True vision requires a means of forming a true image on the receptor cells, and for this purpose special photoreceptor organs or **eyes** have been developed. There are two basic eye structures, found in vertebrate and invertebrate animals, respectively.

### 9.2.1    The vertebrate eye

The vertebrate eye (fig. 9.1) is a refracting eye with a large single **lens**. Light enters *via* the **cornea** and the **iris** and is focused by the lens onto a sensitive surface, the **retina**. The most sensitive area of the retina, and that most sharply in focus, is called the **fovea**, and this may be protected by a pigmented screening layer, the **macula lutea**.

Fig. 9.1. The vertebrate eye. Schematic representation showing relevant structural features of a typical vertebrate (*e.g.* human) eye.

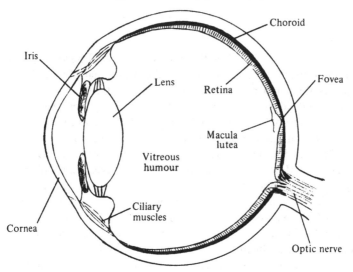

The structure of the photosensitive retina is illustrated in more detail in fig. 9.2. It contains a densely packed array of **photoreceptor cells** which the light reaches after passage through a network of **neural cells**. Retinal receptor cells are of two types, the **rods** which in the human retina are about 28 $\mu$m long and 1.5 $\mu$m in diameter, and the **cones** which are shorter and taper towards the tip. Some animals have more than one rod type and double cone structures are not uncommon. The number of photoreceptor cells is large; the rat retina, for example, is estimated to have at least 15 million rod cells.

The photoreceptor molecules or **visual pigments** are located in the **outer segments** of the receptor cells (fig. 9.3). These outer segments are in contact with the **pigment epithelium–choroid**, which contains stray-light-absorbing **pigment granules**, and in some species a reflecting layer, the **tapetum**. The pigment epithelium is intimately involved in visual pigment regeneration (see §*9.4.3*). The axes of the receptor cells point towards the lens, to maximise light absorption. The visual pigments are located in a stack of **membrane discs** (see §*9.2.3*) perpendicular to the axis.

Fig. 9.2. Schematic representation of the structure of the vertebrate retina (not to scale).

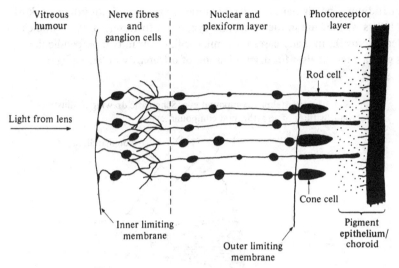

Fig. 9.3. Schematic representation of a vertebrate rod photoreceptor.

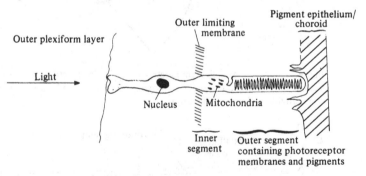

### 9.2.2 The invertebrate eye

Invertebrate animals show a great diversity of eye types. These are generally quite different from those of vertebrates. The **compound eyes** of arthropods are a good illustration of the general features (fig. 9.4). Such compound eyes consist of a large number (up to about 10 000) of small individual photoreceptor units called **ommatidia**. Each **ommatidium** (fig. 9.5a) contains its own lens which focuses a part of the visual field onto its associated receptors. Light passes through the lens into the **crystalline cone** which directs the light into the tube-like **retinular cells** that make up the receptor structure. In the core of the ommatidium the neighbouring edges of the retinular cells form the receptor structure, the **rhabdom**, which contains the visual pigment and is thus analogous to the outer segment layer of the vertebrate retina. Visual pigments with different light absorption maxima can occur in the same ommatidium. The pigment-bearing membranes are formed not into a stack of flat layers, as they are in the vertebrate eye, but into closely packed tubules called **microvilli**. In many cases these microvilli occur in two perpendicular sets giving a mechanism for discrimination of differently polarised light (fig. 9.5*b*).

Fig. 9.4. The invertebrate compound eye. Schematic drawing to illustrate relevant features of an arthropod compound eye.

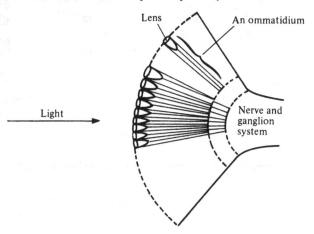

Fig. 9.5. (*a*) Schematic diagram illustrating the main structural features of a single ommatidium. (*b*) Sections showing the perpendicular arrangement of microvilli in alternate rhabdoms.

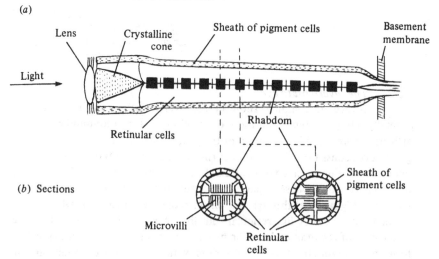

The sheath of the combined retinular cells contains screening pigment granules which can migrate into the ommatidium in conditions of bright light, thereby limiting the amount of light that can reach the receptors. A palisade of long, thin cells surrounding the entire ommatidium contains additional screening pigment which prevents stray light from travelling between the ommatidia.

*The cephalopod eye.* The eye of certain cephalopod molluscs such as the octopus and squid has the appearance of a vertebrate eye rather than the compound eye usually associated with invertebrates. Overall it exhibits features of both types but is in reality a special type of rhabdomeric eye (fig. 9.6). Although a large, single lens forms an image on the receptor layer, as in the vertebrate eye, the receptor layer is in this case composed of rhabdoms, which face towards the light and have neural layers behind them.

Fig. 9.6. Relevant structural features of the cephalopod eye.

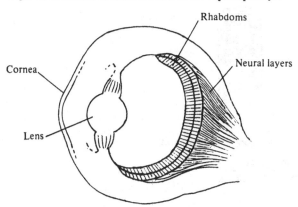

### 9.2.3 Location of visual pigments in receptor membranes

The structure and arrangement of rod outer segment membranes (fig. 9.3), and the location of visual pigment molecules within these structures are now quite well understood, although only a few animal species have been studied. The receptor membranes of vertebrate cone cells and of the microvilli of invertebrate compound eyes have received far less attention. However, all receptor types seem to conform to the same basic pattern of orientated visual pigment molecules in an ordered array of membrane.

Extensive investigations by X-ray diffraction, electron microscopy and other sophisticated techniques have been performed on vertebrate rod outer segments. The picture that emerges (fig. 9.7) is of a series of stacked membrane discs. These discs are in the form of a protein bilayer in which globular protein molecules are centred in the two layers, with lipid – largely phospholipid – filling the spaces inbetween. The visual pigment (*e.g.* rhodopsin) forms a large proportion (*ca* 85%) of the membrane protein. The visual pigment molecules are so orientated in the receptor membrane that absorption of light passing vertically down the receptor is maximised. All the evidence is consistent with a model in which the visual pigment molecules are free to move laterally in the membrane, and to rotate about an axis perpendicular to the membrane but not in any other way.

Specific absorbance measurements reveal that the pigment concentrations in rod and cone cells, and in membranes from a wide variety of animals, are similar. Values of around $10^8$-$10^9$ molecules of visual pigment per segment or rhabdom are generally accepted.

Fig. 9.7. Structure of the vertebrate rod cell outer segment photoreceptor membranes: (*a*) location of stacked membrane discs in the rod outer segment; (*b*) model of the disc membrane ultrastructure.

(*a*)

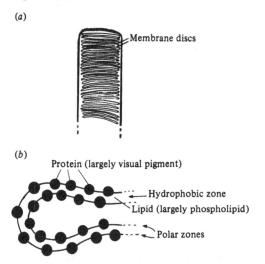

(*b*)

### 9.3     Visual pigments

Within the receptor membranes it is the light-absorbing, photosensitive visual pigments that play the primary role in light detection. In any animal species there will usually be more than one visual pigment (*e.g.* four in man) and different pigments are found in rod and cone cells. The many individual visual pigments are characterised by their particular $\lambda_{max}$ values, which range between 345 and 620 nm, and thus display maximal sensitivity to different wavelength ranges. Although there are so many individual visual pigments, they are all remarkably similar in structure. Each pigment consists of a lipoprotein molecule linked to a smaller chromophore. In the whole of the animal kingdom only two, very similar, chromophoric groups are found.

Slight variations in the structure and conformation of the lipoproteins (opsins) give rise to the considerable variations in $\lambda_{max}$.

### 9.3.1    The chromophore

The two visual pigment chromophores are **retinaldehyde** (**9.1**, also known as **retinal** or **retinene**) and **3,4-didehydroretinaldehyde** (**9.2**; the carotenoid numbering scheme is used). Visual pigments containing the retinaldehyde chromophore appear to have a virtually universal distribution throughout the animal kindom, whereas 3,4-didehydroretinaldehyde is restricted to many fresh-water fish and some amphibian species. The two families of visual pigments derived from these chromophores are termed **rhodopsins** (from retinaldehyde) and **porphyropsins** (from didehydroretinaldehyde).

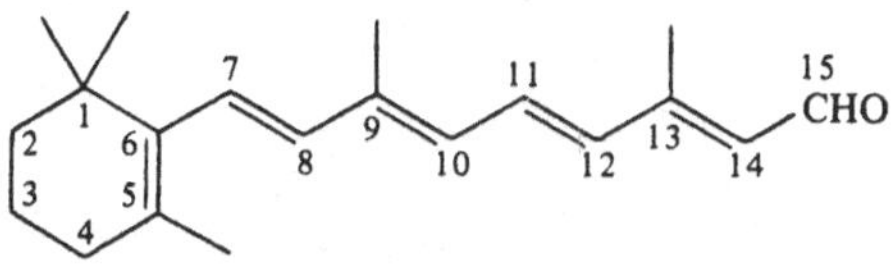

(9.1)  Retinaldehyde

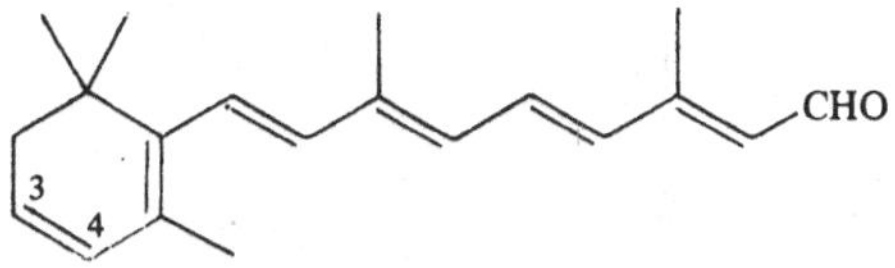

(9.2)  3, 4-Didehydroretinaldehyde

The structural similarity of retinaldehyde to the plant pigment β-carotene (**9.3**, see chapter **2**) is immediately apparent. Animals cannot synthesise retinaldehyde *de novo* but must make it from β-carotene and related carotenoids obtained from the diet, or from dietary vitamin A (retinol, **9.4**). In mammals, an oxygenase enzyme in the intestine cleaves β-carotene to two molecules of retinaldehyde (fig. 9.8). The retinaldehyde is reduced to retinol and stored in the liver as esters (mainly palmitate). A specific lipoprotein, the retinol-binding protein, is used to transport retinol to the retina where it may be converted into retinaldehyde and incorporated into the visual pigments. It is only the sterically hindered 11-*cis*-isomer (**9.5**) of retinaldehyde which is bound in rhodopsin.

### 9.3.2    The protein

The protein to which the retinaldehyde is bound is known as **opsin**. It is not easy to prepare the opsin protein free from lipid and carbohydrate material. This and the instability of the purified free protein have rendered its study difficult. Consequently the term 'opsin' has sometimes been applied to the entire complex of protein, lipid and carbohydrate and sometimes to the protein part alone.

Fig. 9.8. Scheme for the formation of retinaldehyde by cleavage of β-carotene by an oxygenase enzyme from mammalian intestine.

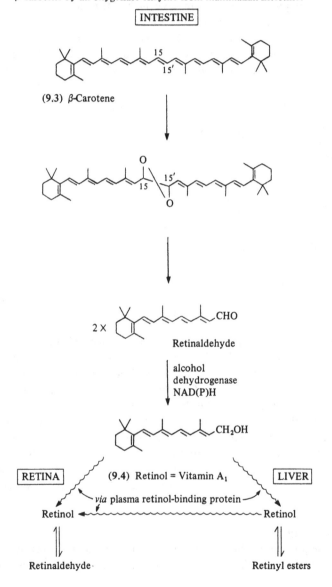

INTESTINE

(9.3) β-Carotene

2 X       CHO

Retinaldehyde

alcohol
dehydrogenase
NAD(P)H

CH₂OH

RETINA       (9.4) Retinol = Vitamin A₁       LIVER

*via* plasma retinol-binding protein

Retinol ◄———————————————► Retinol

Retinaldehyde                     Retinyl esters

All opsins so far examined, from many animal species, are small proteins with relative molecular masses generally in the range 30 000–40 000. Amino acid compositions, but not sequences, have been determined for rod opsins from several animal species. A carbohydrate moiety, consisting of one or more glucosamine and mannose units is firmly bound to an asparagine residue of the protein. A considerable amount of lipid, largely phosphatidyl choline and phosphatidyl ethanolamine, is also strongly associated. Opinions differ about whether this phospholipid is bound as an essential part of the visual pigment molecule or is merely a contaminant from the lipid area of the receptor membrane.

### 9.3.3 Binding of retinaldehyde to opsin

Almost all studies of retinaldehyde–opsin binding have been made with rhodopsin from rod cells of vertebrates, especially cattle. All evidence, however, suggests that other visual pigments are basically similar. Two aspects

Fig. 9.9. Covalent binding of 11-*cis*-retinaldehyde to opsin by Schiff base or aldimine formation with the ε-amino-group of a lysine residue.

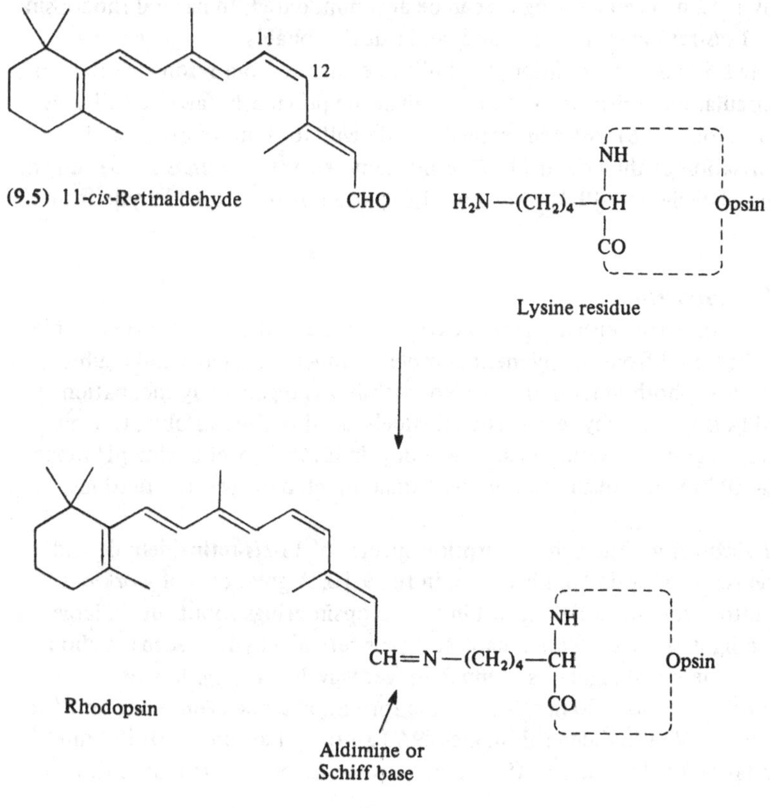

(9.5) 11-*cis*-Retinaldehyde    CHO    $H_2N$—$(CH_2)_4$—CH    Opsin

Lysine residue

Rhodopsin    CH=N—$(CH_2)_4$—CH    Opsin

Aldimine or Schiff base

of the binding must be considered, the primary binding by which retinaldehyde is covalently linked to a functional group of the opsin, and the secondary, non-covalent interactions.

The retinaldehyde is bound covalently by formation of a Schiff base or aldimine between its aldehyde group and an amino-group of opsin. The general opinion is that the ε-amino-group of a lysine residue is involved (fig. 9.9). However, some experimental results have been used to support the alternative theory that the Schiff base may be formed with the amino-group of the ethanolamine residue of phosphatidyl ethanolamine, at least at some stages of the visual process.

Non-covalent interactions between retinaldehyde and opsin are also extremely important, especially interactions between amino acid sidechains and the polyene π-electron system. These factors determine the conformation of the chromophore and also bring about some polarisation of the π-electron system. It is these effects that are responsible for the subtle variations in $\lambda_{max}$ in the different visual pigments.

The conformation that opsin adopts in the visual pigment as a result of the various non-covalent interactions is such that only a narrow range of retinaldehyde isomers and analogues can be accommodated. In natural rhodopsin only 11-*cis*-retinaldehyde is bound, and this is probably maintained in a distorted 6-*s-cis* conformation; the bulk of evidence from resonance Raman and circular dichroism work is now against the previously favoured 12-*s-cis* conformation (**9.6**). All-*trans*-retinaldehyde will not bind to opsin, and isomerisation of the bound 11-*cis*-retinaldehyde to the all-*trans* isomer during the visual cycle (see §**9.4**) results in dissociation of the retinaldehyde from the opsin.

### 9.3.4   Properties

*Stability.* Visual-pigment extracts stored in the dark are very stable. Opsin liberated from the pigment complex is much less stable and readily breaks down both *in vitro* and *in vivo*. Stability is regained by incubation with 11-*cis*-retinaldehyde. (9-*cis*-Retinaldehyde also gives a stable, though artificial, complex, isorhodopsin.) Rhodopsin is stable over a wide pH range (*ca* 5–10) but free opsin can only withstand much narrower fluctuations.

*Light absorption.* The light absorption spectra of 11-*cis*-retinaldehyde and bovine rod rhodopsin are illustrated in fig. 9.10. A great deal of work has gone into trying to explain how binding to opsin brings about such a large shift in light absorption maxima; free 11-*cis*-retinaldehyde absorbs at about 375 nm, the visual pigments at much longer wavelength, *e.g.* bovine rhodopsin at about 500 nm. Schiff base formation alone cannot account for this; simple *N*-retinylidene aldimines (**9.7**) absorb at around 360–380 nm. Protonation of the aldimine (**9.8**) increases the absorption maximum to

(9.6) Twisted 6-*s-cis*, 12-*s-cis* conformation
of 11-*cis*-retinaldehyde

(9.7)  *N*-Retinylidene aldimine

(9.8)  Protonated *N*-retinylidene aldimine

Fig. 9.10. Light absorption spectra of 11-*cis*-retinaldehyde (in hexane) and bovine rod rhodopsin (aqueous solution).

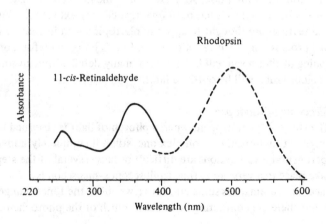

440–450 nm, but this is still well below the 500 nm absorption of the native rhodopsin. The further red shift in $\lambda_{max}$ is probably the result of secondary interactions between the opsin and the polyene chain of the retinaldehyde chromophore. An external point-charge mechanism, in which a protein negative charge above C-13 and a counter-ion near the protonated Schiff base nitrogen cause polarisation of the chromophore, is now considered likely. Presumably it is variation in these interactions that is responsible for the different absorption maxima of the various visual pigments. 3,4-Didehydro-retinaldehyde absorbs at somewhat longer wavelength (393 nm for the 11-*cis* isomer) than retinaldehyde itself, and a porphyropsin, with didehydro-retinaldehyde as its chromophore, will absorb at correspondingly longer wavelengths than the retinaldehyde-containing rhodopsin.

Two u.v.-absorption bands are also usually present in the spectrum of a visual pigment. The $\gamma$-band with $\lambda_{max}$ at about 280 nm is due to the aromatic amino acids tyrosine and tryptophan in the protein, whereas a low-intensity $\beta$-band at about 330 nm is generally considered to be a '*cis*-band' due to the retinaldehyde chromophore's being in the *cis*-configuration (cf. carotenoid '*cis*-peaks', §*2.3.3*). There is evidence that the $\beta$-band absorption is photo-chemically active.

## 9.4    Functioning of the visual pigments – the visual cycles

In the vertebrate retina the rod cells are responsible for **scotopic vision**, or 'night vision', *i.e.* detection of low light intensities. This process is very sensitive, but involves no colour-discrimination mechanism and is there-fore monochromatic, or approximately so. At higher light intensities cone cells are employed for **photopic vision**. The cone cells contain a number of different visual pigments (sometimes called **photopsins**, cf. **scotopsins** in rod cells) which respond maximally to different light wavelengths and thus form the basis of colour discrimination. At present only the mechanism of scotopic vision in vertebrate retinal rods has been investigated very extensively, but the same general features are thought to apply to photopic vision in cone cells and to visual processes in the invertebrate eye. In this book, therefore, only the functioning of rhodopsin will be discussed in any detail, although some aspects of colour vision will be outlined later.

### 9.4.1    Bleaching of rhodopsin

The story of what happens when a photon of light is absorbed by the visual pigment rhodopsin is a complex one, still not completely under-stood. Experimental investigations are difficult because several of the steps involved take place in a very short time (milliseconds–picoseconds). The following account summarises some current views about the functioning of rhodopsin, but there are conflicting ideas about much of the photochemistry.

After a suitable period in the dark the retina becomes dark-adapted, and the visual pigment molecules will be in the fully regenerated form (see below), *i.e.* the chromophore, 11-*cis*-retinaldehyde, is bound to the protein, opsin, in a specific 6-*s-cis*, 12-*s-trans* conformation. The pigment molecules in the receptor membrane behave as individual units. Light entering the eye falls upon the retinal receptor cells and is absorbed by the visual-pigment molecules. Bleaching of the purple visual pigment then occurs by a series of changes which cannot yet be described fully in molecular terms. The overall features, however, are clear (fig. 9.11).

The primary event is the absorption of a photon of light by the 11-*cis*-retinaldehyde chromophore of the rhodopsin. This results in electronic excitation followed by conformational changes and isomerisation to the all-*trans*-retinaldehyde structure. The successive changes in the shape of the retinaldehyde chromophore are accompanied by a series of changes in the opsin conformation, and the specific interactions between the polyene system and the protein are progressively diminished and eventually lost altogether as all-*trans-N*-retinylidene opsin is produced. In this complex only the covalent unprotonated aldimine link remains intact, and this is readily and irreversibly broken to yield free all-*trans*-retinaldehyde and opsin.

Many intermediates in the bleaching sequence have been identified spectroscopically, indicating a series of transformations such as that illustrated in fig. 9.12. However, it is not yet possible to correlate the various spectroscopic intermediates with particular retinaldehyde–opsin configurations and conformations.

Light appears to be required only for the conversion of rhodopsin into bathorhodopsin, possibly only for the generation of a very short-lived (3 ps) intermediate, prebathorhodopsin. Hypsorhodopsin may be involved in an alternative or branch pathway. Subsequent transformations can take place in the dark. There is substantial resonance Raman and circular dichroism evidence to indicate that the isomerisation of the 11-*cis* chromophore to all-*trans* occurs during the photoconversion of rhodopsin into bathorhodopsin probably by the prebathorhodopsin stage; a twisted 11-*trans* chromophore in bathorhodopsin is reasonably well established. In the later products, lumirhodopsin and the metarhodopsins, the retinaldehyde chromophore is certainly in the all-*trans* form, but a considerable degree of interaction remains between the polyene chain and opsin. Metarhodopsin I and all early intermediates are thought to be protonated aldimines. Deprotonation, with its concomitant hypsochromic shift, occurs at the metarhodopsin I–metarhodopsin II stage. The nerve transmitter release also seems to take place at this stage, initiated by protein conformation changes. The bathochromic shift which occurs when pararhodopsin (or metarhodopsin III) is produced from metarhodopsin II *en route* to *N*-retinylidene opsin is unexpected and unexplained.

Fig. 9.11. Main features of the molecular transformations during bleaching of rhodopsin.

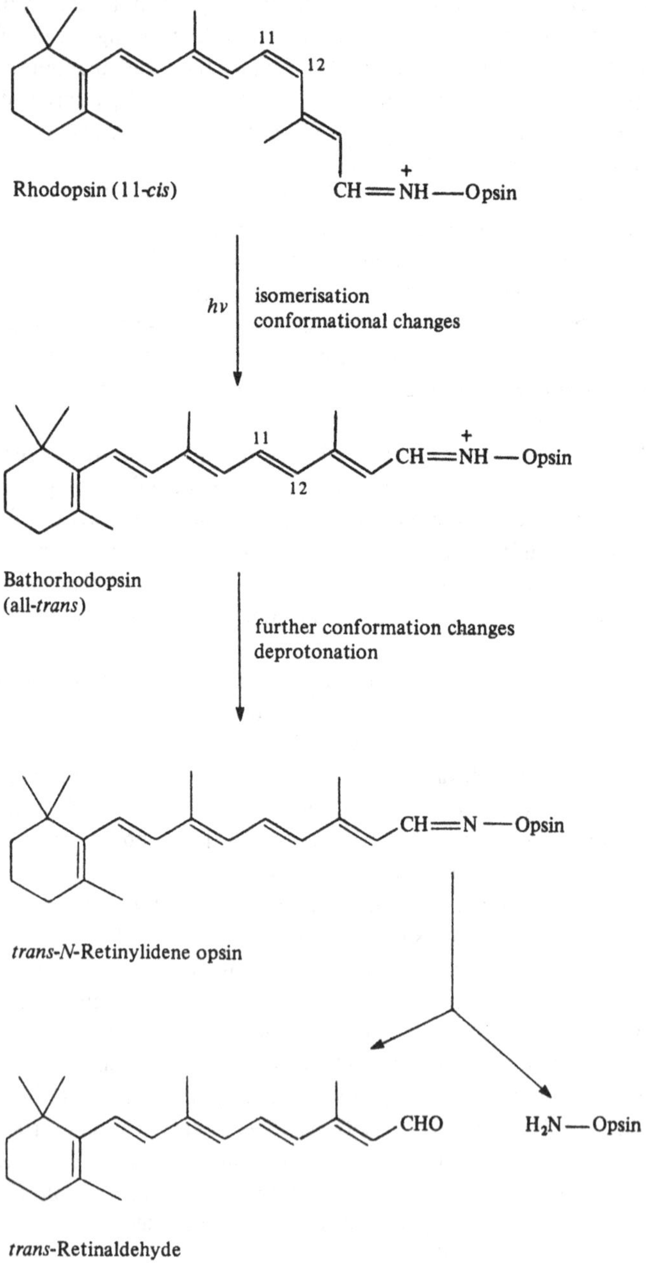
11
12
+
CH═NH—Opsin
Rhodopsin (11-*cis*)
*hν*
isomerisation
conformational changes
11
12
+
CH═NH—Opsin
Bathorhodopsin
(all-*trans*)
further conformation changes
deprotonation
CH═N—Opsin
*trans*-*N*-Retinylidene opsin
CHO
H₂N—Opsin
*trans*-Retinaldehyde

Fig. 9.12. Postulated sequence of intermediates in the bleaching of rhodopsin.

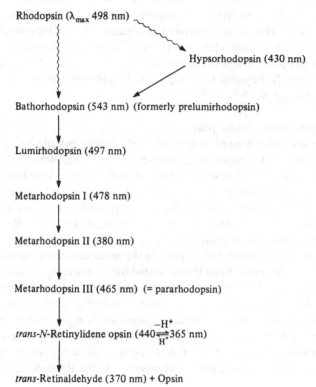

Rhodopsin ($\lambda_{max}$ 498 nm)

Hypsorhodopsin (430 nm)

Bathorhodopsin (543 nm) (formerly prelumirhodopsin)

Lumirhodopsin (497 nm)

Metarhodopsin I (478 nm)

Metarhodopsin II (380 nm)

Metarhodopsin III (465 nm) (= pararhodopsin)

*trans-N*-Retinylidene opsin (440 $\underset{H^+}{\overset{-H^+}{\rightleftharpoons}}$ 365 nm)

*trans*-Retinaldehyde (370 nm) + Opsin

Obviously much more work is needed before the rhodopsin bleaching process can be described completely in molecular terms.

## 9.4.2    *The neural response*

The absorption of light by the visual-pigment molecule, the resulting electronic excitation and the subsequent transformations leading to bleaching of the pigment are clearly related in some way to the neural excitation which leads to a signal being transmitted to the brain. The initial absorption of the light photon is virtually instantaneous, whereas the subsequent molecular transformations occur over a longer time scale. The neural signals are known to appear within a few milliseconds of the absorption of light, and it now seems clear that the metarhodopsin I to metarhodopsin II transition is responsible for generating the impulse. The questions of how photoexcitation of a single pigment molecule can cause the passage of current across the synapse at the distant end of the receptor cell, and how the minute energy of a single photon can bring about the considerable displacement of charge required for a nerve impulse cannot yet be answered. Is is thought that the pigment molecules in some way control ionic channels through the internal membranes of the receptor cells. Photoexcitation of a pigment molecule causes a change in its shape or charge. This opens a channel and releases a flow of ions which polarise the cell and thus stimulate the synapse. Energy must then be provided for ion pumps to restore the normal ion distribution. Such a mechanism could account for the required amplification and rapid response over a relatively long distance.

### 9.4.3    Regeneration of rhodopsin

The photobleaching process results in the breakdown of rhodopsin into its constituent parts, opsin and retinaldehyde. The retinaldehyde liberated is the all-*trans* isomer which must be converted back into the 11-*cis* form before recombination with opsin can be achieved. This process, however, is more than just a simple isomerisation. The several possible transformations are summarised in fig. 9.13. All have been demonstrated experimentally.

The enzymic reduction of *trans*-retinaldehyde to *trans*-retinol by an alcohol dehydrogenase enzyme takes place in the receptor membrane and requires NADH. The *trans*-retinol is transported out of the receptor into the pigment epithelium, where it can be esterified with fatty acids, mainly palmitic and stearic. The retinol and retinyl esters derived from the visual pigment mix with the stores maintained in the pigment epithelium. Isomerisation to the 11-*cis* form at each of the stages, retinaldehyde, retinol and retinyl esters, has been demonstrated. Cleavage of the *cis*-retinyl esters (site unknown) gives 11-*cis*-retinol which can be reoxidised, in the receptor, to the aldehyde. The

Fig. 9.13. Possible routes for the transformation of all-*trans*- into 11-*cis*-retinaldehyde during regeneration of rhodopsin.

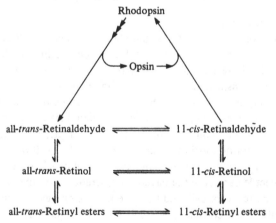

recombination of 11-*cis*-retinaldehyde with opsin is thought to occur spontaneously.

Different animal species may employ different routes for the normal regeneration of their rhodopsin. For example, direct re-isomerisation of all-*trans*- to 11-*cis*-retinaldehyde may be most important in the rat, whereas the more roundabout route *via* retinol and retinyl esters is thought to be the main process in cattle.

Regeneration of rhodopsin *via* retinol and retinyl esters is quite slow compared to the photobleaching process, and may take a few minutes to go to completion. This route is therefore only able to produce sufficient quantities of rhodopsin to keep the visual cycle going in low light intensities. In broad daylight, when light intensity is high, the rhodopsin will remain essentially fully bleached since the rate of regeneration is much slower than that of bleaching. In contrast to this, the cone photopsins which mediate daytime colour vision operate in continuous, high light intensities, so there must be a rapid process for their regeneration. The possibility of a photoregeneration mechanism, involving light-catalysed isomerisation of all-*trans*-retinaldehyde, has been suggested.

The route for rhodopsin regeneration *via* pools of retinol and retinyl esters in the pigment epithelium provides for the efficient renewal of the retin-aldehyde part of all the visual pigments of rod and cone cells. Losses of retinaldehyde are made good from the stores in the pigment epithelium. These stores are replenished as required by transport of retinol and its esters from the extensive supplies maintained in the liver. In extreme cases of vitamin A (retinol) deficiency, liver stores become depleted and, although the supplies in the retina are held most tenaciously, deficiencies there may eventually occur. It then becomes impossible to make good any losses and hence to regenerate the visual pigment. Blindness results, at first reversible, but becoming permanent if the deficiency is prolonged, due to denaturation of the unstable free opsin. The idea that carrots help one to see 'in the dark' (*i.e.* in low-intensity light) has a solid scientific foundation, since carrots are an excellent source of the provitamin A, $\beta$-carotene, and thus help to prevent vitamin A deficiency and the blindness that this deficiency causes.

Some synthesis and turnover of other components of the receptor membrane, *i.e.* protein (opsin), phospholipid and carbohydrate, have been demonstrated.

### 9.4.4    Interconversion of retinol and 3,4-didehydroretinol

Some species of amphibians and fish can transform the retinol in their eyes into 3,4-didehydroretinol (or *vice versa*) so that the pigment in the receptor changes from a rhodopsin to a porphyropsin. The opsin in the receptor has the ability to combine with the 11-*cis* isomers of either retinaldehyde or 3,4-didehydroretinaldehyde, so the type of visual pigment formed depends upon which chromophore is presented to the opsin. Some mechanisms must therefore exist by which the pigment epithelium can bring about the interconversion of the two retinols, and also select the correct retinol required at any particular time.

In other species, particularly of fresh-water fish, it appears that retinol and 3,4-didehydroretinol are formed from different carotenoids, *e.g.* β-carotene is the precursor of retinol, whereas 3,4-didehydroretinol is formed from xanthophylls such as lutein (**9.9**) and perhaps astaxanthin (**9.10**). The mechanisms of these transformations, especially the removal of the ring oxygen substituents, have not been elucidated.

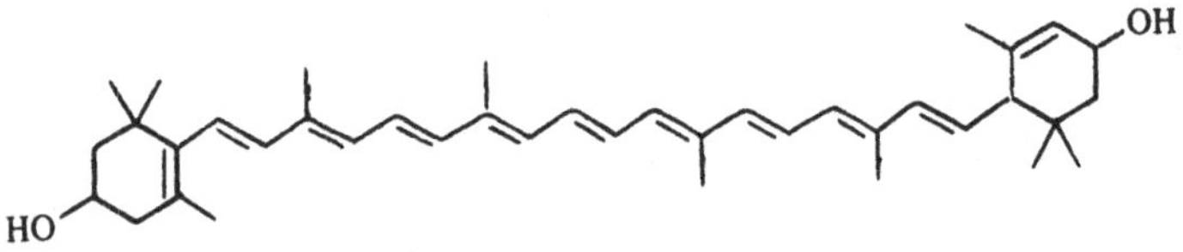

(**9.9**) Lutein

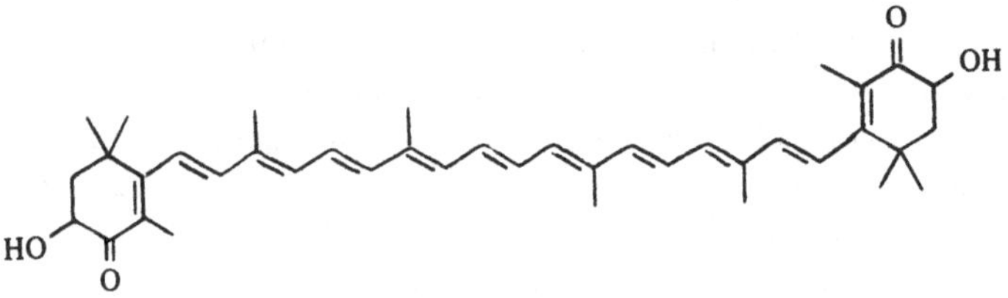

(**9.10**) Astaxanthin

Changes between rhodopsin and porphyropsin production frequently occur at a significant stage in the development of the animal, *e.g.* adult amphibians usually have only, or predominantly, rhodopsins whereas premetamorphic tadpoles utilise porphyropsins. In other cases, fishes for example, such changes either may be seasonal and related to the level of light available or may occur in response to change from a salt-water to a fresh-water environment.

### 9.4.5  Visual cycles in invertebrates

The visual pigments of many invertebrate animals, especially crustaceans and cephalopod molluscs, have been shown to be of the rhodopsin type, similar to those of vertebrates. The visual cycles are similar in principle but differ in detail. For example, the transformations that follow irradiation do not result in the liberation of retinaldehyde and opsin, but terminate at the stage of a metarhodopsin in which *trans*-retinaldehyde remains bound to the opsin. The metarhodopsin can be in the protonated (acid, $\lambda_{max}$ *ca* 500 nm) or the unprotonated (alkaline, $\lambda_{max}$ *ca* 380 nm) form. The absorption maxima of these metarhodopsins, *e.g.* from squid, are at significantly longer wavelengths than those of the corresponding vertebrate *N*-retinylidene opsins, showing that there is still some degree of interaction between the chromophore group and the protein. Circular dichroism studies show that there is no appreciable difference between the degree of protein helical conformation present in the metarhodopsin and in the original rhodopsin. However, the induced circular dichroism, indicative of a twisted conformation of the retinaldehyde chromophore, is given by the rhodopsin but not the metarhodopsin.

Even when the eye contains several visual pigments with different $\lambda_{max}$ the metarhodopsins produced from the different forms have very similar absorption maxima. Those secondary opsin–chromophore interactions which are responsible for the different light absorption properties of the visual pigments are thus lost during the formation of metarhodopsin.

Spectroscopic measurements on eyes and retinas of many invertebrate species suggest that there are many other photobleaching mechanisms and intermediates which remain to be investigated.

*Photoregeneration.* Regeneration of the visual pigment from the metarhodopsin in these animals normally takes place by a mechanism fundamentally different from that in the vertebrates. The relatively stable metarhodopsin is directly converted back into the parent pigment by light. Photoregeneration can take place from either the acid or the alkaline form of metarhodopsin. An intermediate in the regeneration process, termed 'P$_{380}$', has been detected in some species of squid and octopus. It is thought that this intermediate is produced when light isomerises the *trans*-retinaldehyde chromophore of acid metarhodopsin to the 11-*cis* configuration. The small conformational change which is required to complete the secondary chromophore–protein interactions of rhodopsin then occurs rapidly and in the dark.

Regeneration can be brought about biochemically under some circumstances, but the photoinduced process is normally much more important. It appears that the visual pigment level in some insects is controlled by photoregeneration during the day but biochemical regeneration is used at night.

Another route has been described for the photoregeneration of rhodopsins in the eyes of some cephalopods. Retinochrome, a second photosensitive pigment in the layers behind the receptors, apparently acts as a photocatalyst for regeneration *in vivo*. The identity of retinochrome and the mechanism of its action have not been established.

**9.5    Some aspects of colour vision**

Rhodopsin in the retinal rod cells is the pigment responsible for vision in low intensity light. The sensitivity of rod rhodopsin is greatest towards light of wavelengths around 500 nm, the $\lambda_{max}$ of the pigment, but the process is simply one of light detection, not a means of discriminating between different wavelengths, *i.e.* different colours. In most vertebrates, including man, other retinal receptor cells, the cones, are responsible for colour vision and they employ a different set of visual pigments, **photopsins**, for this purpose.

*9.5.1    Human colour vision*

For obvious reasons it is human colour vision that has been studied in greatest detail. This is a trichromatic process which depends upon the relative responses of three colour receptors sensitive in different parts of the visible spectrum. These colour receptors, the cone cells, are most numerous in the central or foveal area of the retina, which therefore shows the greatest colour sensitivity. The three different cone receptors each contain a different visual pigment, which determines the spectral sensitivity of that receptor. The three human pigments have $\lambda_{max}$ at 440, 535 and 575 nm and are therefore sensitive to blue, green and red, respectively. The different forms of human colour blindness are usually caused by the absence of one or more of these cone receptor pigments, so that there is no response to light of the wavelength or colour usually absorbed by that pigment. For example, a person lacking the 575 nm (red-absorbing) pigment can detect only blue and green, and is insensitive to light of longer wavelengths.

Human colour vision is inoperative at low light intensities (*e.g.* moonlight) because the sensitivity of the cone cells is only about one-thousandth of that of rods. However, the pigment concentration is much the same as in rod cells, and the photosensitivity of the isolated cone pigment seems to be no less than that of rod rhodopsin.

Structurally the cone pigments are similar to rod rhodopsin, *i.e.* they are complexes of an 11-*cis*-retinaldehyde chromophore and an opsin. Variations in the conformation of the opsin and the secondary binding of the chromophore are responsible for the variations in light absorption maxima. The cone pigments are more difficult to isolate than rod rhodopsin and consequently details of their structures and of the intermediates in the respective visual cycles are less fully understood. The mechanisms of photobleaching, pigment regeneration and production of neural impulse that occur in cone vision are thought to be similar in principle to those employed in rod photoreceptors, but few supporting data have been obtained.

*9.5.2    Colour vision in other animals*

By no means all animal species possess colour-discrimination mechanisms and cone photoreceptor pigments; many mammals, including

cattle and sheep, are thought to be colour blind. Many vertebrate and invertebrate animals, however, can see and distinguish colours. They too use a range
of photoreceptors maximally sensitive to different light wavelengths, and a
trichromatic system like that of the human eye seems to be most common.
Thus the goldfish (*Carassius auratus*) has three types of cone receptor cell
and three visual pigments with $\lambda_{max}$ at 455, 530 and 625 nm, respectively. In
this case, as in many fresh-water fish, the pigments are porphyropsins, *i.e.* have
3,4-didehydroretinaldehyde as their chromophore.

In many amphibians, the situation is particularly complex. Frogs commonly
have two types of rod receptor, termed 'red' and 'green' rods, which absorb
green and blue light, respectively. In addition there are different cone types,
including double cones, and two or three different cone pigments sensitive
to different spectral ranges. In the adults these are rhodopsins with 11-*cis*-
retinaldehyde as chromophore; in the tadpoles the pigments are very similar
but contain 11-*cis*-3,4-didehydroretinaldehyde (porphyropsins).

Invertebrate animals also may exploit colour vision. Bees have four colour
discriminator pigments with absorption maxima in the ranges 300–340, 400–
480, 480–500 and 500–650 nm. The pigment with maximal sensitivity in the
300–340 nm range allows the insects to see the long-wave u.v. as a colour.
Detailed studies of the biochemistry of colour vision in invertebrates are,
however, sadly lacking.

In the past, trivial names have often been used for visual pigments which
have either been isolated from various species or merely detected spectroscopically. Thus pigments that appear yellow, blue or violet have been called
'chrysopsin', 'cyanopsin' and 'iodopsin', respectively, and the term
'rhodopsin' has been used for red or purple pigments, regardless of their
origin. No information about the structures of these pigments or about
possible correlations between different animal species should be assumed
when these names are encountered.

## 9.6    Accessory pigments in vision
### 9.6.1    *Vertebrates*
*Melanins.* Melanins are present in various tissues of the vertebrate
eye. The choroid complex and pigment epithelium at the back of the eye
contain melanin granules as masking pigments which absorb stray light of all
wavelengths. The choroid–pigment epithelium melanin of several mammalian
species has been isolated and found to be an indolic eumelanin (chapter 7)
bound to protein.

In many mammals, including humans, melanins are also present in the
back of the iris and form a screen which prevents the red colour of the blood
capillaries from being seen. This red colour is revealed in the eyes of albino
animals which lack the melanin layer. Brown and yellow eyes are coloured by
melanin granules in the stroma of the iris, whereas the blue colour of human

and some other eyes is caused by light scattering by minute particles of protein or melanin in the iris. Melanin in the iris is probably not important in the visual process.

*The tapetum lucidum.* A reflecting layer lies behind the receptor layer in the retina of a number of animals which are essentially diurnal but also have good twilight vision. This tapetal layer or **tapetum lucidum** may be in front of, behind, or even part of the pigment epithelium in different species. The function of the tapetum is to return light that is not absorbed by the photoreceptors on its first passage through the retina, back again through the receptor layer. The sensitivity of the eye is thus increased. The tapetal layer usually consists of an array of crystals that act as reflectors and are responsible for the phenomenon of eyeshine. Pteridines and purines such as guanine (**9.11**; see chapter **6**) are commonly used for this purpose. A number of animals, *e.g.* cat, lemur, bushbaby, have a yellow tapetal layer of riboflavin crystals (**9.12**). This provides a mechanism for enhancement of the blue-sensitivity of the eye, for riboflavin absorbs blue and u.v. light of 450 nm and below and re-emits it as fluorescence at 520 nm, a wavelength more efficiently absorbed by rhodopsin in the receptors.

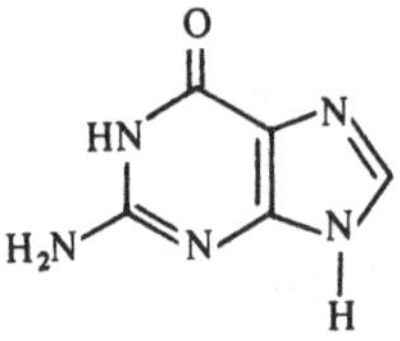

(9.11) Guanine 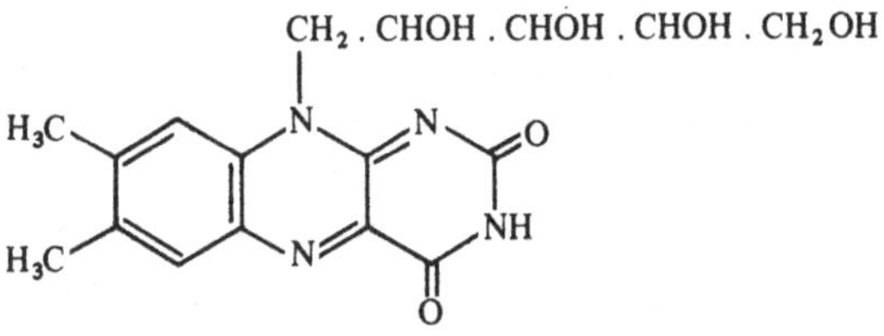        (9.12) Riboflavin

*The macula lutea.* Near the centre of the retina of man and other diurnal primates is a region of high acuity, the foveal region. This has the minimal covering of neural tissue but on the vitreous side it has a yellow filter layer called the **macula lutea**. The probable function of this is to absorb some of the blue light around 450 nm. This selectively lowers the sensitivity of the cone receptors to blue light and serves to reduce blurring of the image due to short-wavelength chromatic aberration of the lens. From its absorption spectrum the human macular pigment appears to be a carotenoid, but it has not been characterised.

*Oil-droplet light filters.* Before reaching the photoreceptor membranes in the vertebrate retina, light must pass through the inner segment of the receptor cell. Some reptiles and birds have coloured oil droplets in these inner segments. The oil droplets are usually as large in diameter as the photosensitive outer segments, so that all light must pass through them before reaching the visual pigment. The chicken retina contains six receptor cell types, and oil droplets are present in the five types of cone cells. Red, orange-yellow, lemon-yellow and even colourless but u.v.-absorbing oil droplets have been identified. The different droplets are associated with different morphological types of cone cell.

The coloured oil droplets have been found to contain extremely high (approx. 1M) concentrations of free carotenoids. In the turkey, astaxanthin (**9.10**), lutein (**9.9**) and galloxanthin – a $C_{27}$ apo-carotenoid ($10'$-apo-$\beta$-carotene-3,10$'$-diol, **9.13**), presumably derived from lutein or zeaxanthin (**9.14**) – have been identified in red, yellow and almost colourless droplets, respectively.

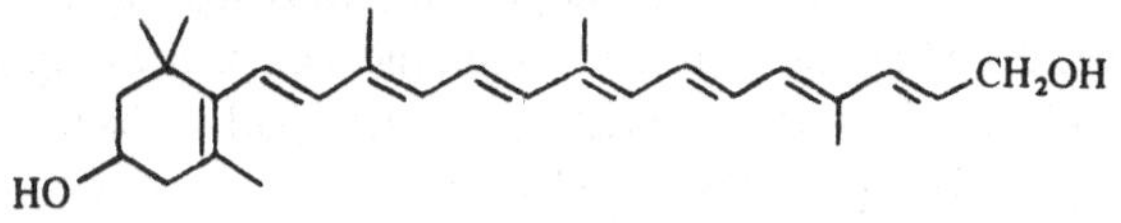

(**9.13**) Galloxanthin

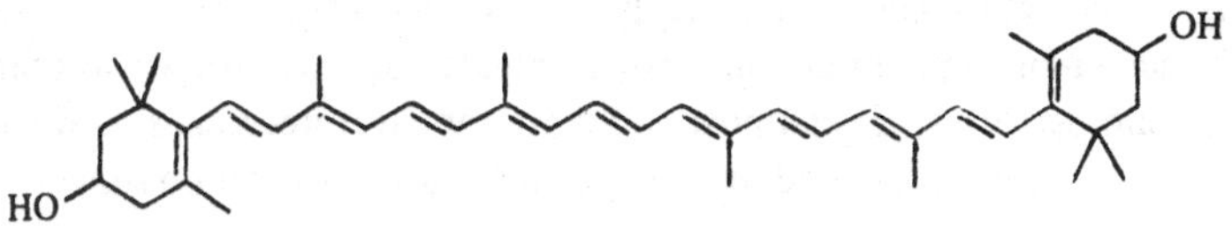

(**9.14**) Zeaxanthin

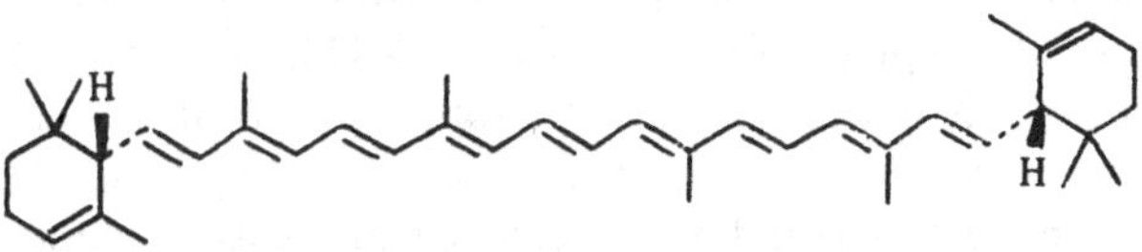

(**9.15**) (6S,6'S)-$\epsilon$,$\epsilon$-Carotene

The most interesting revelation is that the yellow droplet carotene, $\lambda_{max}$ 440 nm, which was previously referred to as 'sarcinene' has now been shown to be the very rare (6S,6'S)-$\epsilon$,$\epsilon$-carotene (**9.15**). The question of the origin of astaxanthin and especially of this isomeric form of $\epsilon$-carotene is an intriguing one, since these pigments are not likely to be present in the birds' diet (see §2.8.1).

Whatever the origin of the pigments may be, the role which these oil droplets play in the avian retina seems clear. They improve wavelength discrimination in colour vision. The differently coloured oil droplets will absorb different light wavelengths, so that a different range of wavelengths will reach the receptor membrane to be absorbed by the visual pigment. This therefore provides a colour-discrimination mechanism in which only a narrow band-width of light reaches the receptor, and the sensitivity ranges of the different receptor cells will be well separated.

Similar coloured oil droplets are present in the retinas of some reptiles. Those in various species of turtle have absorption spectra typical of carotenoids, but they have not been characterised. A colour-discrimination mechanism similar to that of birds is thought to operate.

### 9.6.2    The screening pigments of invertebrate eyes

The outline description of the invertebrate compound eye and ommatidium (see §9.2.2) drew attention to the presence of screening pigments in the outer zones of the receptor units. Static pigmented areas in the palisade of cells surrounding the whole ommatidium form a screen which effectively isolates the ommatidium and prevents stray light from travelling between ommatidia. Only light directed along the axis of the ommatidium is able to stimulate the receptor membranes; all other light is absorbed by the screening pigments. This provides a mechanism for very accurate form and movement discrimination. Other screening cells contain pigment granules which, in bright light conditions, can be dispersed so as to diminish the intensity of the light reaching the deeply situated receptor pigments. In dim light, the screening pigment granules are aggregated and withdrawn so that light from almost any angle can now stimulate the photoreceptors. Sensitivity to light is thus greatly increased, but shape and movement can no longer be discerned so accurately.

The screening pigments used in these structures may be melanin (see chapter **7**) or, particularly in arthropods, ommochromes (*e.g.* xanthommatin (**9.16**), chapter **6**) and pterins (*e.g.* drosopterin (**9.17**), chapter **6**). Pterins and ommochromes have characteristic absorption maxima in the visible region and may also serve to some extent as colour filters. Variations in eye colour, *i.e.* in ommochrome and pterin screening pigments, in mutant strains of *Drosophila* have formed the basis of many of the extensive fundamental genetic studies that have been performed with this fly.

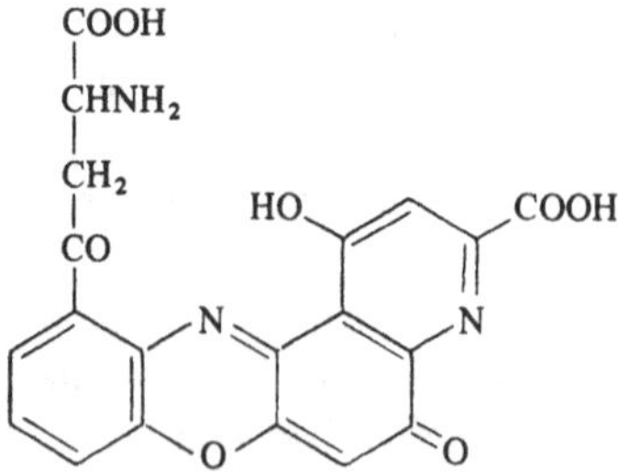

(9.16) Xanthommatin

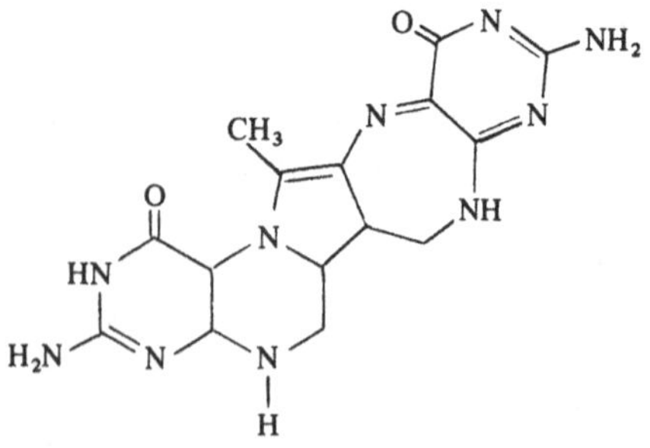

(9.17) Drosopterin

## 9.7.    Conclusions and comments

Many different eye structures may be found in the animal kingdom but, so far as is known, these all use basically the same retinaldehyde–protein visual pigments and mechanisms. There are many refinements to this basic mechanism (*e.g.* pigments with different $\lambda_{max}$, colour filters) to provide for optimal efficiency in low-intensity-light detection or colour discrimination under the prevailing conditions of ambient light, *etc.*, in the natural habitat (*e.g.* terrestrial, aquatic) of a particular animal. Overall though, the similarity of the visual cycles and visual pigments is most striking. The retinaldehyde–opsin complexes seem to be ideal for the purpose of light detection, so much so that it has been suggested that these visual-pigment systems have arisen quite separately at least three times during the course of animal evolution.

With the rapid advances now being made in biochemical and physico-chemical techniques, we can expect, with some confidence, that more and more information will be forthcoming about the visual cycles and their intermediates, and opsin–chromophore interactions, especially for rod rhodopsin. It will, however, be a long time before full structural details are known for some of the transient intermediates, so that the significance of the subtle changes in conformation, protein–chromophore interactions and light absorption properties can be appreciated. Much work is also needed before the mechanism of nerve impulse generation in response to photon absorption by the visual pigments is properly understood. Even when some of these questions can be answered for the functioning of rhodopsin in the few species that are studied most extensively (man, rat, cattle) the amount of work remaining to be done on the biochemistry of mammalian colour vision and the characterisation of the visual pigments and cycles in other animals will be enormous.

## 9.8    Suggested further reading

Vision is of great importance to the efficient functioning of almost all animals, including ourselves, and is therefore a subject which is treated in varying degrees of detail in textbooks of biochemistry and physiology. The number of review articles dealing with a wide range of topics related to vision is enormous. However, for a very comprehensive treatment of all aspects of vision, the reader is referred to a set of volumes entitled *The eye*, edited by Davson (1977). In particular, one volume, *The photobiology of vision* (Knowles and Dartnall, 1977) is most useful to the biochemist, and contains an extensive list of references to earlier work. Ideas about the molecular biology and biophysics of vision are still changing rapidly, and conflicting viewpoints are not uncommon. A very recent review article (Ottolenghi, 1980) presents the most up-to-date information and evaluates the various theories.

## 9.9    Selected bibliography

Davson, H. (ed.) (1977) *The eye*, 2nd edition, 5 volumes. New York, London and San Francisco: Academic Press.

Knowles, A. and Dartnall, H. J. A. (1977) *The photobiology of vision* (*The eye*, 2nd edition, ed. H. Davson, vol. 2B). New York, London and San Francisco: Academic Press.

Ottolenghi, M. (1980) The photochemistry of rhodopsins, *Adv. Photochem.*, **12**, 97.

# 10    Photosynthesis

## 10.1    Introduction

Photosynthesis is the process by which green plants, algae and some bacteria are able to harness solar light energy and convert it into a useful chemical form for the biosynthesis of cell components. Not only is the chemical energy thus produced stored and used by the plant, but also it provides, through the food chains, the primary energy source for non-photosynthetic organisms, especially animals. Plant photosynthesis also supplies the vitally needed oxygen. All life on this planet therefore depends upon photosynthesis.

In plant photosynthesis (higher plants and algae) (fig. 10.1) light energy is harnessed and used to split molecules of water. This simple process, the light reaction, results in the release of oxygen and in the generation of 'reducing power' which is later used, in a series of 'dark reactions', to fix carbon dioxide in a useful form as carbohydrate. This carbohydrate can be used as an energy store and as the source of carbon for all other molecules needed and used by the plant. During photosynthesis ATP production occurs by a coupled photophosphorylation mechanism.

As can be seen from fig. 10.1, four quanta or photons of light are used in the splitting of two molecules of water to release one molecule of oxygen and produce four reducing equivalents. Four further quanta provide the energy to bring about the transfer of the four reducing equivalents, eventually

Fig. 10.1. The overall process of plant photosynthesis: (a), (b) the light-utilising steps or light reactions; (c) the subsequent dark reactions of carbon fixation.

(a)    $2H_2O \xrightarrow{4h\nu} O_2 + 4H^+ + 4e^-$

(b)    $4H^+ + 4e^- + 2NADP^+ \xrightarrow{4h\nu} 2NADPH$

(c)    $2NADPH + CO_2 + 2H^+ \xrightarrow{ATP} [CH_2O] + 2NADP^+ + H_2O$

yielding NADPH which is used, along with ATP, for the reduction of a $CO_2$ molecule and subsequent formation of carbohydrate. It thus requires a minimum of eight quanta or photons to bring about the reduction of one $CO_2$ molecule and the evolution of one molecule of $O_2$.

Fig. 10.2. The overall process of bacterial photosynthesis.

$$CO_2 + 2H_2A \xrightarrow{h\nu} [CH_2O] + 2A + H_2O$$

Some green and purple bacteria which normally live in an anaerobic environment such as mud or stagnant water are also phototrophic and are capable of reducing $CO_2$ to carbohydrate. No $O_2$ is evolved during bacterial photosynthesis. These bacteria are not able to use light energy to split water, but use other hydrogen (electron) donors, $H_2A$ (fig. 10.2). The three major groups of photosynthetic bacteria use different photosynthetic electron donors. The purple non-sulphur bacteria of the Rhodospirillaceae (formerly Athiorhodaceae), *e.g. Rhodospirillum rubrum*, normally utilise simple organic molecules, while the green sulphur bacteria (Chlorobiaceae, formerly Chlorobacteriaceae), *e.g. Chlorobium* spp. normally utilise inorganic sulphur compounds, *e.g.* $H_2S$, or hydrogen. The purple sulphur bacteria of the Chromatiaceae (formerly Thiorhodaceae), *e.g. Chromatium* spp., can use either organic molecules or inorganic sulphur compounds.

Details of the pathways of carbohydrate synthesis and the mechanisms of photophosphorylation are really beyond the scope of this book. However, pigments are fundamental to the harvesting and utilisation of light energy. The light-harvesting role played by chlorophyll in photosynthesis is probably the most obvious demonstration of a specific biological photofunction for a natural pigment. The functioning of carotenoids and phycobilins as accessory pigments is also directly related to their light-absorbing properties. Other coloured molecules, including cytochromes and flavoproteins, are involved in photosynthesis as part of the electron transport systems, although the ability of these compounds to absorb visible light has no relevance to their functioning. The discussion that follows will concentrate mainly on describing how the light-absorbing pigments are disposed within the photosynthetic apparatus of higher plants, algae and photosynthetic bacteria, and the mechanisms by which these pigments are thought to act in the harnessing and utilisation of light energy.

## 10.2    The photosynthetic apparatus of eukaryotes: the chloroplast
### 10.2.1    Chloroplast morphology
The active pigments in photosynthetic cells are situated within lamellar membranes in functionally organised units. In photosynthetic eukaryotes (higher plants and most algae) the pigment-bearing membranes are confined within a specific organelle, the **chloroplast**. Morphological

variation among higher-plant chloroplasts is small, but in the algae consider-able differences are encountered in chloroplast shape and size. *Chlorella*, for example, has a single, cup-shaped chloroplast whereas some species of *Spirogyra* have a long, spirally wound chloroplast which extends over the full length of the cell.

A generalised higher-plant chloroplast structure is illustrated in fig. 10.3(*a*). A plant leaf cell may contain several hundred such chloroplasts, each approxi-mately elliptical or lens-shaped and some 3–10 μm in length. Essentially this organelle consists of an outer double membrane or **envelope** enclosing a matrix, the **stroma**, which contains the internal photosynthetic membranes.

## 10.2.2  *The chloroplast envelope*

The chloroplast envelope is a continuous double membrane which acts as a selective barrier to the transport of metabolites into or out of the organelle. It is also thought that the inner envelope membrane may play some role in the formation of new internal lamellae. In some plant species there

Fig. 10.3. (*a*) A generalised representation of a higher plant chloroplast. (*b*) Schematic enlargement showing the main structural features of the thylakoid membranes.

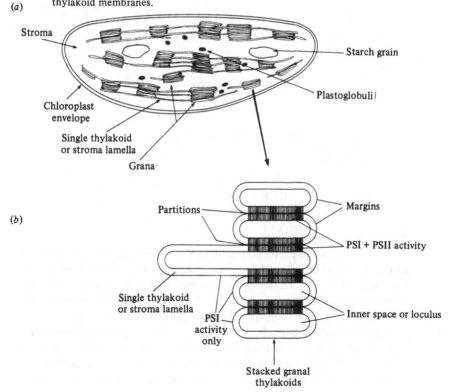

is an extensive system of tubules and vesicles contiguous with the inner chloroplast envelope membrane. This system, known as the **peripheral reticulum**, appears to be a normal feature in plants which possess the $C_4$ carbon fixation pathway (see §10.6), but it has also been detected in some $C_3$ plants, particularly under conditions of environmental stress. The chloroplast envelope contains no chlorophyll, but the presence of carotenoids, notably zeaxanthin (**10.1**), antheraxanthin (**10.2**) and violaxanthin (**10.3**), and their enzymic interconversion within the chloroplast envelope have been reported. There is increasing evidence that the envelope may be an important site of synthesis of chloroplast materials, especially galactolipids, prenylquinones, and perhaps some carotenoids.

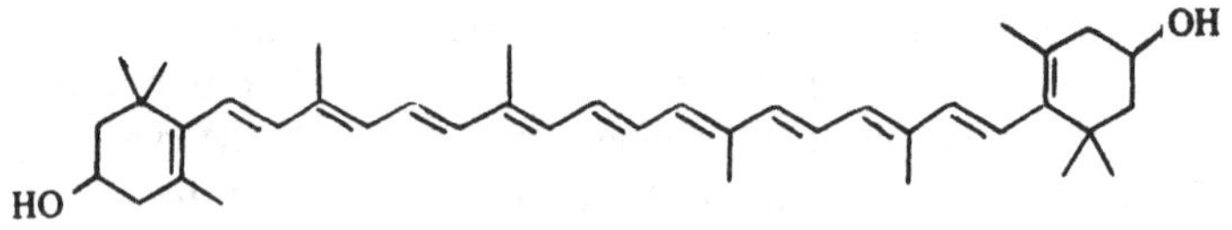

(10.1) Zeaxanthin

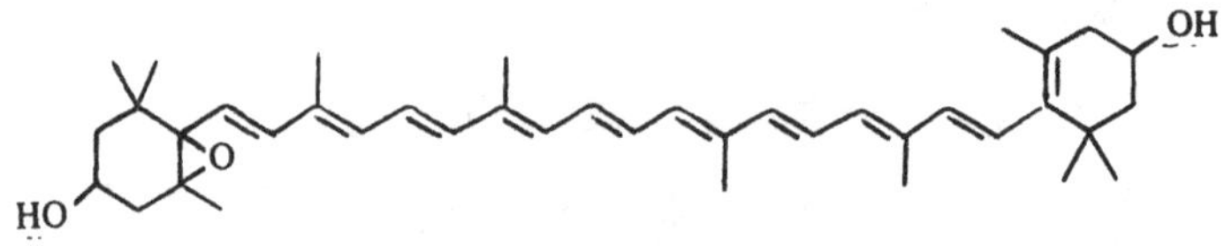

(10.2) Antheraxanthin

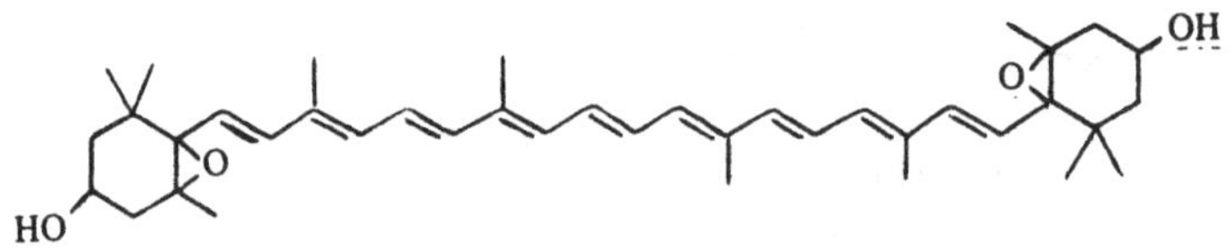

(10.3) Violaxanthin

### 10.2.3    The stroma

The internal proteinaceous matrix of the chloroplast is known as the **stroma**. In addition to the photosynthetic lamellar membranes (see below), electron microscopy also reveals the presence, within the stroma, of various other particulate structures. These include **ribosomes** and strands of DNA, which play a role in chloroplast self-regulation and replication, **grains** of the storage polysaccharide **starch**, osmiophilic globules or **plastoglobuli** which perhaps serve as extra-lamellar pools of membrane lipids, and the **pyrenoids** (in some algae) or related **stroma centres** (higher plants). The exact function of these last two small, dense structures is unknown, but they are largely composed of protein and are known to have ribulose bisphosphate carboxylase activity.

### 10.2.4 The photosynthetic lamellar membrane system

It is the extensive system of internal membranes in the chloroplast that contains the photosynthetic pigments and is thus the site at which photosynthesis takes place. Electron microscopy reveals that among the internal membranes can be recognised a series of sac-like discs, called **thylakoids**, stacked one upon another to form **grana** (fig. 10.3*b*). The interior volume enclosed by one thylakoid is known as the **loculus**, the end portion of the thylakoid, in contact with the stroma, is the **margin**, and a region where two thylakoids are tightly appressed is termed a **partition**. The grana stacks are interconnected by membranes that have been called **single thylakoids, stroma lamellae** or **intergranal lamellae**.

### 10.2.5 Ultrastructure of the chloroplast thylakoid membrane

The structures of both single and granal thylakoid membranes have been studied in detail by electron microscopy and immunological techniques. A great deal is also known about the chemical compositions of these membranes.

The main ultrastructural features are indicated in fig. 10.4. Like most biological membranes the thylakoid is composed largely of protein and lipid, in approximately equal amounts. The membrane structure is asymmetric, and has a basic lipid matrix present in two layers, the outer layer being thicker. Embedded in this lipid matrix are large numbers of proteinaceous particles which are seen on both the outer and inner surfaces of the thylakoid and also on both sides of the hydrophilic middle zone or interior of the membrane where cleavage occurs during sample preparation by the freeze-fracture technique. The particles enjoy quite free movement in the fluid lipid matrix. These particles, up to 6000 per $\mu m^2$, make up about 70% and 50% of the area of the grana and stroma thylakoids, respectively. Photochemical activity,

Fig. 10.4. Schematic representation of the ultrastructure of stacked granal and single thylakoid membranes.

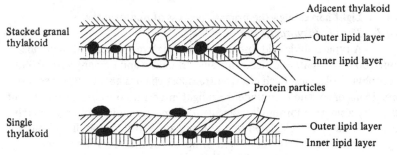

electron transport and metabolic enzyme activity are associated with various types of the particles, and the reactions catalysed are orientated vectorially across the membrane.

There are functional as well as structural differences between the single stromal and the stacked granal thylakoids. As will be explained below, there are two photochemical reactions in photosynthesis, which take place in aggregates known as **photosystems I** and **II (PSI** and **PSII)**. The photosystem and electron transport complexes are located in the membrane structure at specific, though not necessarily regular, intervals in the form of globular clusters. The location of the photosynthetic pigments within the photosystem complexes will be described in more detail below (see §*10.4.2*). However, in order to appreciate fully the significance of the location of the pigments in specific sites or complexes, and the roles which the pigments play, it is first necessary to consider the overall process of photosynthesis.

### 10.3    Plant photosynthesis – a general outline

The thylakoid membrane system of the chloroplast transforms the energy of light into a form which can be used to bring about chemical reactions. The overall process of photosynthesis has been summarised schematically in fig. 10.1. In the discussion which follows, photosynthesis is considered in three phases beginning with the light reactions, that is the primary processes by which light energy is absorbed by light-harvesting pigments and transferred into photochemical **reaction centres**. The second phase deals with the way that the absorbed light energy is used to drive electron transport from water to $NADP^+$. During this electron transport, a charge or proton gradient is established across the functional vesicles of the membrane. The third phase to be considered is that in which NADPH, produced by the electron transport system, and ATP, generated as a result of the electrochemical potential difference of the proton gradient, are used for $CO_2$ fixation and carbohydrate synthesis. Although for sake of simplicity the processes of photosynthesis are described as three phases, it must be remembered that light harvesting, electron transport and electrochemical gradient generation are actually very tightly coupled.

### 10.4    Light harvesting – the primary process of photosynthesis
#### 10.4.1    The photosynthetic unit

A typical thylakoid disc, *e.g.* from a mature spinach leaf, contains about 200 photosynthetic electron transport chains and as many as 100 000 chlorophyll molecules. Each electron transport chain is able to turn over once every 15 ms or so, and even under high light intensities the absorption of light quanta by a single chlorophyll molecule is not nearly rapid enough to match this rate. It has been calculated that as many as 2500 chlorophyll molecules may be involved when a flash of light brings about the liberation of one molecule of $O_2$ and the reduction of one molecule of $CO_2$. This led to the concept of a **photosynthetic unit** in which the primary photochemical reactions of photosynthesis occur, one molecule at a time, in a **reaction centre**, and each reaction centre is associated with a relatively large number of pigment molecules, any one of which may absorb a photon and pass on the energy to the reaction centre. Every photosynthetic electron transport chain includes two distinct photochemical reactions, brought about by the two photosystems, I and II, each of which has its own reaction centre. It is now known that four electrons must be passed through both photosystems (*i.e.* a total of eight photons is required) in order to liberate one molecule of $O_2$ and reduce one molecule of $CO_2$. This leads to a figure of about 300 chlorophyll molecules required per photochemical event. This is, in fact, rather an oversimplification, the number being greatly affected by environmental factors and variations between species. However, although the figure of 300 chlorophyll molecules may be inaccurate, the concept of the photosynthetic unit appears valid. The basic photosynthetic unit for each photosystem is now seen as a complex in which the bulk of the chlorophyll molecules form an antenna of light-harvesting pigments which trap light energy and pass it on to a special chlorophyll dimer in the reaction centre, where the energy is used to drive the photosynthetic electron transport system (fig. 10.5). The most recent figures indicate that the PSI antenna contains 120 chlorophyll *a* molecules, the PSII antenna complex only 60. In addition to these two photosystems, a third light-harvesting pigment complex (LHCP) is also normally present, associated with PSII. Extensive experimental work in recent years has revealed much about the organisation of those three photosynthetic units or particles in the thylakoid membranes.

Fig. 10.5. Diagram representing the photosynthetic unit.

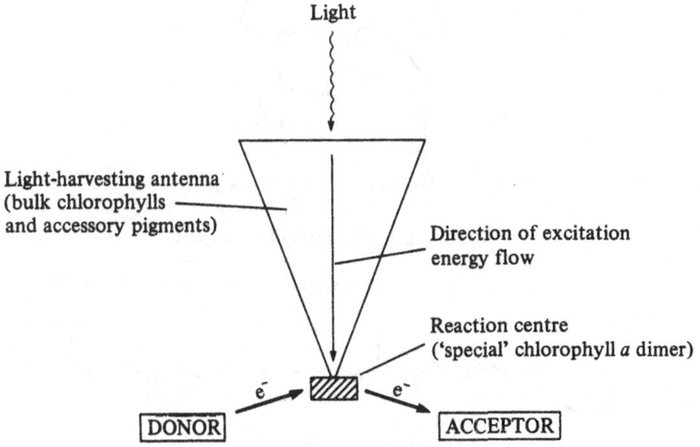

Their distribution is not uniform throughout the membrane. One school of thought believes that the PSI particles are distributed throughout both the single and granal thylakoids, whereas the large aggregates considered to contain PSII and the light-harvesting complex are found only in the stacked thylakoids of the grana. A very recent alternative model places very little PSI in the appressed grana where PSII and LHCP are mainly located, but suggests that PSI is to be found in end grana and outer thylakoids. This model requires that the spatially separated PSI and PSII be linked by a lateral shuttle of reducing equivalents. The huge plastoquinone pool may play this role.

### 10.4.2    The light-harvesting pigments

The three light-harvesting pigment systems can be isolated from detergent-treated chloroplasts, in the form of chlorophyll–protein complexes. Thus PSI can be resolved as a chlorophyll $a$-protein containing some 120 antenna molecules of chlorophyll $a$ (**10.4**) and the P-700 reaction centre (see below, §*10.4.3*). PSII also gives a chlorophyll $a$-protein complex containing the P-680 reaction centre and about 60 chlorophyll $a$ molecules. No chlorophyll $b$ (**10.5**) is present in either PSI or PSII. Both contain some β-carotene (**10.6**) though it is not known whether this forms part of the antenna pigment or is in the reaction centres.

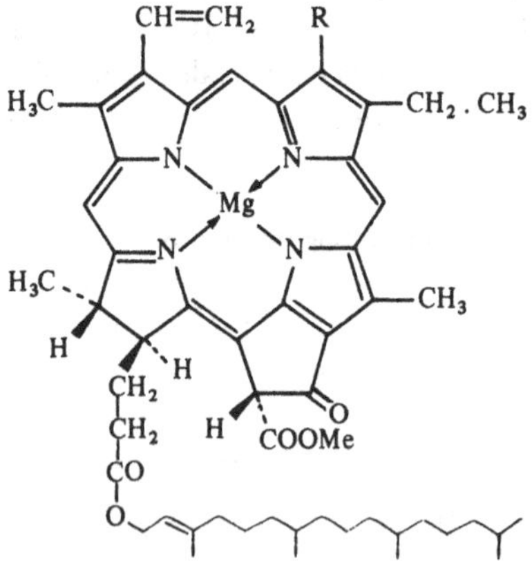

(**10.4**) Chlorophyll $a$: R=CH₃

(**10.5**) Chlorophyll $b$: R=CHO

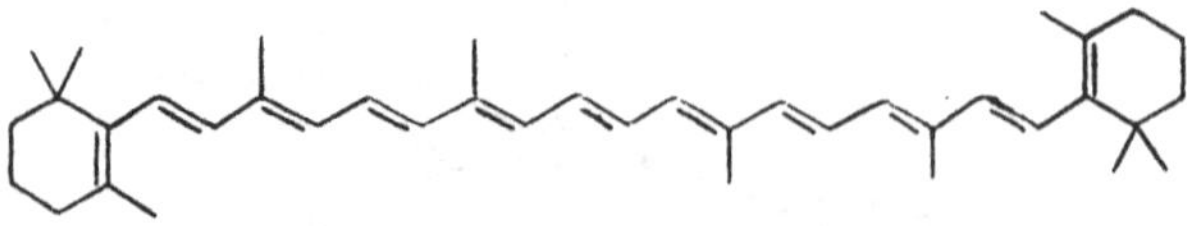

(**10.6**) β-Carotene

The light-harvesting complex can be isolated as a chlorophyll *a/b*-protein, and can be split into three chlorophyll–protein components. This light-harvesting chlorophyll *a/b*-protein (LHCP) appears to be the only site which contains chlorophyll *b*. This functions as an accessory pigment, as does the substantial amount of carotenoid that is also present, consisting mainly of xanthophylls, predominantly lutein (**10.7**). The LHCP is closely associated with PSII. Its role seems to be to increase the light-harvesting ability of PSII, in particular by widening the spectral range of light than can be used by this photosystem. Work with pea seedlings suggests the existence of a 160 Å light-harvesting particle containing the PSII–chlorophyll *a*-protein core, surrounded by four LHCP units (fig. 10.6). The possibility that light energy trapped by the LHCP may also be used to drive the PSI reaction has also been considered.

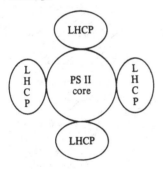

(**10.7**) Lutein

Fig. 10.6. Model of the proposed 160 Å light-harvesting particle of pea seedlings.

Several different forms of chlorophyll *a* with slightly differing $\lambda_{max}$ have been recognised spectroscopically in each light-harvesting system. The antennae of both photosystems have chlorophyll *a*-660, -670, -678 and -685 (named from the long wavelength absorption maximum), but in addition PSI has forms absorbing maximally at 690, 705 and 720 nm. The variation in absorption maxima is considered to be due to differences in the molecular environment of the chlorophyll molecules, *e.g.* association with proteins or interaction with neighbouring chlorophyll molecules. For example, the extension of the chlorophyll *a* absorption *in vivo* into the red spectral region at 720 nm is a consequence of molecular aggregation. However, a recent report suggests that there may be four chemically different forms of both chlorophylls *a* and *b*.

Quanta absorbed by the antenna pigments are passed from molecule to molecule by resonant energy transfer and the energy is funnelled into the reaction centre. The efficiency of the energy-transfer processes is high, so that energy loss by fluorescence emission or wasteful photochemistry is small. The energy transfer through the whole of the antenna system is very rapid, occurring in picoseconds. The resonant energy transfer is directed towards pigments absorbing at longer wavelength, so quanta absorbed by the accessory pigments in the LHCP, *i.e.* carotenoids (400–500 nm) and chlorophyll *b* (640–650 nm), and also by the shorter-wavelength (higher energy) forms of chlorophyll *a* may all be passed on to the longer-wavelength-absorbing chlorophylls and eventually to the reaction centre (fig. 10.7). A larger proportion of the sunlight spectrum can therefore be used.

Fig. 10.7. Diagram illustrating the passage of excitation energy *via* different pigment forms in the light-harvesting antennae of photosystems I and II.

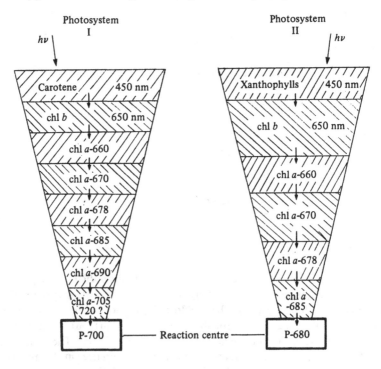

*Protective mechanism against excess light harvesting.* The size of the antenna system is such as to allow electron transport to proceed at full velocity even on a cloudy day. Consequently under bright lighting conditions (sunlight)

many more quanta may be absorbed than can be used by the reaction centre. There are various wasteful pathways by which the excess energy of the excited chlorophyll can be lost. One of these involves intersystem crossing to give the longer-lived but still high-energy triplet state ($^3$Chl). This triplet species can then pass on its excess energy to ground-state molecular oxygen, thereby raising this to the singlet state $^1O_2$. This extremely reactive oxidising species is capable of oxidising any suitable acceptor molecule, such as chlorophyll itself, thus causing extensive and perhaps lethal damage to the photosynthetic membranes. One of the major roles of carotenoids in the antennae is thought to be to give protection against such destruction by interfering in this sequence of events either by reacting preferentially with the oxidising singlet oxygen, or by quenching the excess energy of either the triplet chlorophyll or of singlet oxygen (fig. 10.8).

Fig. 10.8. Loss of excitation energy of chlorophyll (CHL) to yield singlet oxygen ($^1O_2$), and three mechanisms whereby carotenoid (CAR) can afford protection against harmful or lethal oxidation brought about by $^1O_2$.
1. CAR as preferred substrate for oxidation. 2. Quenching of $^3$CHL by $^1$CAR. 3. Quenching of $^1O_2$ by $^1$CAR.

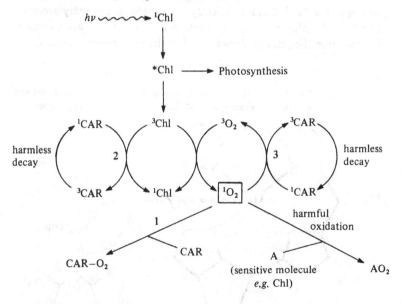

### 10.4.3 The photochemical reaction centres

The antenna and LHCP pigments simply absorb light, undergo electronic excitation and pass the excitation energy on from molecule to molecule until it reaches the reaction centre. It is at the reaction centre that this excitation energy is converted into a chemically usable form. In the reaction centres of both photosystems the active pigment is a special chlorophyll *a* dimer which undergoes reversible oxidation.

*Reaction centre I.* The reaction centre pigment of photosystem I is characterised by light-absorption changes, especially at 700 nm, which are seen after illumination of chloroplasts by a flash. The pigment responsible for this, known as chlorophyll $a_I$ or P-700, is a specific chlorophyll–protein complex containing a dimer of two chlorophyll *a* molecules. The internal structure of the dimer or 'special pair' is extremely important but not yet fully established. The tetrapyrrole rings of the monomers, however, appear to lie approximately flat in the plane of the thylakoid membrane, and models have been suggested in which a water molecule interacts between the central magnesium atom of one chlorophyll molecule and the ring V carbonyl groups of the second (fig. 10.9).

Light absorbed by the antenna system is transferred to the reaction centre and the chlorophyll $a_I$ dimer undergoes electronic excitation. This process takes less than 30 ps. Within the next 20 ns the chlorophyll $a_I$ is oxidised, probably to a radical cation, $(Chl-Chl)^+$, in which the unpaired electron is delocalised over the two porphyrin rings. The electron lost from the chlorophyll $a_I$ in this oxidation is efficiently transferred to a primary acceptor, X (= P-430). Chlorophyll $a_I$ returns to its normal state by accepting an electron from the primary donor, Y. Thus the PSI reaction centre drives this

Fig. 10.9. Model of the possible hydrogen-bonding interactions between water and the two chlorophylls of the P-700 (chlorophyll $a_I$) dimer.

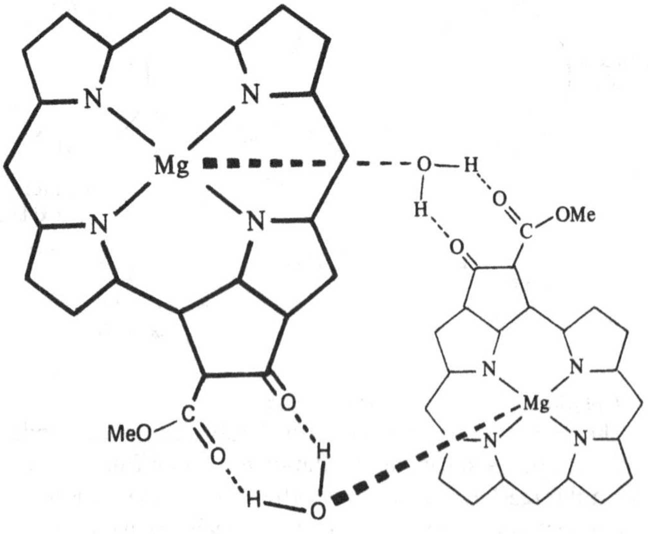

part of the photosynthetic electron transport system by effectively passing on electrons from Y to X, a process which cannot take place spontaneously (fig. 10.10a).

*Reaction centre II.* The pigment of reaction centre II is also a chlorophyll–protein complex containing a chlorophyll *a* dimer, and is known as chlorophyll $a_{II}$ or P-680. Although the different light-absorption change indicates that the chlorophyll *a* molecules are in a different molecular environment or orientation from those of P-700, the processes of light absorption and oxidation that occur in reaction centre II are similar to those of reaction centre I. Again electronic excitation energy is passed from antenna chlorophyll to the chlorophyll $a_{II}$ dimer, which undergoes excitation followed by oxidation to a radical cation with a delocalised, unpaired electron. In this case, the electron lost is passed on to Q (= X-320), the primary electron acceptor of photosystem II. The chlorophyll $a_{II}$ radical cation is then reduced by accepting an electron from a donor, Z. Photosystem II thus effectively passes on electrons from Z to Q (fig. 10.10b).

Fig. 10.10. The primary reactions at the photosynthetic reaction centres; (*a*) photosystem I; (*b*) photosystem II.

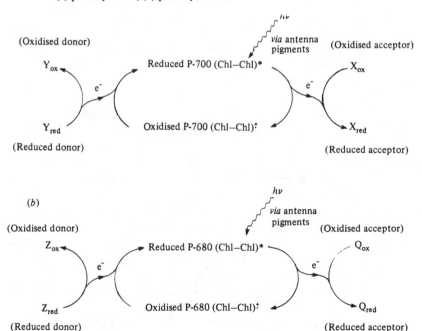

**10.5    Secondary events – the photosynthetic electron transport system**
In eukaryotic cells, the oxidative breakdown of foodstuffs results in the production of reduced nucleotides, especially NADH. A mitochondrial electron transport system is used to bring about the reoxidation of these reduced species by molecular oxygen (fig. 10.11*a*). This process is favoured thermodynamically since the redox potential of $NAD^+/NADH$ ($-0.32\,V$) is more strongly negative than that of $O_2/H_2O$ ($+0.82\,V$), or in other words molecular oxygen is a stronger oxidising agent than $NAD^+$. The reverse process in which $NAD^+$ (or $NADP^+$) is reduced to NAD(P)H by water, which in turn is oxidised to molecular oxygen (fig. 10.11*b*), is obviously energetically unfavourable and will not occur spontaneously. However, this is essentially what happens in photosynthetic electron transport. Energy must be provided to drive the reaction, and it is the light energy harnessed by the primary reactions of photosynthesis that is used for this purpose.

*10.5.1    The Z scheme*
Of the various models proposed for photosynthetic electron transport, that which seems to fit most of the data is a zigzag or Z scheme (fig. 10.12) which illustrates how light provides the energy in two separate photoreactions. In each case the harnessed light energy is used to produce a reducing agent that is strong enough to pass on electrons to (*i.e.* reduce) a series of electron transport intermediates.

Fig. 10.11. (*a*) Mitochondrial oxidation of NADH by molecular $O_2$. (*b*) Photosynthetic reduction of $NADP^+$ to NADPH by water, and liberation of molecular $O_2$.

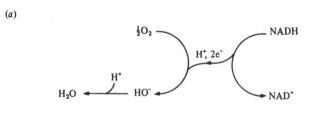

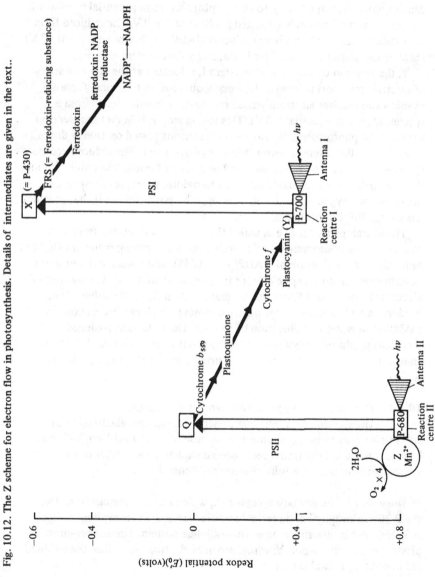

Fig. 10.12. The Z scheme for electron flow in photosynthesis. Details of intermediates are given in the text..

### 10.5.2 Photosystems I and II

Light absorbed by photosystem I provides the energy for electrons to be passed on from a donor, Y (redox potential $+0.4\,V$), *via* reaction centre chlorophyll $a_I$ (P-700), to an acceptor, X (redox potential $\sim -0.6\,V$), *i.e.* against an unfavourable potential gradient of $+1.0\,V$. The reduced form of X then passes on electrons to (reduces) $NADP^+$ (redox potential $-0.32\,V$) *via* an iron–sulphur protein, ferredoxin, and a flavoprotein enzyme.

Y, the electron donor for photosystem I, is located at the end of a series of electron transport intermediates – plastoquinone, cytochrome $f$, plasto-cyanin – and receives electrons passed on *via* these intermediates from a species, Q (redox potential $\sim 0\,V$). This substance, Q, is in fact the electron acceptor for photosystem II, and accepts electrons passed on from a donor, Z, *via* P-680, the reaction centre chlorophyll $a_{II}$ dimer. The reduced acceptor Q passes on electrons to photosystem I as just mentioned. The oxidised donor Z is a sufficiently strong oxidising agent to oxidise water, resulting in the liberation of oxygen. Thus it is light trapped by photosystem II that provides the energy for the splitting of water.

The overall picture is one in which the transfer of electrons from donor to acceptor by photosystem I effectively produces a strong reductant that can bring about the reduction of $NADP^+$ to NADPH, and a weak oxidant that is nevertheless able to accept electrons from plastocyanin, *etc*. The transfer of electrons from donor to acceptor by photosystem II, on the other hand, produces a strong oxidant (the oxidised donor) which can bring about the oxidation of water and liberation of oxygen. The reductant (reduced acceptor) produced, although weaker than that of photosystem I, is strong enough to pass on electrons *via* the plastoquinone–cytochrome $f$–plastocyanin series.

### 10.5.3 The nature of the primary acceptors and donors

The electron transport components that donate electrons to or accept electrons from the photosynthetic reaction centre chlorophylls have been identified mainly from spectroscopic studies. In a number of cases the substances have not been fully characterised chemically.

*Photosystem I.* The primary acceptor, X, which receives the electrons from the P-700 chlorophyll $a_I$ dimer, is identified from light absorption changes as P-430, and is thought to be an iron–sulphur protein. The copper-protein plastocyanin is the donor, Y, which provides electrons to reduce the oxidised chlorophyll $a_I$ radical cation.

*Photosystem II.* The primary acceptor, Q, for photosystem II exhibits a light absorption change at $320\,nm$ and is considered to be a tightly bound form of plastoquinone (**10.8**), distinct from the bulk plastoquinone which follows it

(10.8) Plastoquinone

in the sequence. C-550, an unidentified component with a difference absorption maximum at 550 nm, is closely associated with Q. The donor, Z, which provides the electrons to reduce the chlorophyll $a_{II}$ (P-680) has not been identified. It has been suggested that Z may be either tightly bound manganese or a cytochrome $b_{559}$.

### 10.5.4 The nature of the other components of the photosynthetic electron transport system

Several of the photosynthetic electron transport components are molecules which absorb light of visible wavelengths and are therefore coloured pigments. Thus the cytochromes, like those of mitochondrial electron transport systems, are iron–haem proteins which absorb at around 560 nm and are therefore red. The general properties of haems and cytochromes were outlined in chapter 5. The flavoprotein enzyme ferredoxin-NADP$^+$ oxidoreductase is yellow, having as its prosthetic group riboflavin (see chapter 6). Plastocyanin is a copper–protein, its intense blue colour probably being due to copper charge-transfer absorption. Although not coloured, the important intermediate plastoquinone (10.8) is a benzoquinone with a long isoprenoid sidechain substituent. Properties of benzoquinones have been outlined in chapter 3.

Plastoquinone fulfils several special functions in the electron transport system (fig. 10.13). It is more abundant than the other components in the chain and thus provides an electron 'buffer' which ensures smooth operation of the chain even under large fluctuations in distribution of light quanta between the two photosystems. Also the abundant plastoquinone is able to

Fig. 10.13. Interconversion of several photosynthetic electron transport chains by a large plastoquinone (PQ) pool (cyt: cytochrome; pcy: plastocyanin).

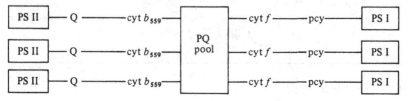

interconnect several electron transport chains and thus improves the reliability of the system. For example, if a reaction centre II fails to operate, then the cross-linking plastoquinone can ensure that the associated reaction centre I will not be starved of electrons but will be served with electrons from another reaction centre II. Another possible role for plastoquinone was mentioned earlier (§ *10.4.1*), when a new model for the distribution of the two photosystems within the thylakoids was outlined. The spatial separation of the different photosystems would require a means of ensuring a flow of electrons between them, and it was suggested that the bulk plastoquinone may be used for this purpose.

However, the ability of all these compounds to absorb visible light has no relevance to their functioning in photosynthesis. They are not involved in the primary processes of light harvesting.

## 10.6    The dark reactions

Although details of these topics are beyond the scope of this book, a brief outline of their significant features is needed to complete the picture of photosynthesis.

### 10.6.1    *Photophosphorylation*

It is now generally accepted that the photosynthetic electron transport chain is arranged vectorially and crosses the thylakoid membrane twice (fig. 10.14). In a single turnover of the chain, each reaction centre transfers one electron from the inner to the outer side of the membrane. At the same time, due to splitting of water and oxidation of dihydroplastoquinone, protons are liberated to the inner side of the membrane. An electrochemical potential gradient is thus produced, with positive charge or low pH on the

Fig. 10.14. Vectorial arrangement of the photosynthetic electron transport chain in the thylakoid membrane.

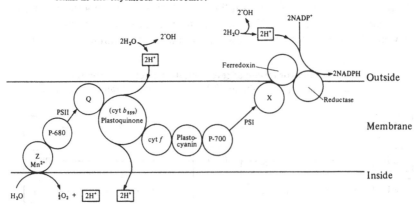

inner side of the membrane and negative charge or high pH on the outer side. According to the Mitchell chemiosmotic hypothesis, the energy of this gradient drives ATP synthesis from ADP by the membrane-bound ATPase enzyme. When both photosystems are working, electron flow is non-cyclic and ATP production is said to occur by non-cyclic photophosphorylation. If only one photosystem (usually PSI) is working, a cyclic electron flow can take place and ATP production is said to occur by cyclic photophosphorylation.

### 10.6.2 Carbon fixation

Chemical energy is trapped in two ways, as the reducing cofactor NADPH and, as in the case of mitochondrial electron transport, as ATP produced by coupled phosphorylation. The NADPH and ATP provide the reducing power and chemical energy for carbon fixation whereby $CO_2$ is effectively reduced to carbohydrate. Although other metabolic sequences are involved in carbohydrate production in various photosynthetic organisms, the Calvin–Benson cycle, summarised in fig. 10.15, is the one mainly applicable, certainly in higher-plant chloroplasts.

In addition to the Calvin–Benson ($C_3$) pathway, mention must be made of a $C_4$ route, the Hatch–Slack pathway (fig. 10.16) which is used by some tropical plants, especially grasses, and allows carbohydrate synthesis from $CO_2$ to proceed in low $CO_2$ concentrations with very little loss of water. Collaboration between two cell types – mesophyll and bundle-sheath cells – is a feature of these $C_4$ plants, which fix $CO_2$ in the mesophyll cell as malate which is then transported into the bundle-sheath cell and decarboxylated. The $CO_2$ released is then fixed by reaction with ribulose 1,5-bisphosphate according to the normal Calvin–Benson pathway.

### 10.7 Photosynthesis in eukaryotic algae

The process of photosynthesis in green algae (Chlorophyceae) very closely resembles that in higher plant chloroplasts; many of the details of photosynthesis have been worked out in studies with green algae, notably *Chlorella*, *Scenedesmus* and *Chlamydomonas*. In general, photosynthesis in all eukaryotic algae seems to be similar to that in higher plants. Oxygen is evolved, two photosystems are utilised and a Z scheme of electron transport operates. There are, however, variations in the photosynthetic pigments in the LHCP of the different algal classes. Some, *e.g.* Chlorophyceae and Euglenophyceae, have chlorophylls *a* and *b* and carotenoid accessory pigments similar to those of higher plants. Others lack chlorophyll *b*, which is replaced as an accessory pigment by chlorophyll *c* or *d* (**10.9, 10.10**) in the Dinophyceae, Chrysophyceae, Bacillariophyceae, Xanthophyceae, Phaeophyceae and some Rhodophyceae. There are also considerable variations in carotenoid compositions (chapter 2). For example, the brown seaweeds

Fig. 10.15. Summary of the Calvin–Benson pathway of photosynthetic carbon fixation and carbohydrate synthesis. (*a*) The essential reactions which fix $CO_2$ and use NADPH and ATP produced by the photosynthetic electron transport chain and phosphorylation. Overall: $6CO_2 + 18ATP + 12NADPH \longrightarrow 1$ fructose 6-phosphate $+ 18ADP + 12NADP^+ + 17P_i$. (*b*) Intermediates in the regeneration of ribulose-1,5-bisphosphate from glyceraldehyde 3-phosphate, dihydroxyacetone phosphate, and fructose 6-phosphate. (*c*) The $CO_2$ fixation reaction, catalysed by the enzyme ribulose 1,5-bisphosphate carboxylase (carboxydismutase).

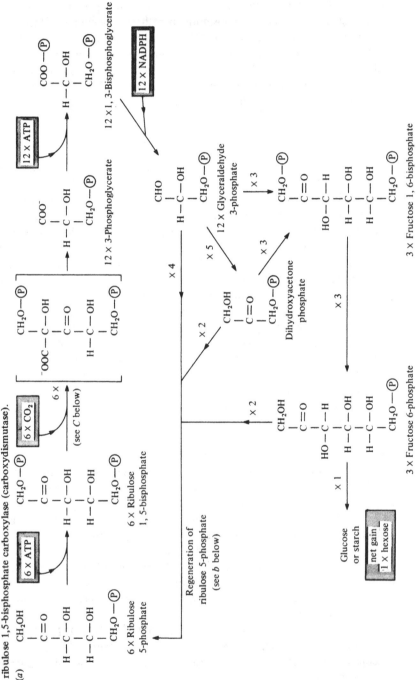

Fig. 10.15. *contd.*

(*b*)

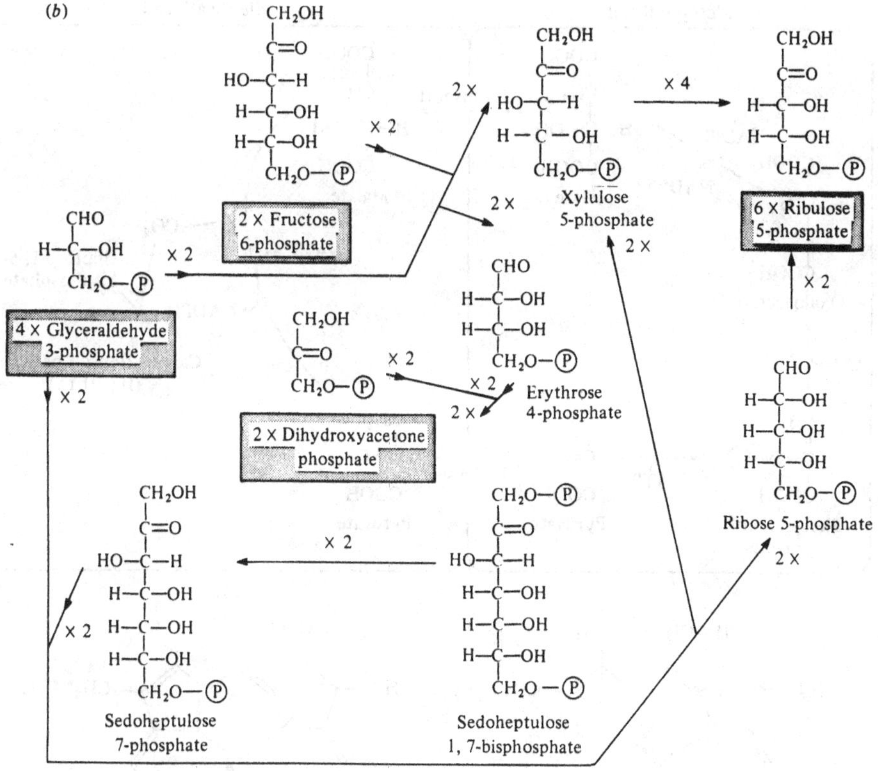
2 × Fructose 6-phosphate
Xylulose 5-phosphate
6 × Ribulose 5-phosphate
4 × Glyceraldehyde 3-phosphate
Erythrose 4-phosphate
2 × Dihydroxyacetone phosphate
Ribose 5-phosphate
Sedoheptulose 7-phosphate
Sedoheptulose 1, 7-bisphosphate

(*c*)

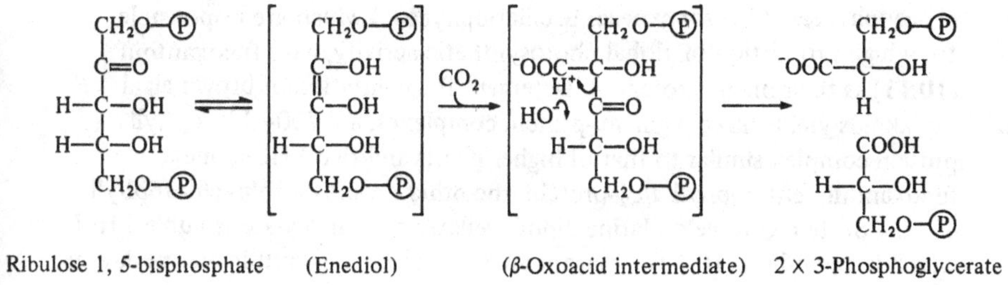
Ribulose 1, 5-bisphosphate
(Enediol)
(β-Oxoacid intermediate)
2 × 3-Phosphoglycerate

Fig. 10.16. The Hatch–Slack ($C_4$) pathway of photosynthetic $CO_2$ fixation.

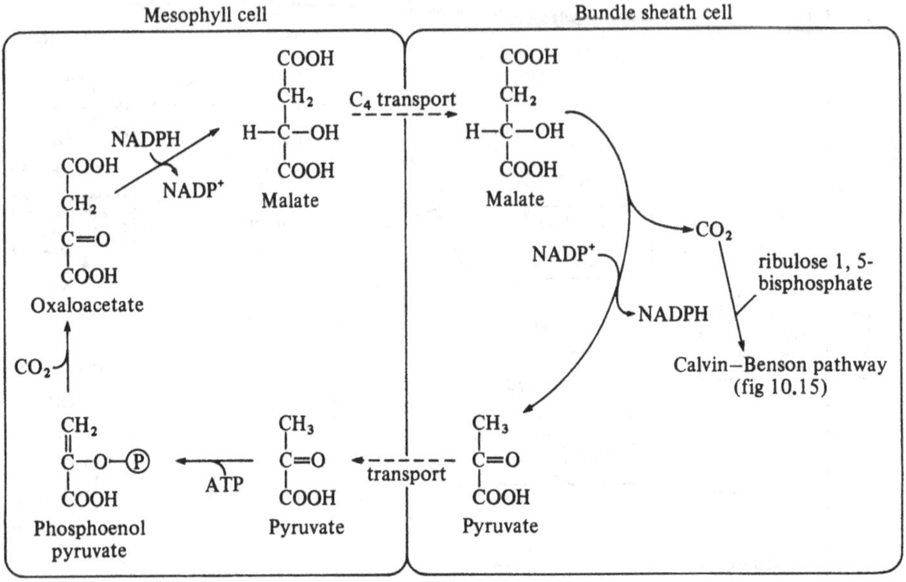

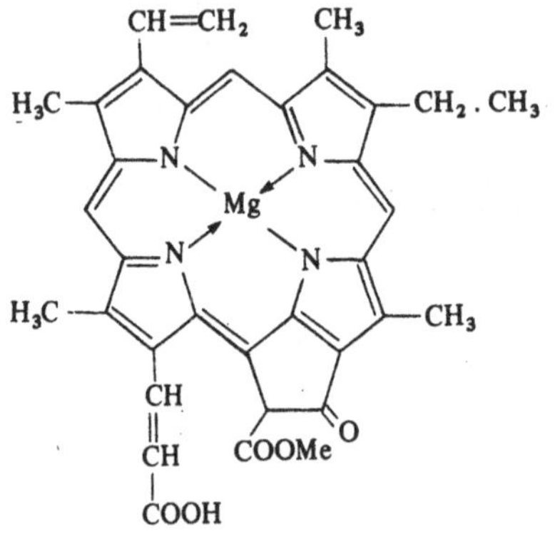

(10.9) Chlorophyll *c*

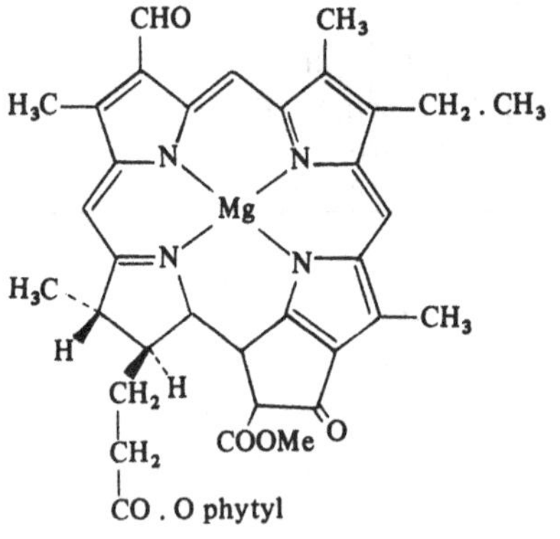

(10.10) Chlorophyll *d*

(Phaeophyceae, Chrysophyceae, Bacillariophyceae) which are responsible for a high proportion of global photosynthetic activity, have fucoxanthin (10.11) as their main carotenoid. Detergent fragmentation of brown algal thylakoids yields three pigment-protein complexes, a P-700–chlorophyll *a*–protein complex similar to that of higher plants and two LHCP, one a fucoxanthin-chlorophyll $a/c_2$-protein, the other a violaxanthin-chlorophyll $a/c_1/c_2$-protein complex. Marine dinoflagellates contain a water-soluble LHCP which is a peridinin(10.12)-chlorophyll *a*-protein. Fucoxanthin and peridinin

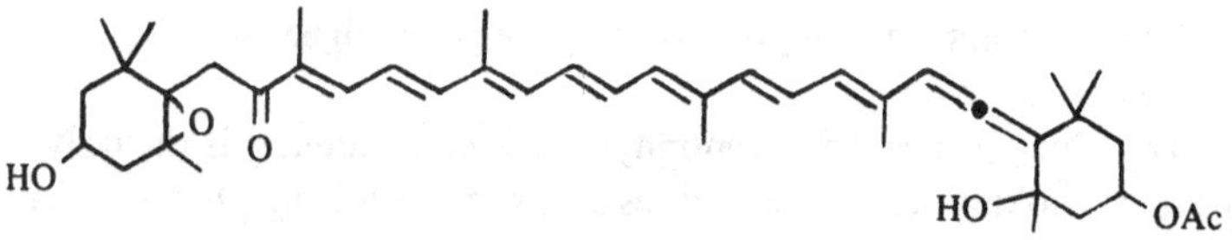

**(10.11)** Fucoxanthin

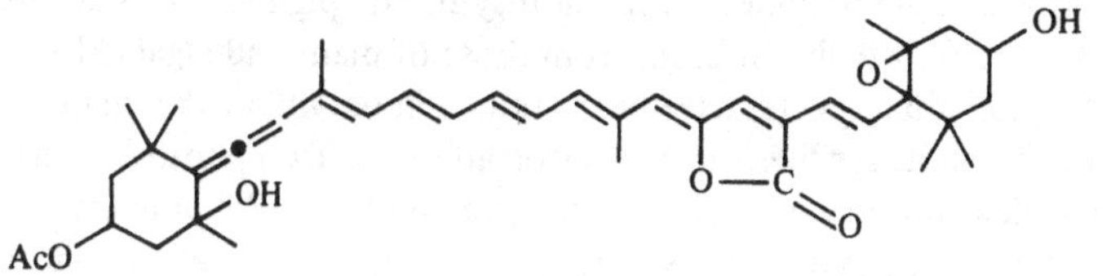

**(10.12)** Peridinin

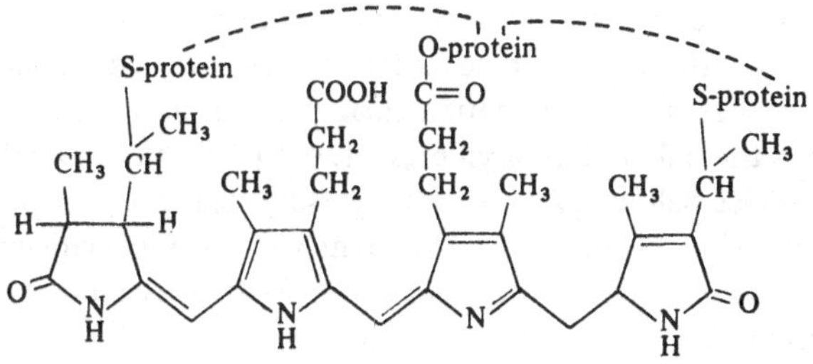

**(10.13)** Phycoerythrin

are important accessory pigments. Energy transfer from them to chlorophyll *a* is very efficient (70%).

Red algae (Rhodophyceae) mostly contain only chlorophyll *a* but have large amounts of phycobilins – complexes between protein and linear tetrapyrroles such as phycoerythrin **(10.13)** (see chapter 5) – which efficiently transfer excitation energy to chlorophyll *a* in photosystem II. The phycobilins are present as protein aggregates in particles, **phycobilisomes**, attached to the chloroplast lamellae (see §**10.8**).

Eukaryotic algae exist in a wide range of habitats and show great variability in growth characteristics, so it is not surprising that, as well as differences in pigmentation, some differences and peculiarities in the nature of the electron carriers, and their sequence in the electron transport chain, have been recognised in some species. Great variation has also been observed in the pathways of carbon metabolism. Few individual species of algae have been studied in detail so many deviations from the higher-plant pattern of photosynthesis may remain to be discovered.

## 10.8     Photosynthesis in the prokaryotic blue-green algae

### *10.8.1     General features*

The blue-green algae (Cyanophyta or Cyanobacteria) is the only large group of prokaryotes which utilises an oxygen-evolving photosynthetic process similar to that of higher plants. In these organisms, however, the thylakoid membranes are not contained in a specific organelle (chloroplast) but are found throughout the cell in the general cytoplasm, being especially prevalent towards the periphery. The photosynthetic pigments of blue-green algae are also substantially different from those of plant and algal chloroplasts. Few individual species have been studied in detail but the main features of the photosynthesis have been established. Two photosystems and a Z-scheme electron transport system are again used, and PSI appears to be similar to that in eukaryotes. However, many of the finer details of photosystem composition and electron transport carriers and sequences in Cyanobacteria probably vary from those of higher plants. In general, the pigments, photosystems and electron transport components seem to be much less firmly associated with the membrane structures than they are in photosynthetic eukaryotes. The pigmentation characteristics have been determined. Chlorophyll *a* is the only chlorophyll present, but phycobilins, located in phycobilisomes (see below, §*10.8.2*) are accessory pigments associated with photosystem II. Of the carotenoids, β-carotene and possibly echinenone (**10.14**) are associated with photosystem I, and perhaps various xanthophylls with photosystem II.

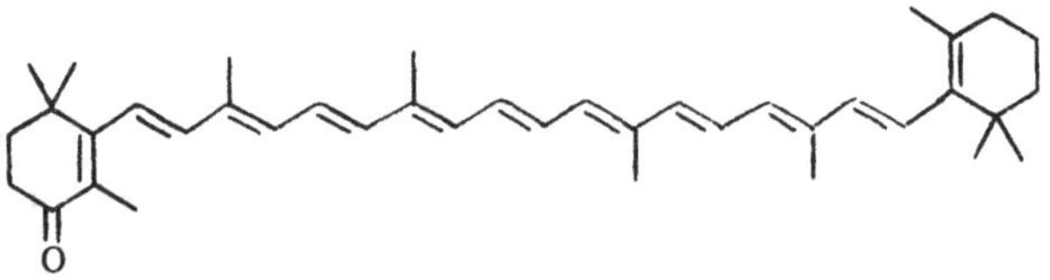

(**10.14**) Echinenone

Again different growth and environmental conditions may cause considerable differences in morphology, pigmentation, and details of the photosynthesis mechanism.

### *10.8.2     Phycobilisomes*

The red algae and the blue-green algae or bacteria differ from all other photosynthetic organisms in that they use as accessory light-harvesting pigments phycobiliproteins (see chapter 5), which are located in specific structures called phycobilisomes. These are macromolecular aggregates, up to about $20 \times 10^6$ daltons, and occur in an ordered array on the stroma side of the thylakoid membrane. The structures seem to be generally similar in all organisms that have been examined. Allophycocyanin, the longest-wavelength-absorbing (and -emitting) phycobiliprotein, constitutes a central core which is closely associated with the thylakoid. Radiating from this core out towards the stroma side are stacked rods containing phycocyanin (inner portion) and phycoerythrin (outer portion) (fig. 10.17). Other, colourless, proteins seem to be present as binders to maintain the phycobilisome structure.

The whole structure forms a light-harvesting complex, with energy transfer occurring from phycoerythrin *via* phycocyanin to the allophycocyanin core. About 95% of the energy trapped can be transferred to PSII, so a close spatial relationship between the phycobilisomes and PSII is assumed. Each phycobilisome appears to serve several PSII reaction centres. PSI in these organisms seems to be similar to that of higher plants and does not use light energy trapped by the phycobilisomes.

Fig. 10.17. Model red algal phycobilisome structure.

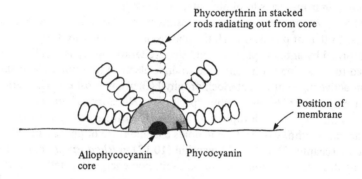

10.9 Bacterial photosynthesis
Three main groups of bacteria, the green and purple sulphur bacteria (Chlorobiaceae and Chromatiaceae) and the purple non-sulphur bacteria or Rhodospirillaceae are able to use light as their primary energy source. In general, photosynthetic bacteria have cell walls similar to those of other Gram-negative bacteria but have a thick cell membrane inside these walls. This membrane may be extensively folded to give lamellar membrane structures or vesicles within the body of the cell. These structures carry the photosynthetic apparatus, though the organisation of this, and the biochemical details of the photosynthetic processes differ greatly from those in the plant chloroplast. Photosynthetic membrane preparations may be obtained from disrupted cells as particles or vesicles termed chromatophores (cf. animal chromatophores, chapter 8) but in general no such discrete structures are present in the intact cells, with the exception of the green sulphur bacteria (Chlorobiaceae). These green bacteria (*e.g. Chlorobium*) differ from the other photosynthetic bacteria in being devoid of internal membranes but possessing flattened, cigar-shaped structures called **chloro-**

**somes** attached to the plasma membrane. The chlorosomes serve as functionally organised light-harvesting antennae for the reaction centres which are
located in the plasma membrane. Energy transfer within the chlorosome
proceeds from bacteriochlorophyll *c* (1000–1500 molecules present) *via*
bacteriochlorophyll *a* (perhaps 100 molecules in a protein complex) to the
reaction centre bacteriochlorophyll *c*.

Bacterial photosynthesis differs from plant photosynthesis in a number
of ways. First, these bacteria are unique among photosynthetic organisms
in being unable to use water as an ultimate reductant; they utilise other
reductants, which may be organic molecules or inorganic sulphur compounds,
and consequently they do not evolve oxygen. Secondly, carbon fixation and
metabolism occur by pathways other than the Calvin–Benson cycle. Thirdly,
the machinery of light harvesting and electron transport is quite different
from that in plants; in particular there is only one light reaction, though this
is similar in many ways to photosystem I of plants.

The *mechanism* used for light harvesting is, however, fundamentally
similar to that of plants, though the photosynthetic units are smaller. Light
is absorbed by antenna pigments and the excitation energy is rapidly transferred to a reaction centre and used to drive electron transport. The main
photoactive pigment is bacteriochlorophyll (Bchl), in most cases bacteriochlorophyll *a* (**10.15**) but in some species (*e.g. Rhodopseudomonas
sphaeroides*) bacteriochlorophyll *b* (**10.16**). The carotenoids present are
the acyclic methoxy or aryl carotenoids characteristic of photosynthetic
bacteria (chapter 2), *e.g.* spirilloxanthin (**10.17**) in *Rhodospirillum rubrum*.
The light-harvesting antennae pass on excitation energy to reaction centre
Bchl which absorbs at 870–875 nm in the Rhodospirillaceae, 890 nm in the
Chromatiaceae. The reaction centre Bchl, P-870 or P-875 is analogous to P-700
or chlorophyll $a_I$ of higher-plant PSI. The bacterial reaction centre has three

(**10.15**) Bacteriochlorophyll *a*
(R = farnesyl or geranylgeranyl)

(**10.16**) Bacteriochlorophyll *b*

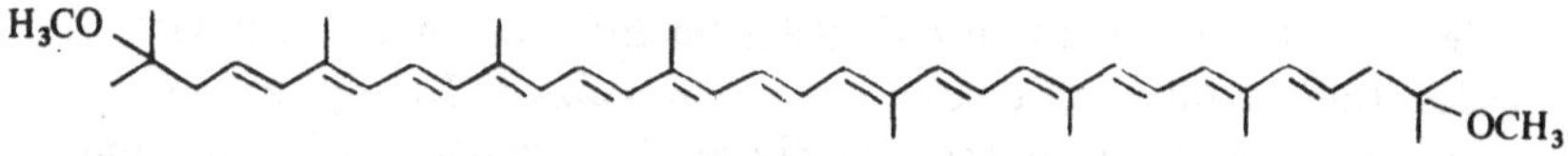

(10.17) Spirilloxanthin

polypeptides (28, 32 and 35 kilodaltons) associated with four bacteriochlorophyll molecules, two molecules of bacteriophaeophytin (10.18), and
one of ubiquinone (10.19), together with non-haem iron. Carotenoid is also
found in small amounts in reaction centre preparations but occurs mainly in
the light-harvesting antenna complex.

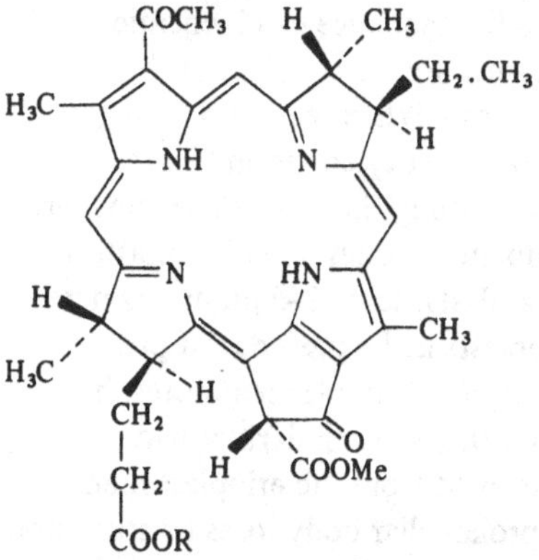

(10.18) Bacteriophaeophytin *a*

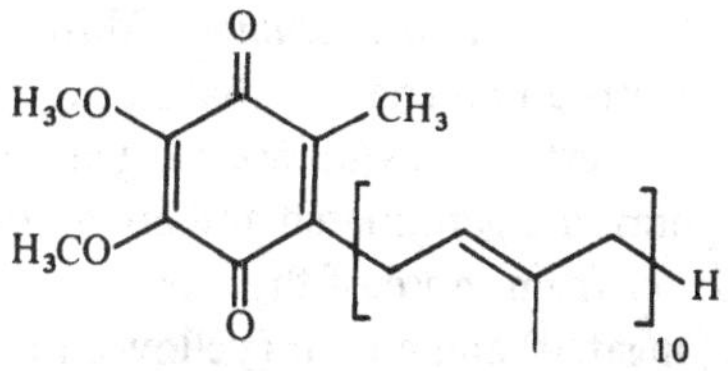

(10.19) Ubiquinone-10

In very close association with the reaction centre is the main light-harvesting antenna, which also absorbs at 875 nm in the Rhodospirillaceae. This
complex contains two polypeptides, and bacteriochlorophyll and carotenoid
in a 1:1 ratio. The Bchl:reaction centre ratio remains constant at about
25:1. In some species, *e.g. Rhodospirillum rubrum*, there is only the one
light-harvesting complex, but in other organisms, especially *Rhodopseudomonas* spp., a second such complex is present absorbing at rather shorter
wavelengths ($\lambda_{max}$ at 800 and 850 nm). It contains two peptides, two forms
of bacteriochlorophyll (two molecules of Bchl-850, one molecule of Bchl-800)
and carotenoid (one molecule). The ratio of this complex to the reaction
centre varies depending on environmental conditions, and there may be
anything from ten to 100 Bchl-800–850 per reaction centre (see §10.11).
   In the primary light reaction, excitation energy is transferred to P-870
and an electron is lost from the bacteriochlorophyll special pair and passed
on to an acceptor. The oxidised P-870 in turn accepts an electron from a
donor molecule. The nature of this donor and of the primary acceptor is
not yet certain.

Details of the electron transport chain are different for different bacterial species. Those studied most extensively are *Rhodospirillum rubrum* and several *Rhodopseudomonas* species (*sphaeroides, capsulata, palustris*) from the Rhodospirillaceae, and several *Chromatium* and *Chlorobium* species from the sulphur bacteria. In all these bacteria several cytochromes are involved in the electron transport chain, but plastocyanin appears not to be used. Ubiquinone may fill the roles that plastoquinone plays in plants. Detailed discussion of bacterial photosynthetic electron transport would be out of place in this book.

## 10.10    Pigment synthesis in relation to chloroplast development
### 10.10.1    *Formation of higher-plant chloroplasts from etioplasts*
*Structural changes.* Most studies of the sequence of biochemical and structural changes that occur during maturation of chloroplasts have been performed with greening etiolated seedlings. When seedlings of angiosperms are germinated and grown in the dark the cotyledons and leaves differ from those of the normal light-grown seedlings in their shape (smaller, elongated) and colour (yellow). In these etiolated tissues no chloroplasts are formed but the **proplastids** develop into **etioplasts**, some 3–5 $\mu$m in diameter. The characteristic structural feature of etioplasts is the presence of **prolamellar bodies**, quasi-crystalline three-dimensional tubular structures, but a few lamellae usually extend outwards from these. When dark-grown seedlings are illuminated, chloroplasts are formed from the etioplast membranes. During this process the crystalline prolamellar body loses its structural regularity and then disperses completely into sheets of perforated membranes which give rise to thylakoids. Recent reports suggest that **prothylakoid** membranes attached to the tubules of the prolamellar body may be extremely important in the synthesis of constituents of the developing photosynthetic membranes, including the chlorophyll and carotenoid pigments.

The time-course of the development of photosynthetic activity depends markedly upon species, age and growth conditions. Some photosynthetic activity usually appears within minutes and is rapid within about 2 h, though the whole process of chloroplast formation from etioplasts takes up to 48 h to complete.

*Formation of active photosystems.* The growth of the thylakoid membrane and the development of functioning photosynthetic apparatus during etioplast to chloroplast differentiation is a multi-step process which involves not only the biosynthesis of structural and functional components but also the integration and assembly of these components into functional units. It is possible to distinguish different steps in the membrane development and to isolate thylakoids containing PSI and PSII units at different developmental

stages. First the cores, including the reaction centres, of the PSI and PSII
units are formed, followed by a simple (monomeric?) form of LHCP. Differentiation of the primary thylakoids into stroma and grana membranes occurs
as the LHCP is synthesised; during this differentiation the PSI and PSII units
increase in size. As the development continues, the pigment–proteins are
gradually organised into the large supramolecular structures of the fully
developed chloroplasts.

*Light regulation of chloroplast development.* The final state of the thylakoid
membranes depends on the environmental conditions, especially light. Chloroplasts developed in high light intensities have relatively small but highly
efficient PSI and PSII, but reduced LHCP. At lower light intensities, when
light harvesting needs to be as efficient as possible, LHCP synthesis is of
great importance and is also related to the stacking of thylakoids into grana.
There are reports that red light and a phytochrome system are the primary
agencies involved in the regulation of chloroplast development. Other
workers, however, dispute this and believe that blue light and an unknown
photoreceptor are important.

(10.20) Protochlorophyllide *a*

*Synthesis of photosynthetic pigments.* The biosynthetic pathway by which
chlorophyll is produced is described in chapter 5. Etiolated seedlings contain
no chlorophyll but do have a small amount of protochlorophyllide (10.20)
bound to the holochrome protein and localised in the prolamellar bodies.
Spectroscopic studies indicate the presence of three different forms of
protochlorophyll(ide), absorption maxima 628, 637, 650 nm, which differ
in their state of aggregation or mode of binding to protein. When dark-grown
seedlings are illuminated, a number of intermediates may be detected spectroscopically ($\lambda_{max}$ 676, 678, 682, 672 nm) as the protochlorophyllide is
converted into chlorophyll *a*. Formation of chlorophyll from protochloro-

phyllide occurs rapidly, but there is then a short lag period before the bulk chlorophyll is synthesised more slowly *de novo*. It has been suggested that the rate of chlorophyll synthesis may be regulated by phytochrome. Although photochemical activity develops rapidly, the light-harvesting assemblies are at first small compared with those in green-leaf chloroplasts. The bulk chlorophyll synthesised during the hours after the lag period serves mainly to increase the size of the light-harvesting antennae.

Etiolated seedlings normally contain no carotenes but do have small amounts of xanthophylls. The various components of the photosynthetic electron transport chain, including cytochromes, plastocyanin and plastoquinone, are also present only in small quantities. On illumination, massive new synthesis of chloroplast carotenoids (see chapter 2) and photosynthetic electron transport components occurs, in parallel with the bulk chlorophyll synthesis, as these molecules are incorporated into the thylakoids. The syntheses of the various components of the functioning chloroplast are greatly interdependent and closely linked genetically. The supply of all structural and functional components must be maintained if chloroplast development is to proceed normally. Interruption of the supply of one component, *e.g.* by an inhibitor, will prevent proper chloroplast development and may lead to the synthesis of other molecules being blocked (see also §*10.10.4*).

### 10.10.2 *Chloroplast development in light-grown plant tissues*
Although the greening of etiolated seedlings provides a convenient, synchronous system with which to study chloroplast development, this greening is not a process which occurs during normal chloroplast formation in plant tissues maintained in the light. In this case the meristematic tissue develops small proplastids which eventually become chloroplasts. Etioplasts are not intermediate in this transformation. The development of chloroplasts from proplastids in normal light-grown plants is inherently much more difficult to study, and virtually no information has been obtained about pigment synthesis and its regulation in such systems. Variation in the size and pigment composition of the LHCP in plants of different species or growing under different light conditions has been noted.

There have been reports that extensive synthesis and breakdown (*i.e.* turnover) of chlorophyll and carotenoid pigments continues in mature, functioning chloroplasts, but further investigation is needed.

### 10.10.3 *Chloroplast development in algae*
There have been very few studies relevant to the development of chloroplasts in most classes of algae. *Euglena gracilis*, however, has proved extremely useful, since it produces normal chloroplasts only in the light. In dark-grown cells, proplastids (structures somewhat similar to plant etio-

plasts) are present and these are converted into functioning chloroplasts on illumination. The first 12 h of illumination is a lag period during which there is a major transfer of energy, small molecules, reducing power and ultimately nuclear-coded proteins to the developing plastid, particularly from the mitochondrion, as a result of light-induced breakdown of the carbohydrate reserve paramylum. From 12 to about 96 h, the plastid itself is highly active and most of the chloroplast components, including chlorophyll and caro-tenoid pigments, are made within it. Two different photoreceptors appear to coordinate the roles of the plastid and non-plastid compartments of the cell. One of these seems to be a blue-red photoreceptor similar to proto-chlorophyll(ide). There is great similarity between dark-grown *Euglena* and **young** etiolated higher plants in the pattern of protochlorophyll(ide) syn-thesis, proplastid organisation and the phototransformations which begin plastid development.

Green algae, especially *Chlorella, Scenedesmus* and *Chlamydomonas*, have been studied extensively. When grown in the light these algae have chloro-plasts similar to those of higher plants and contain a similar collection of photosynthetic pigments. The same pigments are, however, present in dark-grown cultures. No massive synthesis occurs on transfer to the light, although light may be required for proper fixing of the pigments into the photo-synthetic membranes and particles.

Several mutant strains of these algae have been produced which, when grown in the dark, have pigment compositions differing greatly from those of the wild type; chlorophyll may be completely absent and carotenoid biosynthesis may be blocked at an early stage, *e.g.* ʃ-carotene (**10.21**). On illumination of some of these strains normal chloroplast production proceeds, making them very useful for studies of structural changes and pigment transformations, and similar in some ways to greening etioplasts.

(10.21) ʃ-Carotene

Although very few species of blue-green algae (Cyanophyceae or Cyano-bacteria) have been cultured successfully in the dark, one species, *Chlorogloea fritschii*, has given interesting results. In this organism the chlorophyll and carotenoid pigments are synthesised in the dark but are not fixed into the photosynthetic membranes. Light is needed for this deposition to occur.

Red algae (Rhodophyceae) and blue-green algae utilise phycobilins as accessory pigments, but the synthesis of these compounds as the photo-synthetic membranes, and especially the phycobilisomes, develop has not been studied in detail.

When individual phycobilins are considered it seems that, in some species at least, the relative proportion of phycocyanin ($\lambda_{max}$ around 620 nm) and phycoerythrin ($\lambda_{max}$ around 565 nm) produced is controlled to a large extent by the spectral quality of the incident light. Thus in green light conditions synthesis of the green-absorbing red pigment phycoerythrin is favoured, whereas in red lighting conditions synthesis of the red-absorbing blue pigment phycocyanin predominates. This therefore ensures maximum absorption of the available light by these accessory pigments.

### 10.10.4  General regulation of chloroplast and pigment synthesis

There appears to be a general regulatory mechanism which governs the formation of the chloroplast as a whole. The mechanisms of this intimate interrelationship and control are not known, though close genetic links have been identified. All components must be synthesised and made available for incorporation into the thylakoid membrane otherwise the synthetic processes are inhibited. For example, several herbicides act by inhibiting carotenoid biosynthesis. If etiolated seedlings or dark-grown algae (*Euglena*) are treated with the herbicide then the normal chloroplast carotenoids are not made and thus are not available for incorporation into the photosynthetic membranes. The synthesis of other components, notably chlorophyll, is then prevented, so chloroplast development as a whole ceases. Even if this were not so, inhibition of carotenoid production would leave any synthesised chlorophyll, and the incipient photosynthetic membranes, unprotected against photooxidative destruction (see §10.4.2). Herbicides which act against carotenoid biosynthesis in plants are therefore very effective.

### 10.11  Development of the photosynthetic apparatus of photosynthetic bacteria

Many photosynthetic bacteria, including the *Rhodopseudomonas* species *capsulata*, *palustris* and *sphaeroides*, are able to grow both anaerobically in the light and aerobically in the dark. Light intensity and oxygen partial pressure are the major external factors governing the development of the internal membranes to ensure optimal use of the prevailing energy source. Thus in dark, aerobic conditions increased amounts of respiratory electron transport components are synthesised and incorporated into the membranes. On the other hand, cells growing in the light in an oxygen-free atmosphere synthesise photosynthetic membranes and pigments, the greatest synthesis taking place at low light intensities when light harvesting needs to be most efficient.

The main factor regulating the development of photosynthetic membranes and pigment synthesis as a whole appears to be oxygen partial pressure. Above certain oxygen levels, respiration is efficient and no photosynthetic membrane or pigment synthesis takes place. Low oxygen pressure stimulates the production of functioning photosynthetic apparatus and pigments, especially the reaction centre and the main light-harvesting antenna complex B-875. Pigment composition is modified in response to changes in light intensity. Thus in *Rhodopseudomonas* spp., low light intensity causes increased synthesis of bacteriochlorophyll and carotenoids as the secondary light-harvesting antenna complex, B-800–850, is made. High light intensity inhibits formation of this complex; a decreased pigment content results. In the case of *Rhodospirillum rubrum*, which contains no B-800–850 antenna, the pigment content of the main light-harvesting antenna, B-875, is regulated by light intensity. Little is known of the organisation and control of the processes by which the photosynthetic pigments are made and incorporated into the photosynthetic membranes. The genes controlling the synthesis of chlorophyll, carotenoid and perhaps development of the active photo-synthetic apparatus as a whole are located in the chromosome, not on a plasmid, and map very close together. A single large genetic unit may be involved in coding for the photosynthetic apparatus.

## 10.12    Fate of photosynthetic pigments during chloroplast degeneration

Photosynthetic tissues of higher plants may cease their photo-chemical activity for various reasons. Two familiar examples are the senescence of leaves in autumn and the ripening of fruit. In these examples the chloroplast ceases to function and either degenerates or is transformed into a chromoplast.

### 10.12.1 Degeneration of chloroplasts in leaves

In autumn the leaves of most trees change colour from green to yellow, red and brown, and are shed by the tree. This process involves the degeneration of the chloroplast and the destruction of the chlorophyll pigments. The mechanism of chlorophyll degradation is not fully known, but at an early stage phytol and magnesium are lost to give phaeophorbide (**10.22**). The porphyrin ring system is then broken down into colourless low relative molecular mass compounds.

The chloroplast carotenoid is not all lost, hence the yellow colour of many senescing leaves. Considerable oxidation of β-carotene occurs, *via* epoxides and apocarotenals, and the xanthophylls become esterified with fatty acids. The bright-red colours of some autumn leaves are caused by massive synthesis of anthocyanins (chapter 4) during senescence. This is, however, not directly involved with chloroplast degradation.

### 10.12.2 Ripening of fruit

Immature fruits are usually green and contain functioning chloro-plasts. In many cases when the fruits ripen the chloroplasts are transformed into non-photosynthetic chromoplasts. The chlorophyll is destroyed as

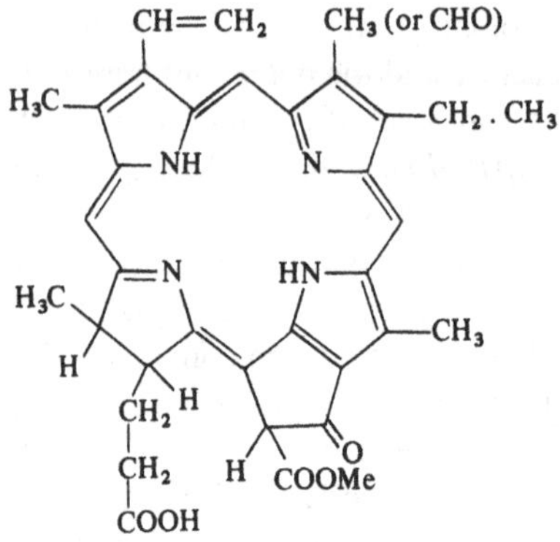

(10.22) Phaeophorbide

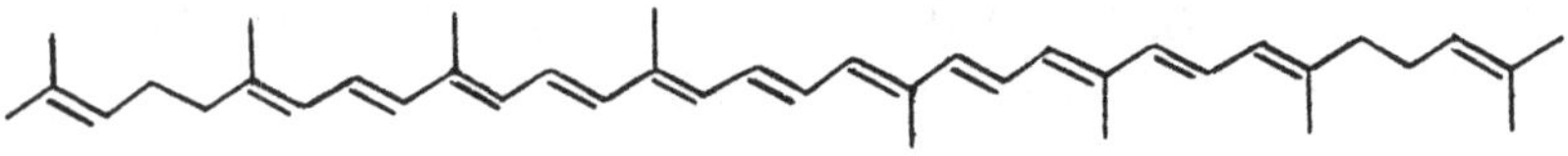

(10.23) Lycopene

photosynthetic activity is lost. Loss of the chloroplast carotenoids may also occur, but these are often replaced by much larger amounts of other carotenoids which give rise to the colour of the ripe fruit. Well-known examples of this include the tomato which becomes red due to extensive synthesis of lycopene (10.23).

## 10.13    Conclusions and comments

The process of photosynthesis is essential to the continued existence of life on this planet, since it brings about the harnessing of the abundant solar energy into a chemically usable form through the fixation of $CO_2$ to carbohydrate. Pigments, especially chlorophyll, play the primary role in photosynthesis, so this will remain a major area for pigment research. Much attention is now being focused on the mechanism and control of pigment synthesis and incorporation into the photosynthetic membranes, the orientation of pigment molecules in the photosynthetic apparatus, and the molecular transformations that occur during the very short time-scale of the primary photoreactions. This increasing fundamental knowledge of the mechanism of photosynthesis will make more likely the development of simpler model systems that could be used for harnessing solar energy. As food and conventional fossil energy supplies (themselves the product of earlier photosynthesis) become less readily available, increased production by photosynthesis becomes a target. Progress is already being made. In Brazil, high-yielding crops such as sugar cane are being used to provide sugar for production of ethanol which is being incorporated into motor fuel. Photosynthetically fixed carbon will probably have to replace coal and oil as a major source of raw material for the chemical industry. Unicellular algae are finding use as a nutritious vitamin- and mineral-rich animal feed. Systems containing chloroplasts with hydrogenase enzyme and other catalysts are being investigated for possible exploitation as hydrogen generators. Greater basic knowledge of the mechanism of light harvesting by photosynthetic pigments is obviously fundamental to the success of such programmes.

## 10.14 Suggested further reading

The volume of literature devoted to photosynthesis is enormous; the number of original papers published each year must run into thousands. Review articles and books also abound, so only a few can be included here. For the non-specialist, introductory books such as those by Rabinowitch and Govindjee (1969) and by Gregory (1977) should be useful. A more detailed general account may be found in the very authoritative book edited by Govindjee (1975). Two volumes in the *Encyclopaedia of plant physiology* series (vols 5 and 6, edited by Trebst and Avron (1977) and by Gibbs and Latzko (1979), respectively) are comprehensive and extremely useful. A recent series edited by Barber contains volumes devoted to chloroplasts (vol. 1, 1976), to photosynthesis (vol. 2, 1977) and to model systems related to photosynthesis and energy production (vol. 3, 1979). All aspects of plant and algal chloroplasts, including structure, ultrastructure and development are covered in an extensive monograph by Kirk and Tilney-Bassett (1978). Details of the development of chloroplasts and photosynthetic apparatus are the subject of another book, edited by Akoyunoglou and Argyroudi-Akoyunoglou (1978). Bacterial photosynthesis and photosynthetic pigments are discussed in detail in a recent monograph on the photosynthetic bacteria (Clayton and Sistrom, 1978). The proceedings of the triennial International Congresses on Photosynthesis (Metzner, 1969; Forti, Avron and Melandri, 1972; Avron, 1975; Hall, Coombs and Goodwin, 1978; Akoyunoglou, 1981) give an up-to-date record of work in progress in hundreds of photosynthesis laboratories throughout the world. Many of the general and most useful methods applied to studies of photosynthesis are given in two volumes of *Methods in enzymology* (San Pietro, 1971, 1972).

## 10.15 Selected bibliography

Akoyunoglou, G. (ed.) (1981) *Proceedings of the 5th International Congress on Photo-synthesis*. Philadelphia: Balaban International Science Services.
Akoyunoglou, G. and Argyroudi-Akoyunoglou, J. H. (eds) (1978) *Chloroplast development*. Amsterdam: Elsevier.
Avron, M. (ed.) (1975) *Proceedings of the 3rd International Congress on Photosynthesis*.. Amsterdam, Oxford and New York: Elsevier.

Barber, J. (ed.) (1976) *The intact chloroplast. (Topics in photosynthesis*, vol. 1). Amsterdam, Oxford and New York: Elsevier–North Holland.

Barber, J. (ed.) (1977) *Primary processes of photosynthesis. (Topics in photosynthesis*, vol. 2). Amsterdam, Oxford and New York: Elsevier–North Holland.

Barber, J. (ed.) (1979) *Topics in relation to model systems. (Topics in photosynthesis*, vol. 3). Amsterdam, Oxford and New York: Elsevier–North Holland.

Clayton, R. K. and Sistrom, W. R. (eds) (1978) *The photosynthetic bacteria*. New York: Plenum.

Forti, G., Avron, M. and Melandri, A. (eds) (1972) *Proceedings of the 2nd International Congress on Photosynthesis Research*. The Hague: Junk N.V.

Gibbs, M. and Latzko, E. (eds) (1979) *Photosynthesis II: Photosynthetic carbon metabolism and related processes. (Encyclopaedia of plant physiology*, vol. 6). Berlin, Heidelberg and New York: Springer-Verlag.

Govindjee (ed.) (1975) *Bioenergetics of photosynthesis*. New York, San Francisco and London: Academic Press.

Gregory, R. P. F. (1977) *Biochemistry of photosynthesis*, 2nd edition. Chichester, New York, Brisbane and Toronto: Wiley.

Hall, D. O., Coombs, J. and Goodwin, T. W. (eds) (1978) *Proceedings of the Fourth International Congress on Photosynthesis*. London: The Biochemical Society.

Kirk, J. T. O. and Tilney-Bassett, R. A. E. (1978) *The plastids: their chemistry, structure, growth and inheritance*, 2nd edition. Amsterdam, New York and Oxford: Elsevier–North Holland.

Metzner, H. (ed.) (1969) *Progress in photosynthesis research*. Tübingen: Laupp.

Rabinowitch, E. and Govindjee (1969) *Photosynthesis*. New York, London, Sydney and Toronto: Wiley.

San Pietro, A. (ed.) (1971) *Photosynthesis, Part A. (Methods in enzymology*, vol. 23). New York and London: Academic Press.

San Pietro, A. (ed.) (1972) *Photosynthesis and nitrogen fixation, Part B. (Methods in enzymology*, vol. 24). New York and London: Academic Press.

Trebst, A. and Avron, M. (eds) (1977) *Photosynthesis I: Photosynthetic electron transport and photophosphorylation. (Encyclopaedia of plant physiology*, vol. 5). Berlin, Heidelberg and New York: Springer-Verlag.

# 11 Other photofunctions of natural pigments

## 11.1 Introduction

The previous chapters in section II of this book have dealt with the main functional areas of natural pigments, in colouring tissues, in the processes of vision, and in photosynthesis. There are, however, many other functions related to the light-absorbing properties of these molecules, although these functions are probably not of such wide occurrence, nor of such fundamental importance, and are certainly not so well understood. The features of several of these photofunctions will be outlined in this chapter.

## 11.2 Phytochrome
### 11.2.1 Introduction

Many aspects of the growth and morphology of plants are regulated by light. Most, if not all, of these photoresponses are mediated by one extremely important photoreceptor pigment, phytochrome, which is present only in minute amounts. Structurally, phytochrome is a protein with a linear tetrapyrrole (bilin) as prosthetic group and chromophore. Details of the structure and properties of phytochrome are given in chapter 5; see especially fig. 5.16.

Two interconvertible forms are involved in the functioning of phytochrome. One form, $P_r$, absorbs red light ($\lambda_{max}$ 660 nm) and is thereby transformed into the other form, $P_{fr}$. $P_{fr}$ absorbs longer wavelength, far-red light ($\lambda_{max}$ 730 nm) and as a result is reconverted into $P_r$. The two forms function as a switching mechanism. $P_{fr}$ is in most cases the active form or 'on-switch', which initiates physiological processes. Thus when red light around 660 nm is absorbed by $P_r$, $P_{fr}$ is produced and will initiate many photoresponses, *e.g.* flower development. However, if the red light is followed by a flash of far-red light around 730 nm, this light is absorbed by the $P_{fr}$ which is converted into the inactive form, $P_r$. The photoprocess will therefore be prevented from taking place.

Whether or not a response occurs is determined by the nature of the final irradiation in a sequence. Thus after a fr–r–fr–r sequence, the phytochrome will be left in the $P_{fr}$ form and will therefore initiate a response. On the other hand a r–fr–r–fr sequence will not, because the final far-red irradiation will leave the phytochrome in the inactive $P_r$ form. The exposure to red light need be only very short (seconds or less) in order to bring about the response, and need be received by only part (*e.g.* one leaf) of the plant. The red light effect can in some cases be cancelled by far-red light treatment after a considerable time (2–3 h) has elapsed.

### 11.2.2    Distribution and localisation

Phytochrome systems are found throughout the plant kingdom. Although the only ones that have been studied in any detail are those of higher plants, similar systems have been recognised in mosses, ferns and algae. In many higher plants, phytochrome is highly localised in some specific tissues. For example, in etiolated oat seedlings, high concentrations of phytochrome are present in parenchyma and epidermal cells just below (0.1– 1.5 mm) the apex of the coleoptile, but the pigment is absent from the tip itself. Within the cell, phytochrome as $P_{fr}$ has been found associated with the nuclear envelope and in organelles such as mitochrondria, amyloplasts, etioplasts and chloroplasts. $P_{fr}$ appears to be associated more firmly than does $P_r$ with membrane structures, where discrete receptor sites have been suggested. The existence of different pools of phytochrome within the cell has also been considered. Thus it may be only a small amount of firmly bound $P_{fr}$ that brings about responses whilst the bulk of the $P_{fr}$ remains in the free form, not associated with the receptor sites, and is not directly involved.

### 11.2.3    Examples of processes controlled by phytochrome

Of the many photoresponses mediated by phytochrome the best known are probably initiation of flowering, seed germination and greening of etiolated tissues. In the first case very brief exposure of even part (one leaf) of a plant to the relevant light wavelength initiates responses that will require weeks for completion. Clearly this involves the expression of new genetic information. In the natural situation it is daylength, or more accurately the length of the dark period, that determines the onset of flowering. Thus flowering is induced in short-day plants by a long dark- and short light-period regime, whereas the converse of this, namely a long light period and short dark period, is required by long-day plants. In both cases phytochrome is the photoreceptor agency which mediates the response.

Seed germination in most cases is promoted by red-light-produced $P_{fr}$ but in some instances light may be inhibitory.

Brief exposure to red light brings about profound changes in etiolated seedlings. Leaf size increases and the tissues become green due to the synthesis of chlorophyll and other chloroplast constituents. Utilisation of starch reserves is stimulated until active photosynthesis becomes efficient.

The phytochrome system is also involved as a photoregulator in dormancy and senescence, root growth, leaf movement, and the general maintenance of growth habit in various plants.

### 11.2.4  Mode of action of phytochrome

Many attempts have been made to explain the action of phytochrome and various models have been proposed. The primary event is clearly the absorption of light, which causes excitation of the phytochrome and brings about the transformation of the $P_r$ chromophore into that of $P_{fr}$, and *vice versa* (see fig. 5.16). Changes in the conformation of the phytochrome protein appear to be only small but differences in association between the two phytochrome forms and membrane structures may be significant.

Recent low temperature and flash photolysis studies have revealed the existence of intermediate forms, with different absorption maxima, between $P_{fr}$ and $P_r$, and a cycle for the interconversions has been proposed (fig. 11.1). This cycle cannot yet be explained in molecular terms, though *cis* isomers of the bilin chromophore are implicated (*e.g.* **11.1**) and conformational changes in the protein have been suggested in addition to the structural changes shown in fig. 5.16. Some interconversion of the two forms of phytochrome can in some cases occur in the dark. Dark reversion of $P_{fr}$ to $P_r$ is direct but any formation of $P_{fr}$ probably occurs from the intermediate $P_{650}$ rather than from $P_r$.

Seed germination very frequently occurs in the dark. Its control by the photoreceptor phytochrome therefore seems paradoxical. A likely explanation is that the phytochrome is 'trapped' as the $P_{fr}$ form when the seeds mature and ripen.Then in the appropriate season the seeds imbibe water, the tissues become hydrated and the $P_{fr}$ phytochrome can assume its active conformation.

It is thought that the phytochrome effect may be mediated through hormonal agents such as cyclic AMP, acetylcholine or various plant growth

Fig. 11.1. Cycle showing intermediates that have been detected in the inter-conversion of the two forms of phytochrome. It is likely that alternative, parallel sequences also operate. The subscripts in $P_{698}$ *etc.* refer to the light absorption peaks of the intermediates. $P_{bl}$ and $P_x$ are weakly absorbing forms. Reactions labelled 'd' are dark reactions.

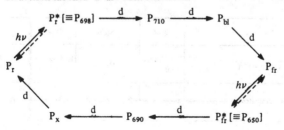

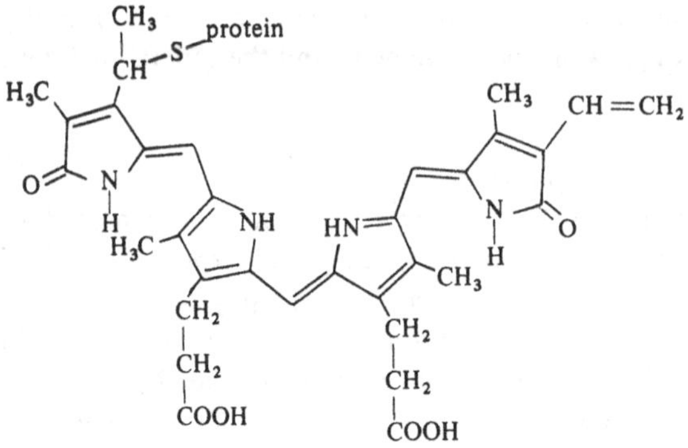

(11.1) '*cis*-form' of phytochrome

substances. There are, however, many different responses to phytochrome in different tissues. Some of these responses occur extremely rapidly (seconds or less), others over a much longer period (hours, days or weeks). There may therefore be several different mechanisms by which phytochrome exerts its action. Among the many that have been considered are effects on ion fluxes, bioelectric potentials, membrane permeability, or respiratory activity, but little evidence is yet available to support any of these hypotheses.

Fig. 11.2. Predominant forms of phytochrome under different daylength conditions: (*a*) short-day/long-night; (*b*) long-day/short-night.

(*a*) Short day

| Day | Dusk | Night | Dawn | Day |
|---|---|---|---|---|
| Sunlight Red: far red ~ 3 : 1 | Red: far red ~ 0.7 : 1 | Dark | Red: far red ~ 0.7 : 1 | Red: far red ~ 3 : 1 |
| ◄─$P_{fr}$ predominates─► | | ──────$P_r$ predominates────── | | $P_{fr}$ predominates |

(*b*) Long day

| Day | Dusk | Night | Dawn | Day |
|---|---|---|---|---|
| Sunlight Red: far red ~ 3 : 1 | Red: far red ~ 0.7 : 1 | Dark | Red: far red ~ 0.7 : 1 | Red: far red ~ 3 : 1 |
| ───$P_{fr}$ predominates─── | | ─$P_r$ predominates─ | | $P_{fr}$ predominates |

Although details of the action of phytochrome have been elucidated mainly by studying the effects of irradiation with either red or far-red light, it must be remembered that under natural conditions plants or plant tissues will either be in darkness or exposed to white light. Light of wavelengths between about 500 and 700 nm produces an equilibrium in which $P_{fr}$ predominates (approx. 80% of the total phytochrome) whereas light outside this wavelength range results in an equilibrium in which $P_r$ is the predominant species. Diurnal variation in the quality of incident light is therefore important. For example, during most of the day, the ratio of red : far-red light energy in sunlight is about 3:1 so the $P_r \rightarrow P_{fr}$ conversion is favoured.However, at dusk and dawn this ratio may be as low as 0.7:1, *i.e.* far-red predominates and the phytochrome equilibrium is in favour of $P_r$. Thus in long-day conditions the $P_{fr}$ form will be in excess for a longer period of time, whereas in short-day, long-night conditions the $P_r$ produced at dusk will be the dominant species for the longer time (fig. 11.2). This provides an explanation of the well-known effects of daylength on plant growth processes. Effects such as this, and other possible diurnal rhythms must be taken into account when assessing phytochrome effects in plants in their natural habitat.

## 11.3 Phototaxis

Phototaxis is light-induced directional motion or change of motion of a living organism, or of cells or organelles within a living organism. The movement induced is usually related to the direction of the incident light and may be positive (towards the light) or negative (away from the light). Changes in light intensity can also influence the rate of motion; this related phenomenon has been termed photokinesis.

Many examples of phototaxis have been recognised in algae, dinoflagellates, fungi and bacteria, and photoactive movement of chloroplasts within algal cells has been described. Several pigments and pigment groups have been suggested to participate in photoactive responses in various organisms, largely on the basis of action spectra. These pigments include chlorophyll, bacteriochlorophyll, carotenoids, biliproteins, phytochrome, and riboflavin, but no definite identification has yet been achieved.

Phototaxis has been studied in many species, particularly unicellular algae and dinoflagellates, but it is *Euglena gracilis* that has received the most attention. In constant, uniform illumination or in darkness no photoresponse occurs, but photostimulation of *Euglena* is brought about by variation of the light intensity above or below a certain threshold level. The normal response of *Euglena* is positive, *i.e.* the cells move towards a source of increased light intensity so that energy production by photosynthesis can be greatest. However, very high light intensities which could be harmful produce a negative phototactic effect, a characteristic movement away from the light.

The *Euglena* cell is represented schematically in fig. 11.3. Two areas are concerned in the photomotile response, the photoreceptor area itself (the paraflagellar body) and also the stigma ('eyespot'). Maximum light is received by the photoreceptor when the cell is orientated so that the incident light is directed staight down the opening or gullet. The cells move or orientate themselves accordingly. The primary photoreceptor almost certainly consists of an ordered array of flavin molecules in the paraflagellar body, but the mechanism of the photoreaction, and the way that the light absorption is translated into a movement response are unknown.

The stigma also plays an important, though not a primary role in the phototactic response. This area contains a large concentration of carotenoids which serve as screening pigments. Since the motion of *Euglena* is complex and rotational rather than directly towards the light source, the pigmented stigma will periodically shade the photoreceptor. This is thought to control the erection and movement of the flagellum and hence control the direction of motion.

Fig. 11.3. Schematic drawing of a *Euglena gracilis* cell (length 40–50 μm).

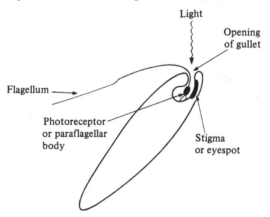

## 11.4    Phototropism

Phototropism is the light-induced growth or curvature of plants and fungi, usually towards the light source (positive phototropism) but occasionally away from the light (negative phototropism). The absorption of light by a photoreceptor pigment is clearly implied. The action spectrum determined for many phototropic responses has a maximum at around 450–460 nm and closely resembles the absorption spectra of both β-carotene (**11.2**) and riboflavin (**11.3**). After much argument, it now seems clear that riboflavin fulfils the requirements for the photoreceptor pigment much better than does β-carotene, and is almost certainly the photoreceptor used.

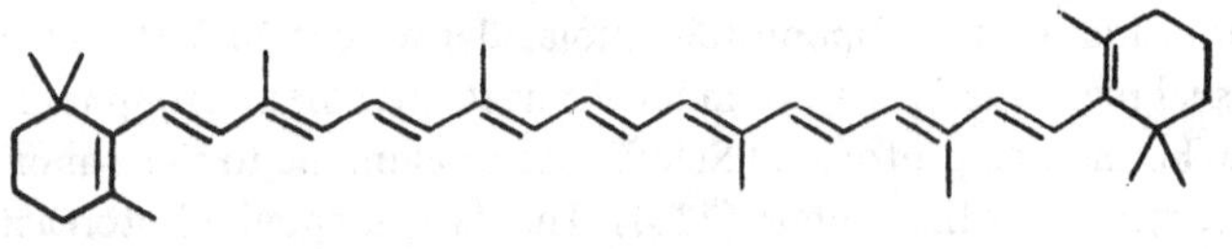

(11.2) β-Carotene

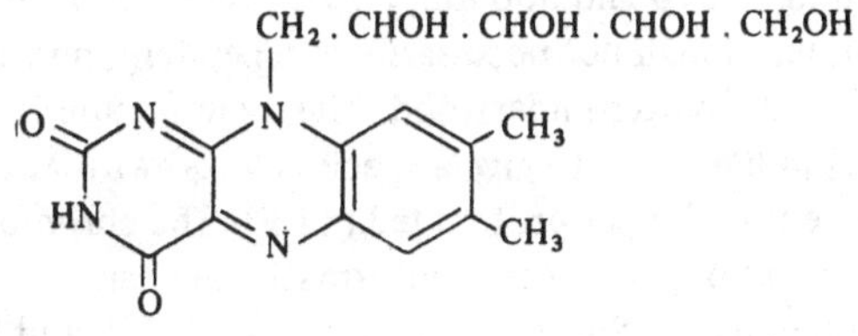

$CH_2 . CHOH . CHOH . CHOH . CH_2OH$

(11.3)  Riboflavin

In young tissues of higher plants it is the pattern of growth that is affected by the phototropic responses. The light is absorbed by riboflavin which thereby photosensitises either the destruction of auxin (indole-3-acetic acid) in the light-exposed side of the plant, or the movement of this hormone away from the illuminated area. The shaded side consequently experiences greater elongation, and the shoot or stem curves towards the light. In mature plant tissues, light-induced turgor changes are responsible for the phototropic reactions.

Riboflavin is similarly the photoreceptor pigment in the movement or growth of fungal hyphae towards light, but auxin is not involved. However, some very recent reports suggest that the riboflavin primary photoreceptor may be complexed with protein and with an antenna of carotenoid molecules, which serve to increase greatly the efficiency of light trapping.

## 11.5    Bacteriorhodopsin

Some salt-loving bacteria are able to harness light energy for ATP production by a process that is not related to the photosynthesis of plants or photosynthetic bacteria. These *Halobacterium* species, in particular *H. halobium*, use normal aerobic respiration to generate ATP when sufficient oxygen is available to them. Under conditions of oxygen deprivation, however, the bacteria develop special purple patches as an integral part of the cell membrane. The pigment responsible for the purple colour is a protein, **bacteriorhodopsin**, which acts as the photoreceptor for a process which converts light energy into a proton gradient which drives the synthesis of ATP by a chemiosmotic mechanism. The photosensitive purple membrane consists of a lipoprotein matrix, and X-ray diffraction shows that the bacteriorhodopsin is organised within this membrane as a rigid, two-dimensional lattice.

Bacteriorhodopsin is a lipoprotein, molecular weight 26 000, and is similar to the visual pigments of the eye in having as its prosthetic group retinaldehyde, bound as a protonated Schiff base or aldimine to the ε-amino group of a lysine residue of the protein (**11.4**). The dark-adapted, photoreactive purple form has $\lambda_{max}$ at 560 nm, and apparently consists of a 1:1 mixture of two species containing an all-*trans*- and a 13-*cis*-retinaldehyde chromophore and absorbing at 570 and 550 nm, respectively. These two forms take part in two separate, though not necessarily independent, photocycles. That of the *trans* form is the better understood. The main feature is that light absorption results in bleaching to give a species ($M_{140}$) with $\lambda_{max}$ 410 nm in which the aldimine is no longer protonated (**11.5**). The chromophore probably has a 13-*cis* configuration, and other conformational changes about a single or double bond seem likely. Species absorbing at 625, 610 and 550 nm have been shown to be short lived (ps-$\mu$s) intermediates in $M_{410}$ production, and interconversion with other species, absorbing at 520 and 640 nm has been suggested. The formation of $M_{410}$ from *trans*-bacteriorhodopsin involves loss of a proton from the protonated aldimine, and the regeneration of bacteriorhodopsin from $M_{410}$ requires proton uptake. It is the directionality of the alternate proton loss and uptake that produces the proton gradient.

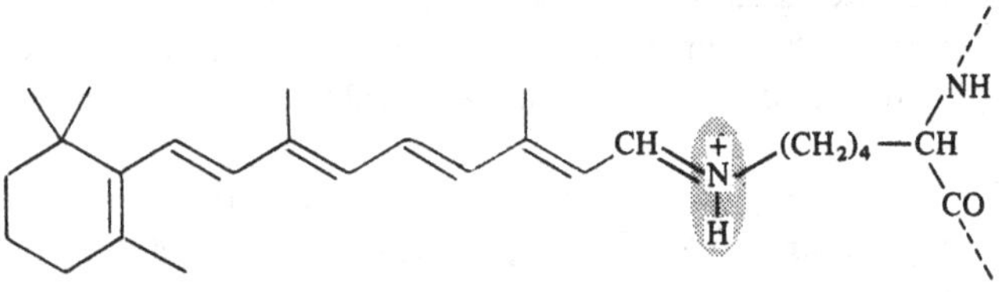

(11.4)  Protonated form of retinaldehyde–lysine aldimine

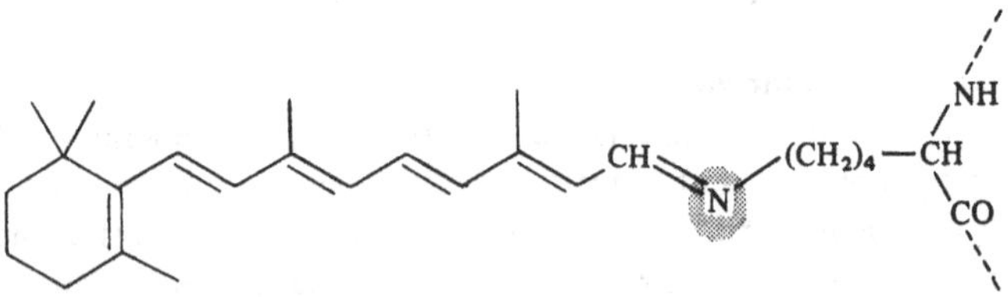

(11.5)  Unprotonated or neutral form of retinaldehyde–lysine aldimine

It is not yet possible to define, even approximately, the sequence of events in the 13-*cis*-bacteriorhodopsin photocycle, though *cis* to *trans* isomerisation certainly seems to occur. Also there seem to be cross-linking points between the two cycles, though details are not available.

There are two fundamental differences between the bacteriorhodopsin photocycles and that of the visual pigment rhodopsin (chapter **9**). In the case of bacteriorhodopsin, the retinaldehyde chromophore never becomes detached from the protein; the aldimine link remains intact. Secondly, the photocycle kinetics of a *trans*-bacteriorhodopsin molecule depend on whether its nearest neighbours are also photocycling.

The generation of ATP *via* the bacteriorhodopsin cycles provides one of the best demonstrations of the chemiosmotic mechanism. The essential feature of the action of bacteriorhodopsin is the establishment of a proton gradient. Light induces the deprotonation of the protonated aldimine group of bacteriorhodopsin and the bleached pigment thus produced is later reprotonated by a dark reaction. The ejection and uptake of the protons are stictly directional, the proton being lost to the outside of the membrane and proton uptake being from the inside of the membrane. A proton and charge gradient is thus established across the membrane and used to drive ATP synthesis from ADP by an ATPase enzyme (cf. photosynthetic phosphorylation, chapter 10).

The functioning of the bacterial purple membrane is currently arousing considerable interest both as a relatively simple mechanism for harnessing light energy without the involvement of chlorophyll, and also because of the similarity of the receptor pigment, bacteriorhodopsin, to the visual pigments of animals.

## 11.6     Extra-ocular and extra-retinal photoreceptors in animals
### 11.6.1    Introduction
It is now clear that in many animals the retina of the eye is not the only light-sensitive tissue. Extra-retinal and extra-ocular photoreceptors and photosensitive areas are now recognised in many species, both vertebrate and invertebrate. These photoreceptors do not allow the animal to 'see', in the sense that vision implies image formation and the rapid detection and recognition of form, position and movement, but they are involved in mediating longer-term effects which depend upon variations in general light intensity. Examples of processes that are regulated by light detected by extra-ocular tissues include the maintenance of light-entrained diurnal and longer-period (*e.g.* lunar) rhythms, colour changes in response to background illumination (*e.g.* skin lightening and darkening), and light modulation of metamorphic changes (*e.g.* diapause in some insects). The fact that the light intensities are monitored by some apparatus other than the eye can readily be demonstrated because the responses are not inhibited or diminished in blinded animals. The photosensitive tissues may be specific organs, *e.g.* the ocellus and the pineal gland, or the integument as a whole may exhibit photosensitivity. Some examples are outlined briefly below. Very few cases, however, have been studied experimentally, and in no instance has the photoreceptor pigment been identified conclusively. There are probably several different mechanisms, but it is unlikely that complex regeneration cycles such as that outlined in chapter 9 for rhodopsin are involved.

### 11.6.2    The invertebrate 'median eye' or ocellus

One of the most highly developed of the primitive photoreceptors is the median ocellus or 'median eye' of many arthropods. As an example, the horseshoe crab *Limulus polyphemus* has a pair of these primitive receptors on the centre-line of the body. Although primitive, these ocelli are equipped with lens structures.

The sensitivity of the ocelli is almost an order of magnitude lower than that of the normal or lateral eye, but the wavelength of maximal sensitivity (520–530 nm) is similar. The ocellus has a second receptor pigment with high u.v.-sensitivity (370 nm). It is thought that the pigments are of the rhodopsin type.

### 11.6.3    The pineal gland and related tissues

In many species of fish and amphibians, it has been shown that the pineal or a closely associated diencephalic area is sensitive to light and is influential in controlling melanophore responses in colour-change mechanisms. In most cases the pineal is responsible for the body-blanching reaction; pineal-ectomised animals remain dark. It is in fact the absence of light which brings about the response of the pineal. The gland then produces a hormonal substance which mediates melanin dispersal and brings about darkening. Light, detected by the pineal, inhibits this effect and lightening or body blanching is observed.

Photoreceptor structures have been found in the amphibian pineal and also in the associated or related 'front organ' or 'Stirnorgan', as well as in the 'parietal eye' of some reptiles. These photoreceptor cells are, superficially at least, similar to the photoreceptors of the retina and are connected to the brain by a nerve tract. The nature of the photoreceptor pigment(s) has not been ascertained.

A somewhat similar case is provided by the Harderian gland of new-born rodents, which is thought to be the extra-retinal photoreceptor controlling their circadian rhythm. This gland contains an unusual tricarboxylic porphyrin (**11.6**) together with a considerable amount of protoporphyrin IX (**11.7**). The obvious implication that these porphyrins are the photoreceptor pigments has not been verified.

### 11.6.4    The animal integument as a general photoreceptor

Many groups of animals are known to exhibit general photosensitivity in the integument, and sometimes also in deeper tissues. Some specific areas may be particularly sensitive, *e.g.* the siphons of bivalves and ascidians. The phenomenon may be demonstrated in animals from which eyes, pineal and ocelli are absent, but which still maintain the appropriate response; this is abolished by covering the general body surface with opaque paint. Commonly the integument of these animals contains considerable amounts of carotenoids,

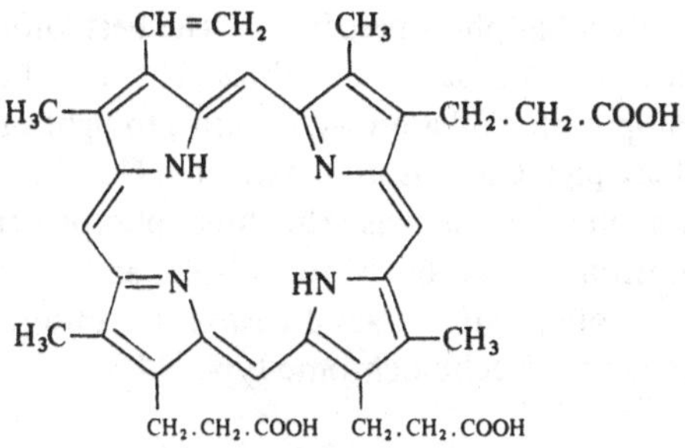

(11.6) Harderoporphyrin

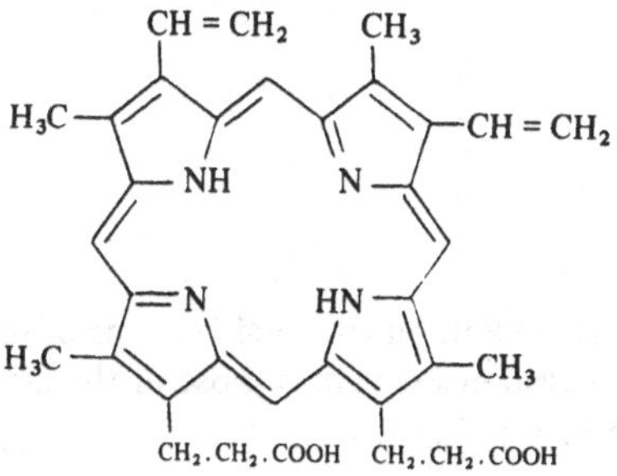

(11.7) Protoporphyrin IX

and pigments of other classes are also frequently present. In general, however, no specific photoreceptor cells have been recognised and the photoreceptor pigment cannot therefore be located and identified. The fact that a photo-receptor pigment of this kind need be present in only minute quantities makes its identification extremely difficult since it could easily be masked by much larger amounts of general integumental pigments. Behavioural studies and action spectra have in some cases given clues to the class of pigment that may be concerned.

The phenomenon is shown by some members of all major invertebrate phyla. It appears commonly in aquatic animals and is also considered very important in controlling the circadian 'clocks' of insects.

Most of the evidence, although not strong, is in favour of carotenoids or retinaldehyde–proteins as the dermal perceptor substances. Recorded action spectra, for example the wavelength sensitivity maxima for some clams, barnacles and fish, are frequently in the rhodopsin–porphyropsin range.

Porphyrins may be the active pigments in some asteroids, holothurians and insects, and in many worms. Riboflavin has also been suggested to have a useful photosensitising function in holothurian integuments and in the dorsal integument of crustaceans.

In some instances nerve cells act as photoperceptors. The best-known
example is in the genital ganglion of the sea hare *Aplysia*. The neural cells contain a carotenoid and a haemoprotein and are very sensitive to light of precisely
the two wavebands which these pigments absorb maximally. Other classes of
pigments have been implicated in other animals. The direct photosensitivity
of the nerve cord of *Branchiostoma* is attributed to melanin, and in an
echinoid, *Diadema*, and the crinoid (feather star) *Antedon*, the observed
action spectra imply a quinone of the echinochrome type (11.8).

(11.8) Echinochrome A

These primitive photoperceptions mediate in two main groups of responses, overt movements and slower responses such as those of the circadian
cycle. The mechanisms of action remain unknown.

## 11.7     Photoprotection
### 11.7.1    Introduction
Light is used by living organisms in so many beneficial ways that it is
easy to overlook the fact that light energy can also be damaging to living
tissues. When light is absorbed by a natural pigment the molecule is raised to
an unstable and short-lived higher energy excited state. The excess energy is
commonly used for the benefit of the organism, *e.g.* in photosynthetic electron transport leading to ATP synthesis or in the production of neural signals
that initiate responses. The various photofunctions of natural pigments outlined in this book are all examples of such activities.

However, the energy trapped in the excited molecules can also cause
damage by bringing about undesirable chemical reactions which can lead to
the destruction of vital tissues. Various natural pigments, but most particularly
porphyrins, can act as sensitisers for light-catalysed damage, especially in the
presence of oxygen, unless there is some suitable protective mechanism. The
harmful effects are not brought about only by visible light. Ultraviolet light
of high intensity is absorbed by proteins and nucleic acids and is potentially
capable of damaging these vital molecules.

One of the most important roles of many natural pigments, especially
carotenoids and melanins, is the protection of vital tissues from photodynamic, especially photooxidative, damage. Three instances of this photo-

protection are well documented. Two of these, the role of carotenoids in the protection of bacteria against photooxidation and the prevention of photo-induced damage in animals, will be described below. The third, photoprotection in photosynthetic tissues, was dealt with in chapter **10**.

### 11.7.2   Repair of DNA damaged by u.v. irradiation

The DNA of a cell encodes the genetic information which determines virtually all aspects of the structure and functioning of that cell, and of the organism to which the cell belongs. Damage to the DNA molecule results in alteration of the information encoded in the DNA sequence, and is likely, therefore, to have dire consequences if it cannot be reversed. Because of the small number of DNA molecules present in the cell and the enormous size of these molecules, there is great risk of damage being brought about by chemicals, radiation, *etc*. Particularly well known are the effects of u.v. irradiation. Nucleic acids absorb u.v. radiation, with $\lambda_{max}$ about 260 nm, and the absorption of this energy can cause structural changes. Perhaps the main result (fig. 11.4) is the dimerisation of adjacent pyrimidine bases (usually thymine) in the DNA sequence to give cyclobutane products **(11.9)**. This dimerisation affects the regular hydrogen bonding between bases on the two chains, and thus disrupts the double-helix structure over a portion of the macromolecule.

Fortunately cells have a range of defence mechanisms whereby they can repair damage to DNA strands. Excision–repair processes, in which a small piece of the defective DNA strand is removed and replaced by a newly-synthesised undamaged section, are used to overcome damage caused by many agents, including u.v. irradiation. Many cells, especially bacteria but also human cells, can use a photoreactivation mechanism, by which the damage is corrected without the need for excision of any of the DNA mole-

Fig. 11.4. Dimerisation of adjacent thymine bases in DNA, and reversal of the process by photoactivated repair enzyme.

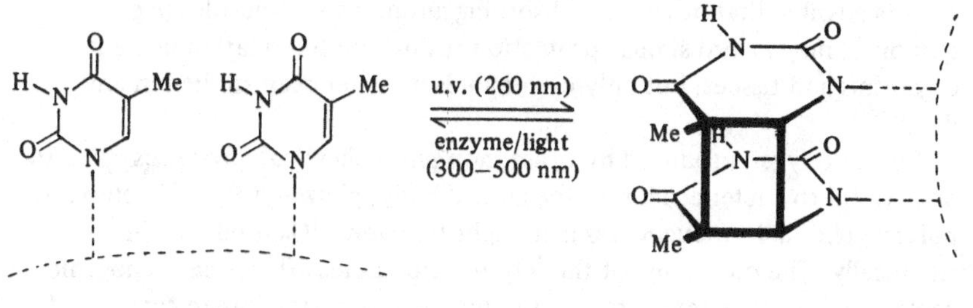

Thymine bases                          (11.9) Thymine 'cyclobutane dimer'

cule. In this process, an enzyme binds to the DNA that contains the pyrimidine dimer. The enzyme is activated when light (300–500 nm) is absorbed by the enzyme–DNA complex, and reverses the damage by splitting the dimers into the normal pyrimidine bases.

### 11.7.3   Protection against photooxidation in bacteria

Many instances are known of carotenoidless mutants of normally carotenogenic bacteria being killed by the combined effects of light and oxygen, whereas the coloured wild-type organisms are nor harmed. This protective effect of carotenoids has been demonstrated in both photosynthetic and non-photosynthetic bacteria.

Only those carotenoids that absorb in the visible region, above 400 nm, are effective in this photoprotection, *i.e.* a chromophore of at least eight conjugated double bonds is required. Although it is possible that in some cases carotenoid in the cell envelope protects by filtering out harmful light wavelengths, the carotenoid is usually involved more directly in the molecular events of photosensitisation.

Photodynamic killing of bacteria requires oxygen and an endogenous or exogenous photosensitiser molecule as well as light. Light is absorbed by the sensitiser pigment, which undergoes intersystem crossing to give the longer-lived but still high-energy triplet state. This reactive species can then pass on its excess energy to molecular oxygen, raising it to the highly reactive singlet state, $^1O_2$. The singlet oxygen can then oxidise any suitable acceptor molecule, such as unsaturated fatty acid, producing peroxides and thus causing extensive and perhaps lethal damage to sensitive cellular processes.

Carotenoids can interfere in this sequence of events by either serving as preferred substrates for oxidation or by quenching the excess energy of the triplet sensitiser or of $^1O_2$. The triplet carotenoid produced decays harmlessly. These protective mechanisms are very similar to those outlined for the protection of photosynthetic membranes (§*10.4.2*) and illustrated in fig. 10.8.

This protection against photooxidation is probably the main function of carotenoids in non-photosynthetic microorganisms, and similar mechanisms may apply in some primitive animal species.

It is possible that other light-absorbing antioxidant molecules (*e.g.* quinones) may afford similar protection against photooxidation in some organisms and tissues, but only carotenoids have yet been studied in any detail.

Singlet oxygen produced by other, non-photochemical, processes can also be harmful to bacteria. Thus in the animal body, phagocytosis of bacteria by polymorphonuclear leucocytes is thought to involve $^1O_2$ produced biochemically. The quenching of the $^1O_2$ by carotenoids affords carotenogenic bacteria a much greater degree of protection against attack than carotenoidless organisms enjoy.

## 11.7.4    *Photoprotection in animals*

In animals, protection against light irradiation is usually afforded by a screening layer of pigment which either absorbs light of all wavelengths or filters out the particularly harmful rays. For screening purposes, dark pigments such as melanin are obviously best since they absorb strongly throughout the visible and u.v. ranges. The most familiar example is the pigmentation of human skin. The so-called white skin of human beings is rather transparent, though the keratin of the stratum corneum does absorb u.v. radiation considerably. In response to prolonged exposure to sunlight increased production of keratin, and more especially of melanin, occurs in the process of suntanning. Suntanned white skin transmits only 5% of u.v. radiation at 300 nm compared with 25% for the untanned skin. Almost all u.v. radiation is absorbed by the substantial amounts of melanin in the skin of the dark human races, thus affording adequate protection against the higher levels of radiant energy encountered in those areas of the earth where the dark races are indigenous.

Melanins are used for protection by lower animals too. The black slug *Arion ater* accumulates an amount of melanin proportional to the amount of photodynamic free porphyrin in the integument. Similar examples of correlation between melanin and free porphyrin content have been described in marine animals.

Free porphyrins can have detrimental, even fatal effects as photosensitisers in mammals, including man. Disorders of haem metabolism can result in the accumulation of free porphyrin intermediates. Such conditions, or porphyrias (§*5.10.1*) are characterised by extreme photosensitivity in the sufferer, who may not be able to withstand exposure to sunlight. Although there is not sufficient endogenous screening pigment present to protect against these effects, pigment screens can be produced artificially. Injected $\beta$-carotene can in suitable cases be deposited in the dermal tissues where it absorbs light of harmful wavelengths and prevents it from reaching the porphyrin photosensitiser.

In other, lower, animals pigments of other classes are thought to act as filters or screening pigments, *e.g.* naphthaquinones in echinoderms. Carotenoids, widespread integumental pigments, are antioxidants as well as light absorbers. It is likely that they afford protection against photooxidation in animals as they do in bacteria (§*11.7.3*). The high concentrations of carotenoids in the eggs of terrestrial and shallow water animals may reflect this photoprotective role.

## 11.8    Bioluminescence

### 11.8.1    *Introduction*

The preceding parts of this chapter, and indeed the whole of this book so far, have been concerned with the absorption of light by various molecules. At this late stage it is appropriate to mention, albeit briefly, the reverse process, bioluminescence, by which biochemical energy is used by living organisms to generate visible light.

### 11.8.2   The incidence of bioluminescence

Bioluminescence has been recognised in marine bacteria and dino-flagellates and in a small number of fungi. The phenomenon is, however, known best in the animal kingdom. Two of the most familiar examples are the firefly and the glow-worm, terrestrial insects which at night use flashing or continuous emitted light to attract mates, but luminescence is most commonly associated with marine animals, including fishes, crustaceans, molluscs, annelids and coelenterates. Some luminous animals, especially fishes and some cephalopods, utilise colonies of symbiotic light-emitting bacteria. In a wide variety of other animals the luminescence is intrinsic and does not depend on symbionts.

There are two genera of light-emitting bacteria. All symbiotic strains are from the genus *Photobacterium*, whereas *Beneckea* includes only free-living strains. Those animals that utilise *Photobacterium* confine the microbes to specific light organs, the most extravagant example perhaps being the lures used by angler fishes. The maintenance of bacterial symbionts presents problems. The bacteria must be restricted to the light organs and not allowed to spread indiscriminately, and they must be passed on, without contamination, into the light organs of successive generations. Bacterial luminescence is continuous, so the animal needs to have some means of controlling light emission from the organ. This may be done by movement of the organ itself, by the use of chromatophores to obscure the emitted light, or by various shutter mechanisms. Luminescence is also restricted to specific light organs in those animals which do not employ bacteria. The light organs are usually greater in number than in the symbiotic species, and usually arranged in definite patterns.

### 11.8.3   Mechanisms

The classical explanation of the phenomenon of bioluminescence describes the oxidation of a substrate, **luciferin**, by an enzyme, **luciferase**, with a photon of light being emitted at some stage of the process. Recent work has revealed some details of the reactions in a few individual cases.

Ideally it would be useful if the various examples of bioluminescence could be classified into a relatively small number of reaction types, but at present our knowledge is too fragmentary for this goal to be achieved.

The continuous light emission by bacteria requires a bacterial luciferase enzyme, a long-chain fatty aldehyde or related substance as a luciferin, molecular oxygen and reduced flavin mononucleotide, $FMNH_2$. It is thought that the light is emitted as fluorescence from an excited state of the oxidised flavin nucleotide produced in the reaction.

The most intensively studied example of bioluminescence is that of an insect, the firefly. The luciferin, lampyrine (**11.10**) first reacts with ATP to give an enzyme-bound luciferin–AMP derivative. This is then oxidised to the oxyluciferin (**11.11**) by molecular oxygen in a several-step reaction. The enzyme undergoes a series of large conformational changes, and light is emitted at some stage in the sequence.

Other examples are similar in principle but differ in the degree of association or complexity of the luciferin–luciferase system. In the hydrozoan jelly-fish *Aequorea*, the luciferase and the luciferin coelenterazine (**11.12**) remain strongly associated as a stable photoprotein complex (**11.13**). In the case of the anthozoan sea-pansy *Renilla*, the same luciferin (**11.12**) is stored as a sulphate derivative and is liberated on removal of the sulphate group by 3′,5′-diphosphoadenosine and the enzyme luciferin sulphokinase. In the presence of molecular oxygen, the unstable luciferin is oxidised to the oxyluciferin and light emission occurs.

The small ostracod crustacean *Cypridina* produces its luciferin (**11.14**) and luciferase in separate glands. The two components are squirted into the sea water where the luminous reaction occurs. Some fishes apparently rely on *Cypridina* in their diet to provide them with luciferin and luciferase.

Although bioluminescence in aquatic animals is usually associated with a marine environment, one fresh-water species, the limpet *Latia*, is well known.

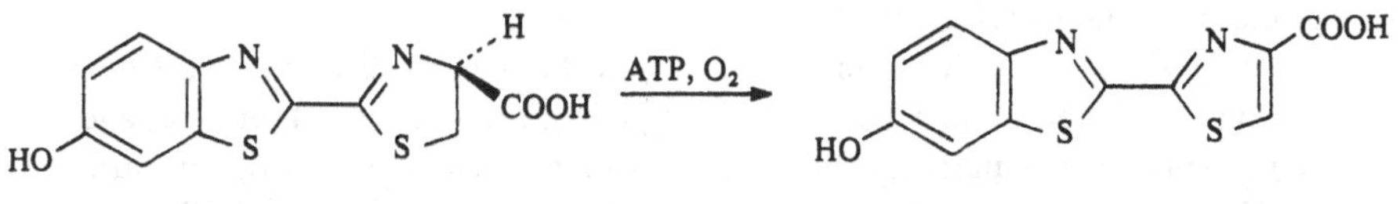

(11.10) Lampyrine (firefly luciferin)

ATP, $O_2$

(11.11) Dehydrolampyrine (oxyluciferin)

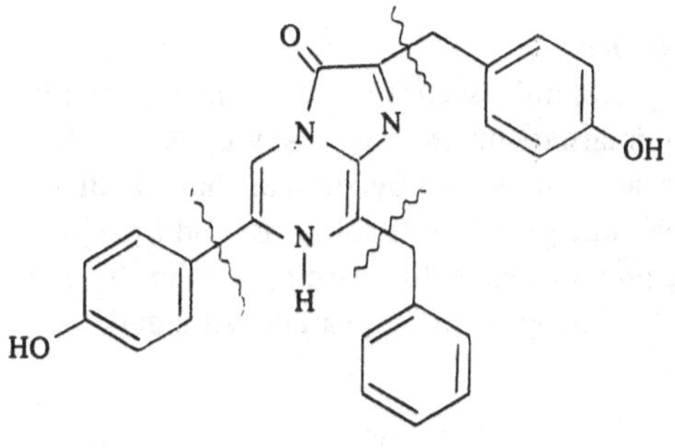

(11.12) Coelenterazine

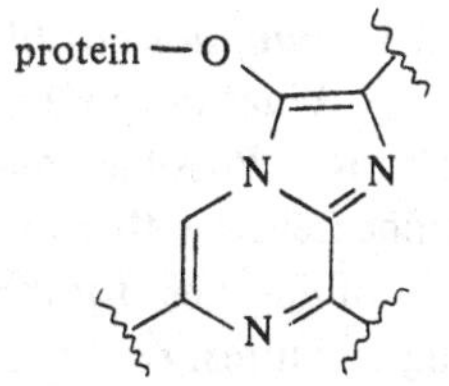

(11.13) Coelenterazine
(protein-bound form
in aequorin)

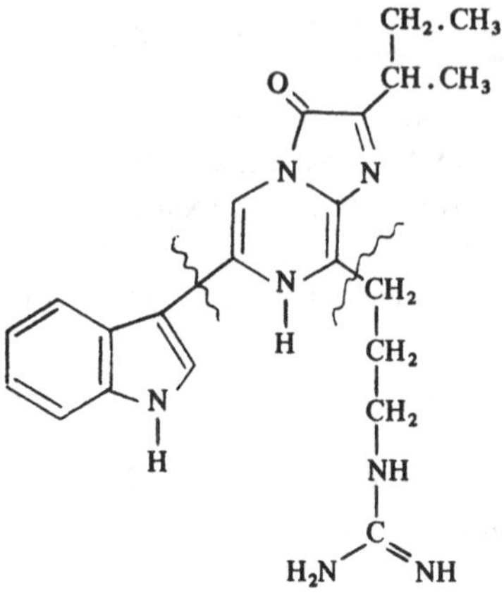

**(11.14)** *Cypridina* luciferin

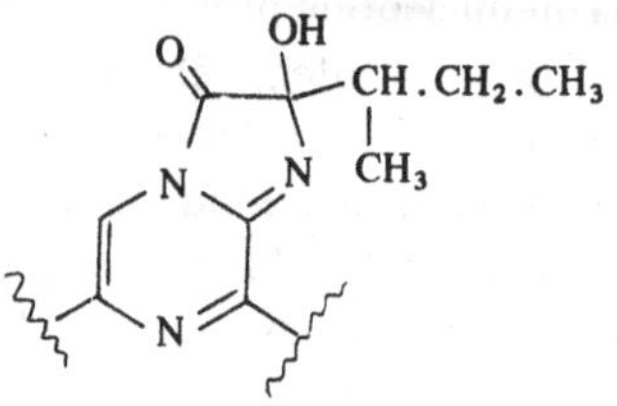

Oxyluciferin

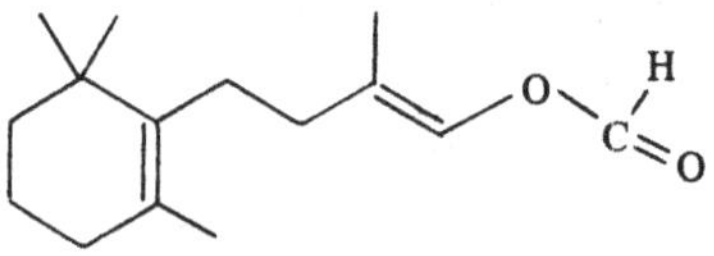

**(11.15)** *Latia* luciferin

The luciferin of *Latia* (**11.15**) bears a close resemblance to the carotenoid group of pigments (chapter 2).

The colour, *i.e.* the wavelength, of the emitted light is different in different examples, *e.g.* red in dinoflagellates, blue-green in *Renilla* and *Aequorea*, and is influenced particularly by the conformation of the enzyme, which in turn is affected by factors such as temperature, pH, pressure, salts and ATP concentration.

### 11.8.4  Functions of bioluminescence

What purpose is served by bioluminescence in free-living bacteria is not known. Possible benefits to animals are much more easily understood. Luminescence, either intrinsic or bacterial, is used by animals that live in a world of darkness for many of the same purposes as colour is used by light-living creatures, *i.e.* for attracting prey or mates, for warning and, in the case of some fishes, for concealment by destroying the appearance of a dark silhouette when viewed from below.

### 11.9    Conclusions and comments

In the first section of this book, the chief characteristics of the main groups of natural pigments were described. Then in the preceding chapters

of the second section, the best known and understood biological functions of
these pigments were discussed, namely, providing colour, detecting light and
recognising colour (vision), and harnessing light energy in photosynthesis.
This final chapter has been used to collect together a variety of other aspects
of photobiology in which natural pigments play an important role. In plants
and microorganisms, photoreceptors such as phytochrome and flavins have
been considered and the remarkable use of bacteriorhodopsin for ATP
production in *Halobacteria* has been outlined. The occurrence, in animals, of
extra-ocular and extra-retinal photoreceptors, as yet unidentified, has been
discussed. Although much less exciting than the above examples, the protection of living organisms and tissues from harmful effects of u.v. and visible
irradiation is extremely important for survival. Finally, brief mention has
been made of bioluminescence, by which living organisms use energy to
generate light for various purposes.

There is great scope for biochemical work on all of these topics, to identify
the actual photoreceptors, to describe the molecular changes which occur
during their functioning, and to show how these changes result in the recognised responses. Such work is extremely difficult; the photoreceptors are
usually present only in minute amounts, and the changes may be very subtle.
This therefore provides a real challenge to the ingenuity of the biochemist. In
addition to the phenomena described in this chapter, there may be many
other forms or examples of interactions between light and living organisms
(*via* natural pigments), which await discovery and will provide further systems
for investigation.

## 11.10  Suggested further reading

As might be expected for a chapter which includes such a wide
variety of topics, a very large number of references could be given. The list
below is by no means exhaustive. The references quoted here are all extremely
useful, but time spent in a good library will surely reveal other books and
reviews on these topics which the reader may find equally useful.

Several books deal with photobiology in general, including most or all of
the topics included in this chapter. Clayton (1971) gives a useful introductory
account. Castellani (1977), Checucci and Weale (1973), Smith (1977) and
Wolken (1975) deal with various aspects in more detail. The annual series of
*Photochemical and photobiological Reviews* (Smith, 1976-9) contains up-to-
date review articles on a range of photobiological topics.

For information on phototaxis and phototropism, the reader is referred to
the books by Clayton (1971) and Wolken (1975), and to an article by Hand
(1977). Phytochrome is the subject of many publications; some which
include a biochemical approach are a book by Mitrakos and Shropshire (1972)
and reviews by Lee (1977), Pratt (1979), Satter and Galston (1976),

Shropshire (1977) and Smith and Kendrick (1976). Of the multitude of references to bacteriorhodopsin and the purple membrane of *Halobacteria*, the book edited by Caplan and Ginzburg (1978) is comprehensive and the review by Ottolenghi (1980) contains the most recent ideas about the mechanism of the bacteriorhodopsin photocycles.

Photoprotection by pigments, especially carotenoids, is described by Krinsky (1971) and information on DNA photodamage and photorepair can be obtained from the book by Wang (1976) and a recent essay by Lehmann and Bridges (1977). Several books and reviews (*e.g.* Millott, 1968; Eakin, 1974; Menaker, 1977; Bennett, 1979; Yoshida, 1979) give accounts of extra-ocular and extra-retinal photoreceptors, but these rather emphasise the lack of biochemical knowledge on the subject. Finally, bioluminescence is covered in reviews by Hastings (1968), Johnson and Haneda (1966) and Ward (1979) and a recent book edited by Herring (1978). Relevant methodology is collected in a volume of *Methods in enzymology* (De Luca, 1978).

## 11.11    Selected bibliography

Bennett, M. F. (1979) Extraocular light receptors and circadian rhythms, in *Handbook of sensory physiology*, vol. VII/6A. *Vision in invertebrates*, ed. H. Autrum, p. 641. Heidelberg, Berlin and New York: Springer-Verlag.

Caplan, S. R. and Ginzburg, M. (eds) (1978) *Energetics and structure of halophilic micro-organisms*. New York: Elsevier–North Holland.

Castellani, A. (ed.) (1977) *Research in photobiology*. New York and London: Plenum.

Checcucci, A. and Weale, R. A. (eds) (1973) *Primary molecular events in photobiology*. Amsterdam: Elsevier.

Clayton, R. K. (1971) *Light and living matter*, vol. 2, *The biological part*. New York: McGraw-Hill.

De Luca, M. A. (ed.) (1978) *Bioluminescence and chemiluminescence*. (*Methods in enzymology*, vol. 57). New York: Academic Press.

Eakin, R. M. (1974) *The third eye*. Berkeley: University of California Press.

Hand, W. G. (1977) Photomovement, in *The science of photobiology*, ed. K. C. Smith, p. 313. New York and London: Plenum.

Hastings, J. W. (1968) Bioluminescence, *Ann. Rev. Biochem.*, **37**, 597.

Herring, P. J. (ed.) (1978) *Bioluminescence in action*. London, New York and San Francisco: Academic Press.

Johnson, F. H. and Haneda, Y. (eds) (1966) *Bioluminescence in progress*. New Jersey: Princeton University Press.

Krinsky, N. I. (1971) Function, in *Carotenoids*, ed. O. Isler, p. 669. Basel and Stuttgart: Birkhäuser.

Lee, J. (1977) Bioluminescence, in *The science of photobiology*, ed. K. C. Smith, p. 371. New York and London: Plenum.

Lehmann, A. R. and Bridges, B. A. (1977) DNA repair, *Essays in Biochemistry*, **13**, 71. London: The Biochemical Society – Academic Press.

Menaker, M. (1977) Extraretinal photoreception, in *The science of photobiology*, ed. K. C. Smith, p. 227. New York and London: Plenum.

Millott, N. (1968) The dermal light sense, *Symp. Zool. Soc. London*, **23**, 1.

Mitrakos, K. and Shropshire, W., Jr. (eds) (1972) *Phytochrome*. London and New York: Academic Press.

Ottolenghi, M. (1980) The photochemistry of rhodopsins, *Adv. Photochem.*, **12**, 97.

Pratt, L. H. (1979) Phytochrome, in *Photochemical and photobiological reviews*, vol. 4, ed. K. C. Smith, p. 59. New York: Plenum.

Satter, R. L. and Galston, A. W. (1976) The physiological functions of phytochrome, in *Chemistry and biochemistry of plant pigments*, 2nd edition, vol. 1, ed. T. W. Goodwin, p. 681. London, New York and San Francisco: Academic Press.

Shropshire, W., Jr. (1977) Photomorphogenesis, in *The science of photobiology*, ed. K. C. Smith, p. 281. New York and London: Plenum.

Smith, H. and Kendrick, R. E. (1976) The structure and properties of phytochrome, in *Chemistry and biochemistry of plant pigments*, 2nd edition, vol. 1, ed. T. W. Goodwin, p. 378. London, New York and San Francisco: Academic Press.

Smith, K. C. (ed.) (1976–1979) *Photochemical and photobiological reviews*, vols 1–4. New York: Plenum.

Smith, K. C. (ed.) (1977) *The science of photobiology*. New York and London: Plenum.

Wang, S. Y. (ed.) (1976) *Photochemistry and photobiology of nucleic acids*, vol. 2, *Biology*. New York: Academic Press.

Ward, W. W. (1979) Energetics of bioluminescence, in *Photochemical and photobiological Reviews*, vol. 4, ed. K. C. Smith, p. 1. New York: Plenum.

Wolken, J. J. (1975) *Photoprocesses, photoreceptors and evolution*. New York, San Francisco and London: Academic Press.

Yoshida, M. (1979) Extraocular photoreception, in *Handbook of sensory physiology*, vol. VII/6A, *Vision in invertebrates*. Heidelberg, Berlin and New York: Springer-Verlag.

## Problems

---

1      Most natural white 'colours' are considered to be structural in origin. Give some examples and suggest what structural elements and physical phenomena may be responsible.

2      Define bathochromic, hypsochromic and hypochromic effects in terms of energy changes and transition probabilities.

3      A simple colorimeter is frequently used to assay biochemical substances by measuring the amount of light transmitted by a coloured solution. It is common to interpose a simple colour filter between the white-light source and the sample. Explain why the results obtained should be most accurate when the filter used is of the colour complementary to that of the sample under examination.

4      A sample (0.026 mg) of a compound $M$ was dissolved in 10.0 ml of a suitable solvent. Given an absorbance coefficient $A_{1\,cm}^{1\%}$ of 1600 for $M$, calculate:
   (i) the absorbance of the solution in a 10 ml cell of 1 cm pathlength
  (ii) the absorbance of the solution in a 10 ml cell of 3 cm pathlength
 (iii) the absorbance of the solution in a 3 ml cell of 1 cm pathlength
 (iv) the absorbance, in a 10 ml cell, 3 cm path length, of a solution prepared by diluting 3 ml of the original solution to 10 ml.

5      A leaf extract, in aqueous 80% acetone, had absorbances of 0.50 and 1.00 at 645 and 663 nm, respectively, in a 1 cm pathlength cell. Calculate the concentrations (in mg/l) of chlorophylls $a$ and $b$ in the extract given the following absorbance coefficients for chlorophyll $a$: $A_{1\,cm}^{1\%} = 820$ (at 663 nm), 170 (at 645 nm); for chlorophyll $b$: $A_{1\,cm}^{1\%} = 100$ (at 663 nm), 450 (at 645 nm).
     A simplified procedure for calculating chlorophyll concentrations uses the equations:
Total chlorophyll (mg/l) $= 20.2 A_{645} + 8.02 A_{663}$
Chlorophyll $a$ (mg/l) $= 12.7 A_{663} - 2.69 A_{645}$

Chlorophyll $b$ (mg/l) $= 22.9 A_{645} - 4.68 A_{663}$

where $A_{645}$ and $A_{663}$ are the recorded absorbances at 645 and 663 nm, respectively. Does the application of these equations give accurate values in this case?

6 (a)   A pigment $X$ (3.43 mg) in solution in hexane (500 ml) had an absorbance, after 5-fold dilution, of 0.346 at 450 nm. Calculate the molar absorbance coefficient of $X$ (relative molecular mass 536).

(b)   The solvent was evaporated and a suspension of $X$ was prepared and incubated aerobically with an intestinal enzyme preparation, in the presence of NADPH. After incubation the products were extracted into hexane (800 ml). The hexane solution had absorbance 0.577 at 450 nm, and a new absorption peak (absorbance 0.272) was present at 366 nm. Given a specific absorbance coefficient, $A_{1cm}^{1\%}$ of 1740 and a molar absorbance coefficient of 49 500 for the product $Y$ ($\lambda_{max}$ 366 nm) calculate the total amount of $X$ and $Y$ recovered (in mg and $\mu$moles), and draw conclusions about the enzymic reaction. (*N.B.* $X$ does not absorb at 366 nm, $Y$ does not absorb at 450 nm.)

(c)   The purified reaction product, $Y$ (1.25 mg) on incubation with a specific protein gave, in quantitative yield, a water-soluble complex, $\lambda_{max}$ 500 nm. The total volume of the aqueous solution was 100 ml, and this gave an absorbance of 1.76. Given a value of 9.8 for $A_{1cm}^{1\%}$ (at 500 nm) for the complex and assuming an equimolar ratio of $Y$:protein, calculate the relative molecular mass of the protein. (*N.B.* In all cases, cuvettes of total volume 3 ml and light pathlength 1 cm were used.)

7   The table below lists the millimolar absorbance coefficients for the oxidised ($\epsilon_{ox}$) and reduced ($\epsilon_{red}$) forms of a cytochrome $c$. Plot the (reduced minus oxidised) difference spectrum. Would it be practicable to try to differentiate between the oxidised and reduced forms simply from the absorption spectra of the two forms?

| $\lambda$ (nm) | $\epsilon_{ox}$ | $\epsilon_{red}$ | $\lambda$ | $\epsilon_{ox}$ | $\epsilon_{red}$ | $\lambda$ | $\epsilon_{ox}$ | $\epsilon_{red}$ |
|---|---|---|---|---|---|---|---|---|
| 370 | 28.0 | 14.0 | 440 | 20.0 | 8.0 | 520 | 8.0 | 13.5 |
| 380 | 34.5 | 20.0 | 450 | 13.0 | 4.5 | 530 | 9.0 | 9.0 |
| 390 | 45.5 | 32.0 | 460 | 9.5 | 2.0 | 540 | 9.5 | 8.0 |
| 400 | 73.0 | 53.0 | 470 | 6.5 | 1.0 | 550 | 8.0 | 24.0 |
| 410 | 105.0 | 110.0 | 480 | 5.5 | 1.5 | 560 | 6.5 | 2.0 |
| 415 | 88.0 | 128.0 | 490 | 5.5 | 2.0 | 570 | 5.5 | 1.5 |
| 420 | 60.0 | 94.0 | 500 | 6.0 | 4.0 | 580 | 4.5 | 1.0 |
| 430 | 32.0 | 32.0 | 510 | 6.5 | 8.0 | 590 | 2.0 | 0.5 |

8   Estimate the positions of the main light absorption maxima of the following carotenoids: rhodopin, $\epsilon$-carotene, 4-oxo-$\gamma$-carotene, 5,6-

epoxy-5,6-dihydro-$\psi,\psi$-carotene, $\gamma$-carotene-5,6-epoxide, zeaxanthin, 4-oxo-$\beta$-zeacarotene, loroxanthin ($\beta,\epsilon$-carotene-3,19,3'-triol), $\beta,\beta$-carotene-2,2'-dione; given that neurosporene, lycopene, $\gamma$-carotene, $\beta$-carotene and echinenone have $\lambda_{max}$ at 440, 470, 460, 450, 458 nm, respectively. (Structures and semisystematic nomenclature are given in chapter **2**.)

9    Explain the following observations:

   (i) A dilute solution of chlorophyll in diethyl ether is green, whereas a concentrated solution may appear to be red.

   (ii) Chlorophyll is red when viewed in ultraviolet light.

   (iii) When a tree, in sunlight, is viewed through a filter transparent to a narrow range of red light wavelengths, the leaves appear bright pink against an otherwise nearly black background.

10    A new animal species was discovered which was blue in colour. How would you determine whether the colour was structural or pigmentary in origin and, if pigmentary, identify the class of compound responsible?

11    A carotenogenic bacterium was incubated with [2-$^{14}$C, (5$R$)-5-$^3$H$_1$]-mevalonic acid and the labelled carotenoids phytoene, $\beta$-carotene and zeaxanthin were extracted and purified and assayed for radioactivity. The following $^{14}$C:$^3$H ratios were obtained:

   [2-$^{14}$C-(5$R$)-5-$^3$H$_1$]-mevalonic acid    1:1
   phytoene    1:1
   $\beta$-carotene    1:0.5
   (3$R$,3'$R$)-zeaxanthin    1:0.25

   What conclusions can be drawn about the biosynthetic sequence and reactions?

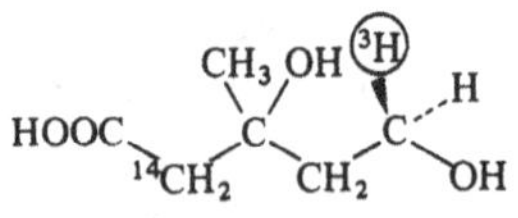

[2-$^{14}$C-(5$R$)-5-$^3$H$_1$]-Mevalonic acid

12    The biosynthesis of the naphthaquinone mollisin occurs from two polyketides of three and four C$_2$ chains, respectively. Devise a scheme for the construction of mollisin in this way, and explain how this pathway could be distinguished experimentally from an alternative one involving only a single heptaketide chain.

   Also devise plausible polyketide folding patterns for the biosynthesis of the anthraquinones solorinic acid (a lichen pigment) and laccaic acid D (from insects).

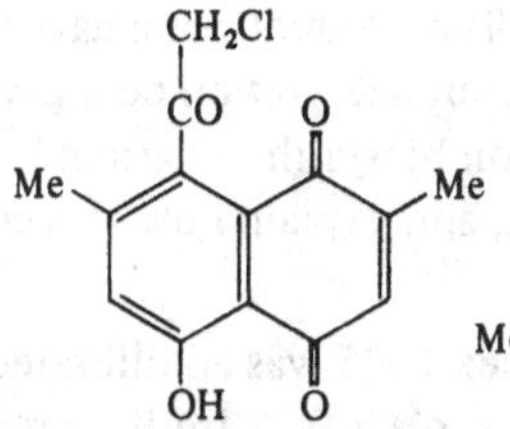

Mollisin

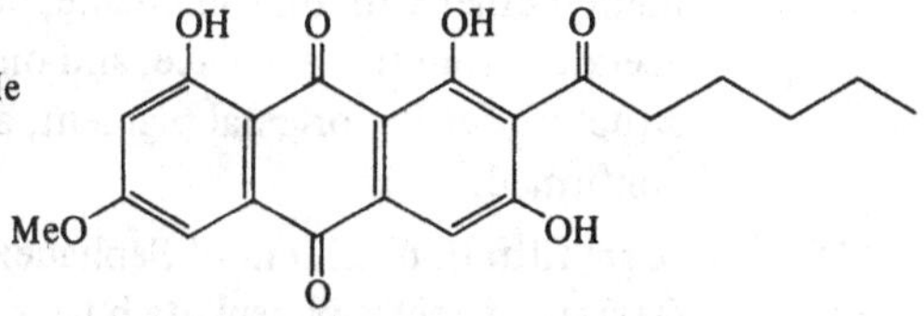

Solorinic acid

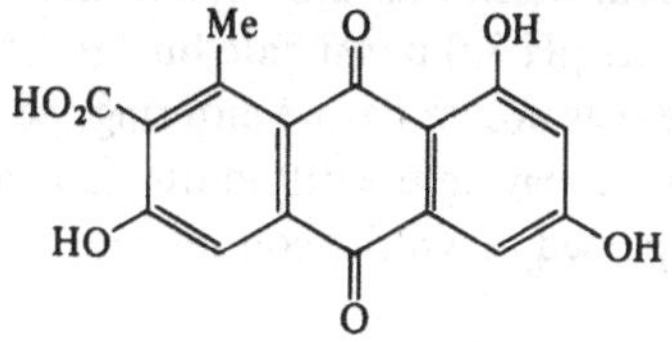

Laccaic acid D

13 The anthraquinone alizarin (**3.27**) is thought to be biosynthesised by a combination of the shikimate pathway and the mevalonate pathway, *via* the prenylated intermediate

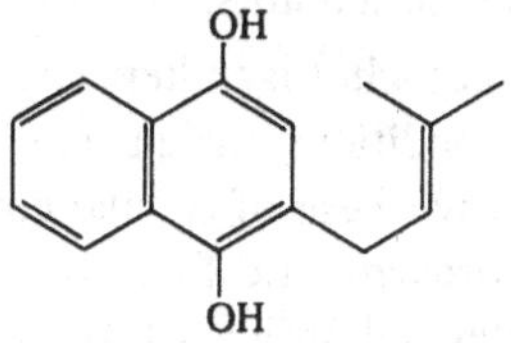

Suggest a mechanism for the formation of alizarin by this route and also by alternative routes *via* a polyketide mechanism, either alone or in combination with the shikimate pathway. Would studies with isotopically labelled acetate allow you to distinguish between these alternatives?

14 The cream-coloured petals of a common garden flower yielded substantial amounts of a water-soluble 'pigment' which could be labelled by incubation with either [$^{14}$C]-acetate or [$^{14}$C]-phenylalanine. The compound had relative molecular mass 480 and $\lambda_{max}$ (in methanol) at 365 nm, shifting to 425 nm in the presence of alkali. With diazomethane the compound gave a pentamethyl ether, whereas a nona-acetate was afforded on treatment with acetic anhydride and pyridine. Hydrolysis with β-glucosidase gave a product, $C_{15}H_{10}O_8$, which had an absorption spectrum only slightly shifted from that of the

original pigment ($\lambda_{max}$ 378 nm). The product again gave a penta-methyl ether with diazomethane, but acetylation now gave a hexa-acetate. From this evidence, and on biosynthetic grounds, deduce a structure for the original pigment, and explain how this could be confirmed.

15     A gel filtration column of Sephadex G-25 was equilibrated with 0.05 M potassium phosphate buffer, pH 7.0. A freshly prepared solution of sodium dithionite was applied to the column and allowed to enter the gel. Then a mixture of haemoglobin and potassium ferricyanide was added. When this had entered the column, elution was continued with the pH 7.0 phosphate buffer. The haemoglobin–ferricyanide mixture was brown before entering the dithionite zone, but the substance leaving the dithionite zone was purple, and became scarlet as it passed down the column and was eluted. Explain these findings.

16     Explain and account for the effects of the following changes in the oxygen affinity of haemoglobin A:
    (i) a 4-fold increase in $CO_2$ partial pressure
    (ii) a 4-fold increase in 2,3-bisphosphoglycerate (DPG) concentration
    (iii) a decrease in pH from 7.4 to 7.2
    (iv) dissociation into monomeric subunits.

17     A unicellular alga can grow non-photosynthetically on glucose as carbon source. Under these conditions, the glucose is metabolised to acetate, which is used for biosynthesis of cellular components. If the organism is cultured non-photosynthetically in 100% deuterium oxide ($^2H_2O$) as its only water and ordinary [$^1H$]-glucose as the only carbon source, predict the distribution of $^1H$ and $^2H$ in carotene and chlorophyll molecules biosynthesised. If the culture, still in deuterium oxide, were then transferred to photosynthetic growth (light + $CO_2$ as carbon source), how would the labelling patterns in the pigments change with time?

18     Deduce the labelling pattern for a phycobilin chromophore (*e.g.* 5.41) biosynthesised from [5-$^{14}C$]-ALA ($\delta$-aminolaevulinic acid). Could this labelling allow you to distinguish between the two alternatives, formation from haem or chlorophyll?

19     In the rat, cleavage of haem gives biliverdin which in turn is con-verted into bilirubin. Two oxygen functions (keto-groups) are intro-duced during the cleavage, and three alternative mechanisms have been proposed: (i) both oxygens introduced come from the same oxygen molecule; (ii) both oxygens come from molecular $O_2$, but

from two different molecules; (iii) one oxygen function comes from molecular oxygen, the other from water.

The relative molecular mass of normal ($^{16}O$) bilirubin is 584. For each mechanism, predict the labelling pattern and relative molecular mass of the bilirubin that would be produced if the reaction were performed in $^{18}O_2$ (100%) and $H_2^{16}O$ (100%). If the cleavage were performed in an atmosphere containing a mixture of the oxygen species $^{18}O_2$ (20%) and $^{16}O_2$ (80%), could the ratios of the 584, 586 and 588 molecular species of bilirubin, as determined by mass spectrometry, distinguish between the three mechanisms?

20  Drosopterin (**6.14**) is a dimeric pterin. What likely monomeric form(s) give rise to this structure? Suggest a possible mechanism for the dimerisation.

The absorption spectrum of drosopterin was determined in neutral solution, in 0.1M HCl and in 0.1M NaOH. In neutral solution $\lambda_{max}$ were at 265 and 485 nm. How would the spectra in acid and alkaline solution differ from this?

21  The bacterium *Pseudomonas phenazinium* produces and secretes large amounts of the phenazine pigment iodinin (**6.50**). Two mutant strains have been produced. One of these, F 11, accumulates 6-hydroxyphenazine-1-carboxylic acid (see fig. 6.10), the other, 13 Z, produces no pigment. Cultures of these two strains were streaked on to an agar plate. Examination of the plate after 7 days revealed the formation of iodinin around the points of intersection of the two strains. Comment on this result.

22  A model partial structure for sepiomelanin is shown in fig. 7.2. Give the structures of the monomeric units from which this structure is derived. Draw an alternative partial structure for melanin constructed from these monomeric units and deduce a scheme for its biosynthesis.

23  The caterpillars of a hawkmoth feed upon leaves of either poplar (*Populus* spp.) or willow (*Salix* spp.). Field studies showed that caterpillars found on the white underside of poplar leaves were grey, whereas those on bright-green willow leaves were themselves green. A research programme was begun to investigate the ecological and nutritional factors which determine and maintain this colour dimorphism. Among the experiments planned were (i) to analyse the pigment contents and compositions of both types of caterpillars and leaves, and (ii) to investigate the effects of rearing the caterpillars in either light or permanent darkness on leaves of the different species. What useful information should these experiments give? Suggest

possible explanations of the dimorphism and design other experiments for testing these hypotheses.

24      Explain how a trichromatic system, based upon three photoreceptor pigments absorbing maximally at different wavelengths, can provide a sensitive colour discrimination mechanism.

25      The retina of the avian eye contains several visual pigment photoreceptors and a range of intensely but differently coloured oil droplets which act as light filters; light reaches a receptor only after passage through an oil droplet. Consider two photoreceptors absorbing in the regions 400–600 nm ($\lambda_{max}$ 500 nm) and 470–670 nm ($\lambda_{max}$ 570 nm) and two oil droplets absorbing at 400–500 nm ($\lambda_{max}$ 450 nm) and 420–520 nm ($\lambda_{max}$ 470 nm), respectively. What effect would each of these filters have on the absorption of light by each photoreceptor, and how would this affect the sensitivity of hue discrimination? (Assume all absorption spectra to be symmetrical in shape.)

26      Sunlight can penetrate clear oceanic water to a considerable depth. Some absorption occurs, however, and the range of wavelengths remaining becomes narrower with increasing depth, so that at 1000 m only a very narrow wavelength band centred at 470 nm remains and can just be detected by the eye. In more turbid, coastal waters light is detectable only to a depth of 50 m, and wavelengths around 570 nm penetrate furthest. Discuss the strategies of pigmentation and photoreceptor systems that would be most advantageous to fish and other animals (a) that live at different, though approximately constant depths in oceanic and coastal waters, respectively, (b) that live in oceanic waters but move freely between surface and sea bed, and (c) that move freely at all depths in both coastal and oceanic waters.

27      The prokaryotic blue-green algae or cyanobacteria are the simplest oxygen-evolving photosynthetic organisms. They contain no chloroplasts, but have more or less extensive internal photosynthetic membranes containing chlorophyll *a*, carotenoids and phycobilins. Describe any changes in morphology, ultrastructure and pigmentation which might be expected to occur when a cyanobacterium is transferred from (i) low light intensity to high light intensity and high oxygen concentration, (ii) high to low light intensity, (iii) saturating white light to green or red light.

28      Explain the following phenomena:
(a)     Leaves in autumn change from green to yellow to red to brown.
(b)     The floribunda rose 'Masquerade' has green flower buds which open to give yellow flowers which later turn red.

(c) Leaves of a normally green plant become chlorotic (yellow)
   (i) if the plant is maintained in the dark for several days
   (ii) if the plant is suffering from magnesium or iron deficiency.

29 The action spectrum for a phototropic response in a fungus was very similar to the absorption spectra of both β-carotene and riboflavin. Suggest experimental approaches to determine which of these pigments might be the photoreceptor.

30 Long-day plants require a long-day/short-night regime for induction of flowering. Short-day plants, on the other hand, flower only under short-day/long-night conditions. Explain how the same photoreceptor, phytochrome, can be responsible for the induction of flowering in both plant types.

## Answers

1      See §1.3.4. Structural whites are usually due to light being reflected from minute crystals, particles or droplets.

2      Definitions of these expressions – though not in terms of transition probabilities and energy changes – are given in chapter 1. The energy change in a transition determines the wavelength of maximal absorption, transition probabilities determine the intensity of absorption.

3      The colorimeter photocell detects transmitted light over the entire visible range. The chosen filter will transmit only the wavelength range where absorption by the sample occurs, so that small differences in concentration will give the largest differences in absorbance.

4      (i) 0.416    (ii) 1.248    (iii) 0.416    (iv) 0.374

Remember the laws of Beer and Lambert: absorbance is proportional to concentration and pathlength.

5      The simplified equations give the concentrations of chlorophylls $a$ and $b$ as 11.35 and 6.77 mg $l^{-1}$, respectively. These figures are virtually identical to the values obtained by the long method requiring the multiple equations

Absorbance at 645 nm

$$= (A_{1\,cm}^{1\%} \text{ for chl}a \text{ at 645 nm} \times \text{chl}a \text{ concentration})$$

$$+ (A_{1\,cm}^{1\%} \text{ for chl}b \text{ at 645 nm} \times \text{chl}b \text{ concentration})$$

Absorbance at 663 nm

$$= (A_{1\,cm}^{1\%} \text{ for chl}a \text{ at 663 nm} \times \text{chl}a \text{ concentration})$$

$$+ (A_{1\,cm}^{1\%} \text{ for chl}b \text{ at 663 nm} \times \text{chl}b \text{ concentration})$$

6    (a) $135 \times 10^3$

(b) 1.83 mg $X$ (3.415 $\mu$moles); 1.25 mg $Y$ (4.40 $\mu$moles) recovered. This gives the relative molecular mass of $Y$ as 284 in agreement with the reaction

$$X + O_2 \rightarrow 2Y \ [\text{i.e. } Y = \tfrac{1}{2}X + O]$$

(c) 180 mg complex is 4.4 $\mu$moles $\therefore$ relative molecular mass
$$= 41 \times 10^3$$

7    The plot is straightforward; $(\epsilon_{red} - \epsilon_{ox})$ as ordinate (can be positive or negative) against wavelength as abscissa. The oxidised and reduced forms have their respective $\lambda_{max}$ at approximately 410 and 415 nm, and it would not be practicable to try to differentiate between them or to attempt to define the composition of a mixture of the two forms simply from the absorption spectra.

8    $\lambda_{max}$ at 470, 440, 468, 455, 455, 450, 438, 445, 450 nm, respectively. Neurosporene, chromophore with 9 conjugated acyclic double bonds, has $\lambda_{max}$ 440 nm; each additional acyclic double bond in conjugation adds 15 nm. A $\beta$-ring has one double bond in conjugation with the acyclic polyene, but contributes only 5 nm to $\lambda_{max}$. A C-4 oxo-group in conjugation adds 8 nm. Epoxidation of a double bond breaks conjugation at that point; other substituents, e.g. hydroxy and non-conjugated oxo-groups, do not affect the chromophore. The double bond of an $\epsilon$-ring is not in conjugation and does not contribute to the chromophore or to $\lambda_{max}$.

9    The key to all these answers is that chlorophyll not only absorbs strongly light of several parts of the visible spectrum but also exhibits very strong red fluorescence.

10    Structural blue is seen as blue only by reflected, not transmitted light. The colour may also depend on the incident and viewing angles. The colour is likely to be lost (though not extracted) on soaking in water or other solvent, then regained when the solvent evaporates. If the colour is pigmentary it should be possible to extract it from the tissues with water or solvent. Resonance Raman spectroscopy may give information about the chromophore. Study of the solubility, general and spectroscopic properties of the isolated pigment (see the individual chapters) should allow the class of compound to be identified.

11    See chapter 2 for the overall pathway of carotenoid biosynthesis. Phytoene is made from eight molecules of mevalonic acid (MVA)

and this phytoene sample therefore retains all eight labelled carbon atoms and eight tritium atoms from the doubly labelled MVA substrate. The formation of phytoene is therefore stereospecific; the hydrogen lost from C-1 of geranylgeranyl pyrophosphate during the introduction of the C-15,15' double bond of phytoene must be the unlabelled hydrogen from C-5 of MVA. β-Carotene retains eight [14]C atoms but only four tritium atoms. The tritium must have been lost stereospecifically during desaturation. Similarly a tritium substituent must be lost from C-3 of each ring of β-carotene during hydroxylation to zeaxanthin.

12    The most likely polyketide patterns for those compounds are illustrated below:

Mollisin

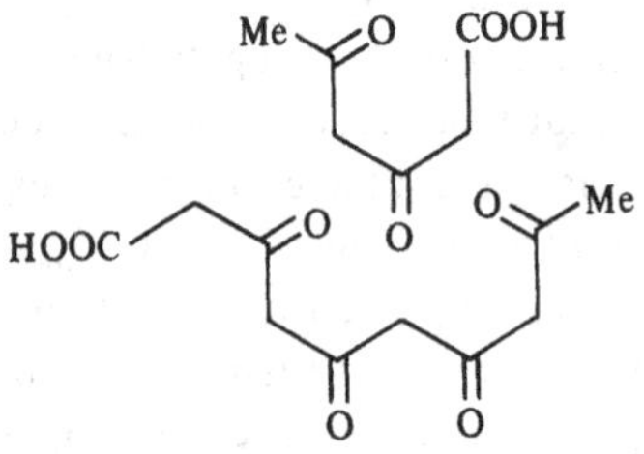

Solorinic acid

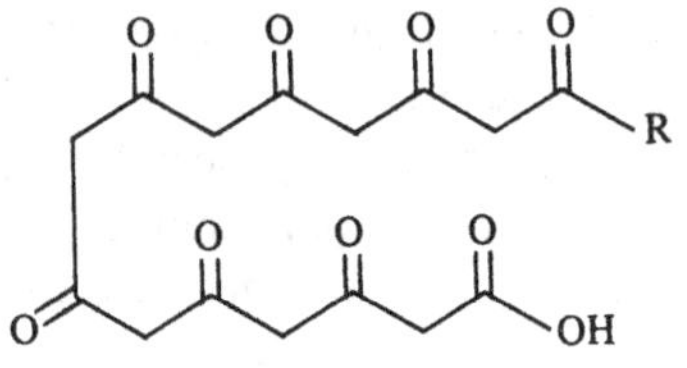

Laccaic acid D

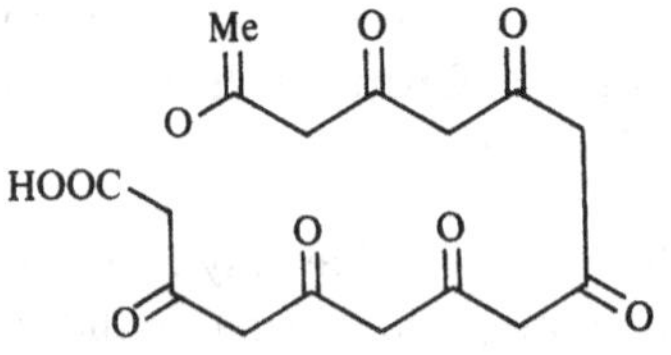

The best method for verifying such schemes is the use of the doubly labelled $[1,2\text{-}^{13}C_2]$-acetate. If this labelled substrate is incorporated efficiently enough into the product (not less than about 1% incorporation or enrichment), those carbon atoms in the product which arise from an intact acetate molecule will exhibit $^{13}C$–$^{13}C$ coupling in the n.m.r. spectrum. Unless the level of incorporation is greater than 10%, no appreciable coupling will be observed between other carbon atoms. The position in the final molecule of the acetate units that form the polyketide chain can therefore be established.

13    This intermediate would be formed by addition of a $C_5$ isoprene unit to *o*-succinylbenzoate or a naphthol/naphthoquinone intermediate (see fig. 3.12). Four of the five carbon atoms of this isoprene unit would then be used to construct ring C of the alizarin. The labelling procedure with $[^{13}C_2]$-acetate, mentioned in the answer to question 12 would distinguish between most of the possible alternative mechanisms – any part of the molecule derived from acetate would not be labelled.

14    The incorporation of labelled acetate and phenylalanine suggests a flavonoid structure and the other data are consistent with its being the β-D-glucoside of mytricetin (**4.19**). Diazomethane methylates only the phenolic hydroxy-groups, whereas all primary and secondary hydroxy-groups are acetylated.

15    Ferricyanide oxidises haemoglobin in methaemoglobin, giving a brown colour. The column will retard small molecules and ions such as dithionite and ferricyanide much more than the larger haemoglobin. Once on the column, the methaemoglobin encounters dithionite and is reduced to haemoglobin which, as it passes through the column and is eluted, is oxidised to oxyhaemoglobin by air dissolved in the eluting buffer.

16    The effects of $CO_2$, $H^+$ and DPG on the oxygen affinity of haemoglobin and the cooperativity between the associated subunits are all outlined in chapter 5 (§5.5.2).

17    Acetate formed from $[^1H]$-glucose will retain $^1H$ in its methyl group (C-2). When this acetate is metabolised further in $^2H_2O$ medium, any hydrogen introduced by reduction of the C-1 carboxyl-group of acetate will be entirely in the form of deuterium. Thus molecules such as carotene and chlorophyll will contain only deuterium at those positions derived from C-1 of acetate but will retain $^1H$ at positions derived from C-2 of the acetate. Under photosynthetic growth, with $CO_2$ at the carbon source supplied, the only hydrogen available will be deuterium from the $^2H_2O$ of the medium, so all positions in the molecules biosynthesised will be fully deuterated.

18    Schemes for the conversion of ALA into haem and chlorophyll and for the formation of the phycobilin structure from the porphyrins are given in chapter 5, and show that the labelling pattern would be the same whether chlorophyll or haem were the intermediate.

19    If the oxygen present were a mixture of $^{18}O_2$ and $^{16}O_2$, then if mechanism (i) applies both oxygen atoms of bilirubin would come

from the same $O_2$ molecule, whether this were $^{18}O_2$ or $^{16}O_2$. In mechanism (ii) each oxygen atom could come from either $^{18}O_2$ or $^{16}O_2$. Mechanism (iii) would require only one of the two oxygen atoms to come from $O_2$, either $^{18}O_2$ or $^{16}O_2$. If the oxygen used contained $^{18}O_2$ (20%) and $^{16}O_2$ (80%), then the bilirubin sample obtained in these three cases would have the composition:

(i)   584 (80%), 586 (0%), 588 (20%)
(ii)  584 (64%), 586 (32%), 588 (4%)
(iii) 584 (80%), 586 (20%), 588 (0%)

and these would readily be distinguished by mass spectrometry. If the oxygen used in the experiment were entirely the $^{18}O_2$ species, the bilirubin formed as a result of mechanism (iii) would be entirely the singly labelled species with relative molecular mass 586, whereas mechanisms (i) and (ii) would each give only the relative molecular mass 588 species and could therefore not be distinguished.

20    Sepiapterin (6.15) is the likely monomeric precursor with the $C_3$ side chains being used to construct the linking five-membered ring. See fig. 6.2 for the absorption spectra and §6.2.4 for an outline of the relevant properties. Also consider the ionisation properties of the amphoteric pterin molecules to explain the spectral changes.

21    The two strains each have incomplete pathways. The intermediate normally accumulated by F 11 can be metabolised to iodinin by 13 Z.

22    The likely main monomeric units are illustrated below. Many alternative melanin structures are possible, all produced by similar phenolic oxidative coupling.

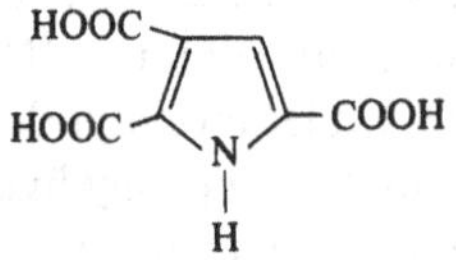

HOOC, HOOC —COOH   and other pyrrole carboxylic acids

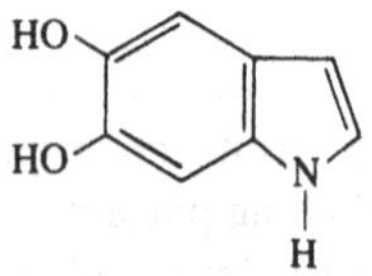

HO, HO   and its *o*-quinone

23    Consider the general strategies of camouflage colouring outlined in chapter 8. Experiment (i) would show if the caterpillars simply

absorb and deposit the different collection of pigments that may be present in the different leaf species. However it is likely that the difference in colour in the leaves is due to structural effects rather than pigment differences. Experiment (ii) would investigate whether, for example, the caterpillars were reacting to background colour via some photoreceptor mechanism or whether the colour adopted were in response to some other stimulus, e.g. some chemical substance present in one leaf species but not the other. It would also take into account the possibility that the different colour forms represent two genetically distinct populations.

24    A trichromatic visual pigment system is similar in principle to the trichromatic system used in colour television or colour photography. Light of three different wavelengths corresponding to the three primary colours can be mixed in infinitely variable proportions to give any colour, hue or intensity required. Likewise if the absorption range of three visual pigments covers virtually the whole of the visible spectrum, with well-separated absorption maxima, light of any colour or wavelength distribution can be resolved into three primary components which are detected by the three visual pigments.

25    Plot qualitative absorption spectra for the photoreceptors in the absence and presence of the oil droplets – note that these droplets are intensely coloured and will prevent all light in their absorption range from reaching the photoreceptor. Consider how this gives sharper and more widely separated absorption spectra for the visual pigments.

26    Although there will inevitably be interspecific variation, the evolutionary tendency will be for the animals to use photoreceptors which absorb maximally the light wavelength which penetrates furthest in their normal environment and to use in their coloration pigments that either absorb or transmit maximally in this optimum wavelength region. Free movement between different environments means that the animals will need photoreceptors for the different ambient light conditions and may use much more complex pigmentation strategies.

27    (i) Light harvesting need not be so efficient, so smaller amounts of accessory light-harvesting pigments will be required, but the photosynthetic apparatus needs to be well protected against photooxidation so more synthesis of carotene may be expected.
(ii) Light harvesting must be made as efficient as possible by the

synthesis of more light-harvesting pigments; effects seen will include
an increase in the size of the phycobilisomes.

(iii) More of the appropriate green- or red-absorbing phycobilin will
be synthesised to take maximum advantage of the light available.

28      (a) As the leaves senesce, chloroplasts degenerate and the green
chlorophyll is destroyed revealing the presence of yellow carotenoids. Synthesis of red anthocyanins then occurs and finally
oxidative coupling of phenolic compounds gives brown polymeric
molecules.

(b) The colours seen are again due to green chlorophyll, then yellow
carotenoid and finally red anthocyanin.

(c) (i) The chloroplasts degenerate and chlorophyll is destroyed. The
resulting yellow etiolated plants contain not chloroplasts but etioplasts which are coloured by carotenoids. The green colour is
restored and chloroplasts and chlorophyll made again on exposure to
light.

(ii) Chlorophyll synthesis is inhibited by iron or magnesium deficiency (see §5.9.7 and 5.9.15).

29      This is very difficult to do. Such a photoreceptor need be present
only in very small amounts, along with much larger amounts of inactive bulk pigment. Possibly sophisticated very short time-scale
spectroscopic studies may be useful, but perhaps the only way to be
sure is to isolate and characterise the actual photoreceptor pigment.

30      Consider the properties of phytochrome outlined in §11.2, especially
fig. 11.2. Remember that the active form appears to be $P_{fr}$; $P_r$
normally plays only a passive role.

# Index

(Numbers in **bold** type refer to the main
references within a group)